Rust Diseases of Willow and Poplar

We dedicate this book to our children Michael and Jeffrey,
Sarah and Philippa

Rust Diseases of Willow and Poplar

Edited by

Ming Hao Pei

Rothamsted Research, Harpenden, Hertfordshire, UK

and

Alistair R. McCracken

Applied Plant Science Division, Department of Agriculture and Rural Development, Belfast, UK

CABI Publishing

CABI Publishing is a division of CAB International

CABI Publishing
CAB International
Wallingford
Oxfordshire OX10 8DE
UK

Tel: +44 (0)1491 832111
Fax: +44 (0)1491 833508
E-mail: cabi@cabi.org
Web site: www.cabi-publishing.org

CABI Publishing
875 Massachusetts Avenue
7th Floor
Cambridge, MA 02139
USA

Tel: +1 617 395 4056
Fax: +1 617 354 6875
E-mail: cabi-nao@cabi.org

A catalogue record for this book is available from the British Library, London, UK.

Library of Congress Cataloging-in-Publication Data
Rust diseases of willow and poplar / edited by Ming Hao Pei & Alistair R. McCracken.
p. cm.
ISBN 0-85199-999-9 (alk. paper)
1. Willows--Diseases and pests. 2. Poplar--Diseases and pests. 3. Rust diseases. I. Pei, Ming Hao. II. McCracken, Alistair R. III. Title.

SB608.W65R87 2005
634.9′7236--dc22 2004018883

ISBN 0 85199 999 9

Typeset by AMA DataSet Ltd, UK.
Printed and bound in the UK by Biddles Ltd, King's Lynn.

Contents

Contributors

Gemma Assante, *Istituto di Patologia Vegetale, Università di Milano, Via Celoria 2, 20133 Milano, Italy*
Gaddam Bagyanarayana, *Department of Botany, Osmania University, Hyderabad 500 007 (A.P), India*
Carlos Bayon, *Rothamsted Research, Harpenden, Hertfordshire AL5 2JQ, UK*
Diane Carlisle, *Department of Applied Plant Science, Queen's University of Belfast, Newforge Lane, Belfast BT9 5PX, UK*
Irene Dellavalle, *CNR, Istituto per la Protezione delle Piante, Area della Ricerca del CNR di Firenze, Via Madonna del Piano, 50019 – Sesto Fiorentino (FI), Italy*
Malcolm Dawson, *Applied Plant Science Division, Northern Ireland Horticulture and Plant Breeding Station, Department of Agriculture and Rural Development, Loughgall, Co. Armagh, BT61 8JB, UK*
Nicolas Feau, *Centre de Recherche en Biologie Forestière, Université Laval, Sainte-Foy (QC), G1K 7P4, Canada*
Pascal Frey, *UR Pathologie Forestière, INRA, F-54280 Champenoux, France*
Pierre Gérard, *Laboratoire Ecologie, Systématique et Evolution, UMR ENGREF-UPXI-CNRS 8079, Université Paris-Sud, 91405 Orsay, France*
A.K. Gupta, *Department of Mycology and Plant Pathology, Dr Y.S. Parmar University of Horticulture and Forestry, Nauni, Solan – 173 230 Himachal Pradesh, India*
Claude Husson, *UR Pathologie Forestière, INRA, F-54280 Champenoux, France*
Tom Hunter, *4 Wally Court Road, Chew Stoke, Bristol BS40 8XL, UK*
Sergio Hurtado, *Plant Pathology and Biocontrol Unit, Swedish University of Agricultural Sciences, PO Box 7035, S-75007 Uppsala, Sweden*
Marlene Karlsson, *Umeå Plant Science Center, Department of Forest Genetics and Plant Physiology, Swedish University of Agricultural Sciences, SE-901 83 Umeå, Sweden*
Makoto Kakishima, *Institute of Agriculture and Forestry, University of Tsukuba, Tsukuba, Ibaraki 305-8572, Japan*
Mirko Liesebach, *Federal Office and Research Centre for Forests, Department of Forest Genetics, Hauptstrasse 7, A-1140 Vienna, Austria*
Nicola Longo, *Dipartimento di Biologia Vegetale, Università di Firenze, Via La Pira 4, 50121 Firenze, Italy*
Alistair McCracken, *Applied Plant Science Division, Department of Agriculture and Rural Development, Newforge Lane, Belfast BT9 5PX, UK*
Salvatore Moricca, *Dipartimento di Biologia Vegetale, Università di Firenze, Via La Pira 4, 50121 Firenze, Italy*

Chris C. Mundt, *Department of Botany and Plant Pathology, 2082 Cordley Hall, Oregon State University, Corvallis, OR 97331-2902, USA*
Biancamaria Naldini, *Dipartimento di Biologia Vegetale, Università di Firenze, Via La Pira 4, 50121 Firenze, Italy*
Ming Hao Pei, *Rothamsted Research, Harpenden, Hertfordshire AL5 2JQ, UK*
Jean Pinon, *UR Pathologie Forestière, INRA, F-54280 Champenoux, France*
Alessandro Ragazzi, *Dipartimento di Biotecnologie Agrarie, Sezione di Patologia Vegetale, Università di Firenze, Piazzale delle Cascine 28, 50144 Firenze, Italy*
Mauritz Ramstedt, *Plant Pathology and Biocontrol Unit, Swedish University of Agricultural Sciences, PO Box 7035, S-75007 Uppsala, Sweden*
Marisa Rodriguez-Buey, *Umeå Plant Science Center, Department of Plant Physiology, Umeå University, S-901 87 Umeå, Sweden*
Carmen Ruiz, *Rothamsted Research, Harpenden, Hertfordshire AL5 2JQ, UK*
Berit Samils, *Department of Plant Biology and Forest Genetics, Swedish University of Agricultural Sciences, Uppsala, Sweden*
Yan Zhong Shang, *College of Forestry, Inner Mogolia Agricultural University, Huhehot, China*
K.R. Sharma, *Department of Forest Products, Dr Y.S. Parmar University of Horticulture and Forestry, Nauni, Solan – 173 230 Himachal Pradesh, India*
R.C. Sharma, *Department of Mycology and Plant Pathology, Dr Y.S. Parmar University of Horticulture and Forestry, Nauni, Solan – 173 230 Himachal Pradesh, India*
S. Sharma, *Department of Mycology and Plant Pathology, Dr Y.S. Parmar University of Horticulture and Forestry, Nauni, Solan – 173 230 Himachal Pradesh, India*
Cheng-Ming Tian, *College of Resource and Environment, Beijing Forestry University, 35 Tsinghua Eastern Road, Beijing 100083, China*
Mikaela Torp, *Umeå Plant Science Center, Department of Forest Genetics and Plant Physiology, Swedish University of Agricultural Sciences, SE-901 83 Umeå, Sweden*
Gunnar Wingsle, *Umeå Plant Science Center, Department of Forest Genetics and Plant Physiology, Swedish University of Agricultural Sciences, SE-901 83 Umeå, Sweden*
Johanna Witzell, *Umeå Plant Science Center, Department of Forest Genetics and Plant Physiology, Swedish University of Agricultural Sciences, SE-901 83 Umeå, Sweden*
Zhiwen W. Yuan, *Institute of Applied Ecology, Academia Sinica, PO Box 417, Shenyang, China*
Irmtraut Zaspel, *Federal Research Centre for Forestry and Forest Products, Institute for Forest Genetics and Forest Tree Breeding, Eberswalder Chaussee 3A, D-15377 Waldsieversdorf, Germany*

Preface

The plant family *Salicaceae* comprises two major genera, willow (*Salix*) and poplar (*Populus*). Some 300–500 species, according to different authorities, are recognized in *Salix* and 30–100 species in *Populus*. Willows and poplars are among the most common woody plants in the northern hemisphere and, through centuries of human intervention, they have been widely planted in various regions of the world. Traditionally, willows were used for an age-old practice – basket-making. Willows are also common as amenity trees in urban areas and as garden shrubs. More recently, increasing demand for energy from renewable sources has provided a new impetus to growing willows in short-rotation coppice (SRC) plantations for biomass energy production. Poplars, on the other hand, have long been cultivated for timber and pulp production, shelterbelts and amenity purposes.

Rust caused by the members of the genus *Melampsora* is the most widespread and damaging disease of willow and poplar. Conventionally, willow and poplar rusts are regarded as nursery diseases. However, as SRC plantations sustain dense, young canopy during the growing season, the rust can cause serious damage. Up to 40% yield loss due to rust has been recorded in SRC willow. In Europe, poplar rusts did not cause serious damage until the 1980s and 1990s, when the rise of new pathotypes had a serious impact on poplar cultivation. Concerns were raised because poplar rust outbreaks severely affected not only SRC plantations but also single-stem tree plantations. In some cases, the rust resulted in stem dieback and even death of coppice stools.

In Europe, early research on willow and poplar rusts was represented by the series of studies carried out by Klebahn in the late 19th and the early 20th centuries. In the Far East, Japanese mycologists such as Takashi Matsumoto and Naohide Hiratsuka conducted much of the early work. These early studies established the foundation for understanding the morphology, host alternation and host range in willow and poplar rusts. The work of the Canadian mycologist W.G. Ziller, in the mid-20th century, contributed much to the knowledge of host range and life cycle of willow and poplar rusts in North America.

Over the past two decades, *Melampsora* rusts on poplar and willow have been subject to extensive study, partly because of the severe damage caused by the occurrence of new, virulent types of rusts and partly because of the growing interest in SRC for biomass energy. As well as conventional plant pathological and mycological methods, DNA profiling techniques have been used widely in willow and poplar rust research. These studies have had a profound impact on our understanding of willow and poplar rust diseases, especially in the fields of population biology, genetics of resistance and functions of host–genotype mixtures.

Current state-of-the-art knowledge of poplar rusts would not have been achieved without the outstanding contributions over the past two decades by Jean Pinon and others at the Institut National de la Recherche Agronomique, Nancy, France. George Newcombe, now based at the University of Idaho, played a leading role in poplar rust research in North America. George declined our invitation to contribute a chapter on North American poplar rusts, simply because he had just written a review on the subject published elsewhere. Colleagues from India, China and Japan also made valuable contributions to the progress of research on poplar rusts.

Compared to poplar rusts, little work had been done on willow rusts until the late 1980s, when growing willows for biomass gained momentum. Over the years, the Swedish Agricultural University, the Department of Agriculture and Rural Development Northern Ireland and Long Ashton Research Station (the team moved to Rothamsted Research due to the closure of the Long Ashton Research Station in 2003) have played important parts in driving forward the research on willow rusts. Swedish colleagues contributed a great deal to the knowledge of population biology of willow rusts. The outstanding achievements in mixture plantation research in Northern Ireland would not have been possible without dedicated efforts and long-term commitment by the colleagues involved in the past 15 years.

During 2000–2004, the Swedish Agricultural University, Queens University Belfast, Long Ashton Research Station and the Institute of Forestry and Tree Breeding, Germany, jointly carried out an EU project on integrated control of willow rusts in Europe. Although relatively new to the subject, German colleagues forwarded remarkable efforts in the studies of pathogenic and molecular variation of willow rusts and the mycoparasite *Sphaerellopsis filum* in central Europe. In terms of the quality of science and practical usefulness, the project was hugely rewarding. The project was truly memorable for the genuine teamwork and unreserved support among the participants.

The primary purpose of this book is to summarize the progress of research on *Melampsora* rust on willows and poplars, and to provide insights into how this important group of plant pathogens behaves and interacts with their hosts. Such knowledge is essential in developing strategies and means of maintaining healthy crops and reducing disease impacts.

The content of the book is based mainly on the papers presented at the 'International Symposium on *Melampsora* on *Salicaceae*' held in Belfast, Northern Ireland on 11–13 September 2003. The Department of Applied Plant Science, Queen's University Belfast, hosted and chaired the meeting. This meeting was a part of the activities of the EU project QLK5-1999-01585 'Integrated, non-fungicidal control of *Melampsora* rusts in biomass willow plantations' which was conducted during 2000–2004 in England, Northern Ireland, Sweden and Germany. Colleagues from various countries contributed to the meeting on taxonomy and population biology, rust resistance in the hosts, the use of host mixtures as a means of rust control and the potential of biological control using mycoparasites. Inspiring background lectures presented by Professor Chris Mundt, Editor-in-Chief of *Phytopathology*, and Dr Mogens Hovmøller were greatly appreciated by all the participants. The Department of Agriculture and Rural Development, Northern Ireland, and the Society of Irish Plant Pathologists offered organizational and financial support for the meeting. The symposium provided a forum in which participants were able to exchange ideas, highlight common ground and explore new directions for research. All the participants felt that it was a truly useful and enjoyable meeting. Several colleagues who were not able to attend the meeting also kindly agreed to contribute chapters for this book.

The book comprises five parts: Part 1, Taxonomy and overview of rusts; Part 2, Occurrence and population biology of *Melampsora*; Part 3, Rust resistance and infection process; Part 4, Rust management; and Part 5, Rust mycoparasites and their potential for biological control. Few books addressing specific tree diseases are available. This book should give research workers, teachers and students a very comprehensive review on the progress and current knowledge of rust diseases of willow and poplar. Thus, it also presents an overview of a woody plant/rust pathogen system.

We would like to thank all the contributors for completing their chapters in time and at a high standard. M.H.P. would like to express his deep gratitude to Professor Li-Ping Shao, Northeastern Forestry University, China, who guided him into rust disease research in the early days of his career.

Ming Hao Pei
Alistair R. McCracken
July 2004

Abbreviations

AFLP	amplification fragment length polymorphism
AMOVA	analysis of molecular variance
AUDPC	area under the disease progress curve
BCA	biocontrol agent
EBI	ergosterol-biosynthesis inhibiting
EtOAc CEs	ethylacetate crude extracts
GUA	genotype unit area
ITS	internal transcribed spacer
IPF	incubation period for flecking
LD	*M. larici-daphnoides*
LET	*M. larici-epitea typica*
LM	light microscopy
LPU	latent period for eruption of first uredinium
LR	*M. larici-retusae*
LSU	large subunit
NMR	nuclear magnetic resonance
PCR	polymerase chain reaction
RAPD	random amplified polymorphic DNA
rDNA	ribosomal DNA
RFLP	restriction fragment length polymorphism
SEM	scanning electron microscopy
SIF	stem-infecting form
SNP	single nucleotide polymorphism
SRC	short-rotation coppice
SRF	short-rotation forest
SSR	single-sequence repeat
SSU	small subunit
TEM	transmission electron microscopy
ULD	uredinia per leaf disc
UPGMA	unweighted pair group method with arithmetic mean
VAF	% variance accounted for

1 Phylogenetic Position of *Melampsora* in Rust Fungi Inferred from Ribosomal DNA Sequences

Ming Hao Pei, Carlos Bayon and Carmen Ruiz
Rothamsted Research, Harpenden, Hertfordshire AL5 2JQ, UK

Rust Fungi and their Classification

The rust fungi (*Uredinales*) are one of the largest groups of fungi, accounting for one-third of the teleomorphic species of *Basidiomycetes* (Hawksworth *et al.*, 1995). Over 5000 species of rust fungi have been recognized and it is estimated that the order *Uredinales* may contain more than 7000 species (Hawksworth *et al.*, 1995). The species in *Uredinales* are grouped into more than 100 genera (Cummins and Hiratsuka, 1983).

Rust fungi are unique in many aspects. They thrive only on living tissues of plants and produce up to five different types of spores during their life cycle. Rust species are either heteroecious or autoecious. Heteroecious species infect taxonomically very different plants to complete their life cycle. Autoecious species, on the other hand, complete their life cycle on the same hosts. The hosts of rust fungi are immensely diverse, including ferns, conifers, dicotyledonous and monocotyledonous plants. Another important feature of rust fungi is their host specificity, each group being capable of infecting a certain range of plant taxa.

Traditionally, teliospore morphology provided the basis for rust taxonomy. Dietel (1900) proposed four families to group the genera of rust fungi, i.e. *Melampsoraceae*, *Coleosporiaceae* and *Cronartiaceae* to include those forming sessile teliospores, and *Pucciniaceae* for those having pedicelled teliospores. Later, Dietel (1928) reduced the number of families to two, by placing those having sessile teliospores in *Melampsoraceae* and those having pedicelled teliospores in *Pucciniaceae*. The two families were widely accepted for a long time by various workers (Arthur, 1929; Bessey, 1950; Azbukina, 1974). While accepting *Puccinaceae*, Gäumann (1959) added a few more families for those traditionally grouped under *Melampsoraceae*. Wilson and Henderson (1966) adopted three families, *Coleosporiaceae*, *Melampsoraceae* and *Pucciniaceae*. Further modifications were proposed by Leppik (1972) and Savile (1976). More recently, largely based on spermogonial morphology, Cummins and Hiratsuka (1983, 2003) proposed a 13- to 14-family system. In their treatment, only a single genus, *Melampsora*, is included in *Melampsoraceae*.

Genus *Melampsora*

The genus *Melampsora* was established by Castagne in 1843 based on the rust on

Euphorbia, *M. euphorbiae* (Scheb.) Cast. (Castagne, 1843). The main characteristic of the genus is formation of a crust of sessile, laterally adherent single-celled teliospores on the host surface. *Melampsora* species are either heteroecious or autoecious. Heteroecious species of *Melampsora* occur on willows (*Salix*) and poplars (*Populus*), both belonging to *Salicaceae*. The majority of heteroecious *Melampsora* species alternate on conifers, but some have dicotyledonous or monocotyledonous plants as their alternate hosts (Fig. 1.1). Most of the autoecious species of *Melampsora* occur on dicotyledonous plants, including *Euphorbiaceae* and *Linaceae*.

The taxonomy of *Melampsora* is notoriously difficult (Chapter 2). Numerous species have been described under *Melampsora* but with many, it is not certain whether they should be treated as valid species or merely as synonyms. The number of species in *Melampsora* is estimated to be somewhere between 80 (Hawksworth *et al.*, 1995) and 100 (Hiratsuka and Sato, 1982). To date, volume III of Sydow and Sydow's *Monographia Uredinearum* (1915) remains the only monograph available for *Melampsora*. It appears that more than half of *Melampsora* species occur on *Salicaceae*. For example, of 51 species compiled by Sydow and Sydow (1915), ten were described on *Populus*, 22 on *Salix* and others on various dicotyledonous plants, including *Euphorbiaceae* and *Linaceae*.

Some Viewpoints on the Evolution of Rust Fungi

The evolution of rust fungi has long been a fascinating subject of speculation among mycologists. Savile (1955) speculated that the rust fungi have evolved from an ascomycetous *Taphrina*-like ancestor. In contrast, Leppik (1965) supported Jackson's (1935) views that an auriculariaceous (jelly-fungi-like) fungus, which occurred on some primitive ferns, gave rise to rusts. Savile (1976) modified these ideas and suggested that a certain *Taphrina*-like fungus parasitic on plants gave rise to ancestral *Basidiomycetes* which, in turn, produced *Uredinales*, which remained parasitic on plants, and *Auricularis* (jelly fungi), which became saprophytic (see Hennen and Buriticá, 1980).

In regard to the origin of the heteroecism among the rusts, Fischer (1898) suggested that the ancestors of present heteroecious rusts had possessed multi-spore forms and were capable of living indiscriminately on a wide range of hosts (cited by Arthur, 1929).

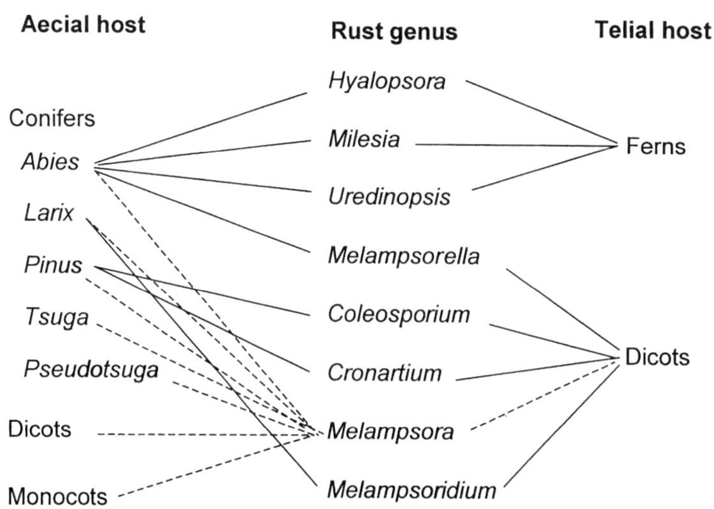

Fig. 1.1. Host specificity of some rust genera related to *Melampsora*.

Each species gradually narrowed its host range and became confined to fewer hosts. During the process, the haploid stage was adapted to one or a group of hosts, and the dikaryotic/diploid stage to another. This early explanation was modified by Klebahn (1904) and Dietel (1904, cited by Arthur, 1929) by assuming that the early forms of rusts were more or less restricted and fixed, and that the change from an autoecious to a heteroecious condition was brought about by some internal change, possibly in the nature of mutation, which led to the acceptance of a very unlike host.

As obligate parasites of plants, rusts are considered to have coevolved with their hosts. The species in *Uredinopsis*, *Milesina* and *Hyalopsora* have uredinial and telial stages on ferns and spermogonial and aecial stages on firs (*Abies*). They have sessile teliospores and produce five spore stages (macrocyclic) during their life cycle. Many species of *Uredinopsis* and *Hyalopsora* also produce an extra kind of urediniospores, called amphispores. These rusts have long been regarded as the most primitive because the two groups of host are considered to be more primitive than other plants serving as hosts to rust fungi (Ando, 1984). Other rust genera producing aecia on conifers also bear sessile teliospores, and predominantly have macrocyclic and heteroecious life cycles. In contrast, autoecious and shortened life cycles become common in the rusts on more advanced hosts (Laundon, 1973; Hiratsuka and Sato, 1982). Therefore, it is generally believed that the heteroecious, macrocyclic life cycle is the primitive feature and that various reduced life cycles are derived from it. However, as Hiratsuka and Sato (1982) pointed out, 'it was a vexing problem to imagine that the most complicated life-cycles (as found in *Uredinopsis*) are the most primitive forms'. A possible explanation is that the extraordinarily complex heteroecious macrocyclic life cycle evolved only once (Mendgen, 1997). Reductions could have easily occurred at any stage of evolution (Laundon, 1973).

Melampsora species form relatively simple aecia (no peridia or bearing only rudimentary peridia) and occur on a diverse range of plants, including the members of *Pinaceae*. Usually, aecial hosts of rust genera capable of infecting *Pinaceae* are confined to a single genus. For example, *Uredinopsis*, *Milesina*, *Hyalopsora* and *Melampsorella* occur on *Abies* at their aecial stage, *Coleosporium* and *Cronartium* on *Pinus*, *Melampsoridium* on *Larix*, and *Chryxomyxa* on *Picea* (Fig. 1.1). In contrast, the species in *Melampsora* occur on several genera of *Pinaceae*, i.e. *Abies*, *Pinus*, *Picea*, *Larix*, *Pseudotsuga* and *Tsuga*. For this reason, Durrieu (1980) suggested that *Melampsora* could have been the first rust to attack *Pinaceae*.

Ribosomal DNA Sequences and Phylogenetic Studies of Fungi

The ribosome is an organelle responsible for protein synthesis, occurring in every living cell. Because of its universal existence and phylogenetically informative nature, ribosomal DNA (rDNA) is widely used to study evolutionary relationships in living organisms, including fungi. The rDNA is a multigene family with nuclear copies arranged in tandem arrays in eukaryotes (organisms having cells with a true nucleus and other organelles). Each unit within a single array consists of the genes coding for the small subunit (SSU) and large subunit (LSU) rRNAs. The internal transcribed spacer (ITS) region (comprises ITS1, 5.8S rRNA gene and ITS2) lies embedded between the SSU and LSU regions (Fig. 1.2). The rDNA sequences encoding the SSU and the LSU rRNAs are highly conserved, while the sequences of ITS region evolve much faster. The ribosomal genes undergo concerted evolution and, therefore, sequence heterogeneity among repeats within a nucleus is rare.

In recent years, phylogenetic studies using SSU rDNA data suggested that *Ascomycetes* and *Basidiomycetes* originated from a common ancestor but evolved separately (Berbee and Taylor, 2001; Tehler *et al.*, 2003). Thus, the molecular data did not support the theory that rust fungi evolved

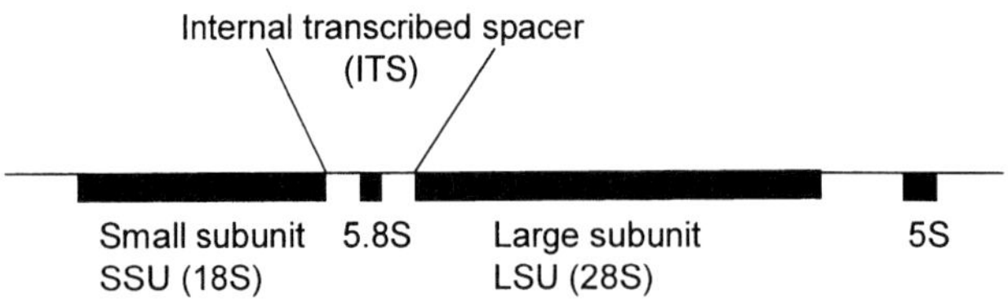

Fig. 1.2. Structure of ribosomal DNA.

from the *Taphrina*-like ancestor. Gene trees derived from the LSU (Berres *et al.*, 1995; Begerow *et al.*, 1997) and SSU (Berbee and Taylor, 2001; Tehler *et al.*, 2003) regions also suggested that rust fungi are monophyletic, forming a distinct group in *Basidiomycota*. Such studies also supported the recognition of the three major groups within *Basidiomycota*, i.e. *Urediniomycetes* (rusts and related fungi), *Ustilaginomycetes* (smuts and related fungi) and *Hymenomycetes* (mushrooms and related fungi). *Auricularia* (jelly fungi) was placed within *Hymenomycetes* in the gene trees, which showed that the split between *Urediniomycetes* and *Hymenomycetes* occurred long before the split between jelly fungi and mushrooms. Such results contradict the theory that there was a close phylogenetic association between rusts and jelly fungi. Additional data from multiple genes are probably needed to resolve the deepest divergences in the *Basidiomycota*.

Phylogenetic Position of *Melampsora* in *Uredinales*

Cullings and Vogler (1998) included *M. epitea* and *M. lini* in their 5.8S rDNA sequence analysis and found that the *Melampsora* spp. formed a clade (bootstrap value = 95%) before the divergence among other genera, such as *Pucciniastrum*, *Cronartium*, *Chrysomyxa*, *Puccinia* and *Uromyces*. Recently, Maier *et al.* (2003) examined partial DNA sequences of the LSU region of rDNA from some 50 rust species belonging to 25 genera. Three *Melampsora* species, *M. helioscopiae*, *M. euphorbiae* and *M. hypericorum* were included in their study. From their results, the genus *Melampsora* also appeared to be somewhat separated (bootstrap value = 62%) from all other genera, which included those having sessile teliospores and those having pedicelled teliospores.

We further examined the phylogenetic position of *Melampsora* among the rust fungi, using rDNA sequence data from willow and poplar rusts and the sequence information available in World Wide Web databases. The partial LSU region of ten species/forms of *Melampsora* on willows, three species on poplar, *Puccinia recondita* f.sp. *tritici* and *Melampsoridium betulinum* was sequenced using the primers described in Berres *et al.* (1995). The rDNA sequence information for other rusts was retrieved from the GenBank database (http://www.ncbi.nlm.nih.gov/).

When the partial LSU sequences (about 1000 bases of the 5′ end of LSU) of 53 species belonging to *Uredinales* (four other species related to *Uredinales* were used as the outgroup) were examined, the neighbour-joining tree placed those having pedicelled teliospores into one group (bootstrap value = 75%) (Fig. 1.3). Among the rest, *Melampsora* formed a distinct clade (bootstrap value = 100%). The remaining rusts having sessile teliospores were placed together, but this clade was not supported by bootstrap value (< 50%). Within *Melampsora*, *M. occidentalis*, which occurs on *Populus* in North America, was more distant from other *Melampsora* species (Fig. 1.3).

When the SSU and LSU regions of 16 *Uredinales* species (see Fig. 1.4; five species related to *Uredinales* were used as an outgroup) were examined, the average *p* distance (percentage difference/100) for the SSU region among the *Uredinales* species was 0.034 (standard error = 0.003) and that for LSU region was 0.098 (standard error = 0.008). The neighbour-joining tree constructed based on the combined SSU and

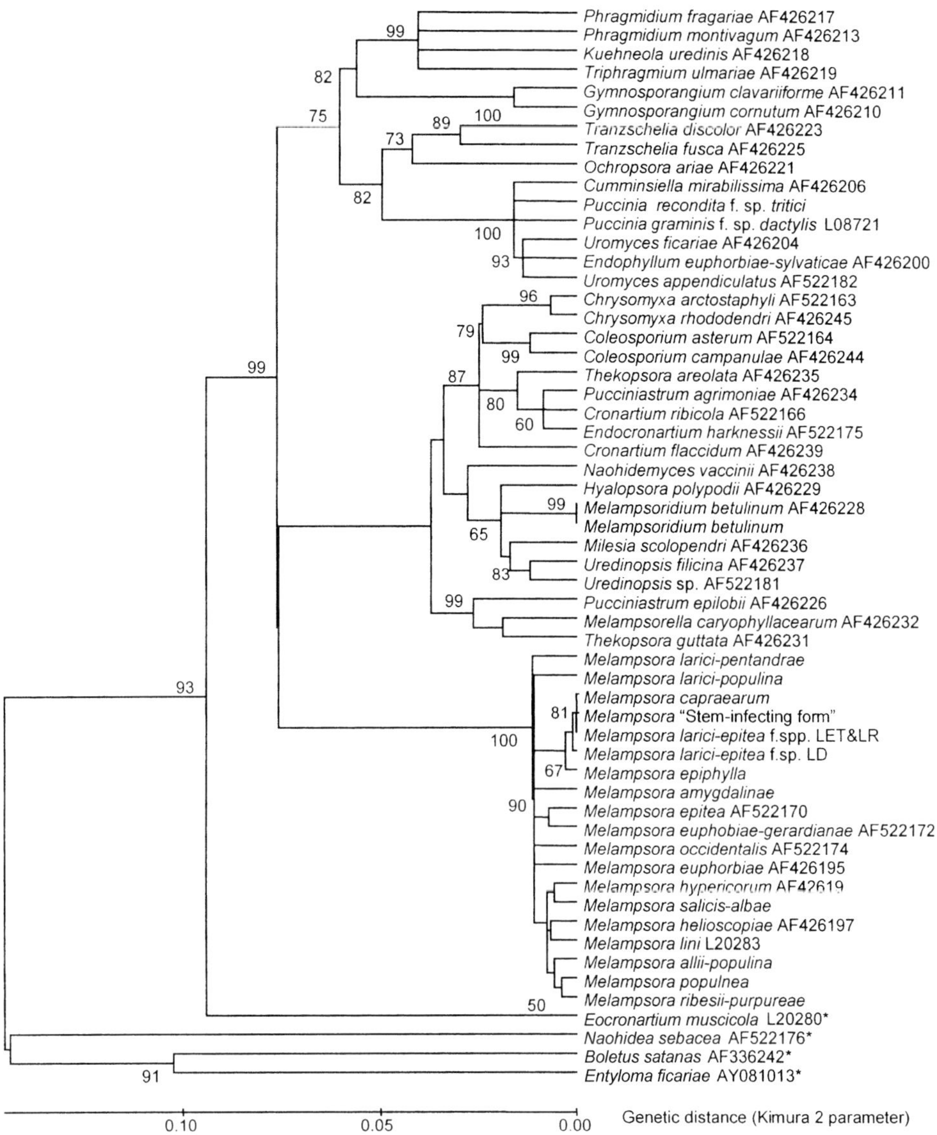

Fig. 1.3. Neighbour-joining tree based on 57 partial large subunit (LSU) sequences. Figures on branches indicate bootstrap values and asterisks (*) indicate the species used as an outgroup.

LSU data placed *Gymnosporangium* and *Kuehneola* (having pedicelled teliospores) into one clade and the rest (having sessile teliospores) into another (Fig. 1.4). There was strong support (bootstrap value = 96%) for the grouping of those having sessile teliospores, including *Melampsora*. This somewhat differed from the results of Cullings and Vogler (1998) and of Maier *et al.* (2003), that the genetic distance between *Melampsora* and other rusts may be greater than that among other rusts.

Implications for the Evolution of the Rust Life Cycle

The molecular evidence from rDNA sequences of rusts shows that the rust fungi

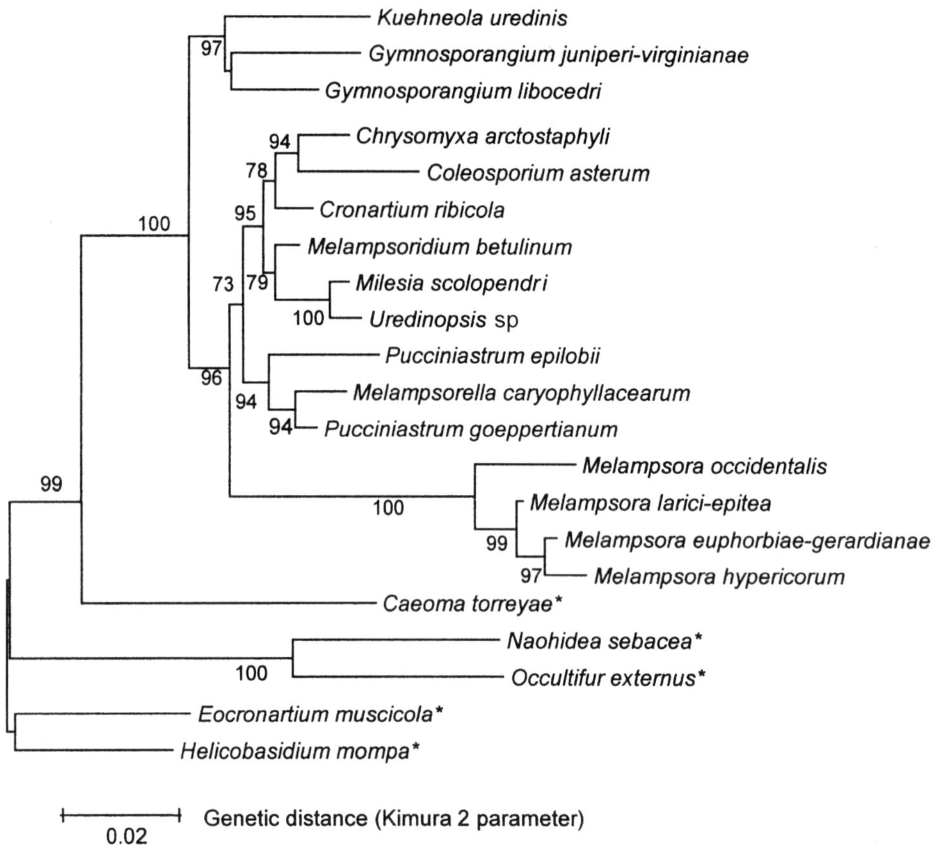

Fig. 1.4. Neighbour-joining tree based on combined small subunit (SSU) and large subunit (LSU) rDNA sequence data. Figures on branches indicate bootstrap values and asterisks (*) indicate the species related to *Uredinales* and used as an outgroup.

are monophyletic (Berbee and Taylor, 2001; Tehler *et al.*, 2003) and that the complex, multi-spore form life cycle may have evolved before the ancestral rust diverged into different groups (having either sessile or pedicelled teliospores) (Figs 1.3 and 1.4). The neighbour-joining trees derived from rDNA sequences placed the rust genera which do not infect conifers at their aecial stage (they also have pedicelled teliospores) into a distinct clade (Maier *et al.*, 2003; Figs 1.3 and 1.4). Gymnosperms have a much older fossil record (~320 million years; Doyle, 1998) than angiosperms (at most 130 million years; Crane *et al.*, 1995). Results from molecular studies suggested that gymnosperms are monophyletic, evolved from a common ancestor (Chaw *et al.*, 1997). Angiosperms are also considered to be monophyletic (Doyle, 1998). One can assume that there had been a close host–parasite association between the ancestral *Pinaceae* and the ancestral rusts before flowering plants–rust fungi relationships were established.

Implications for the Relationships of *Melampsora* with Other Rusts

In explanation of the distant relationships between *Melampsora* and other rusts (based on LSU data), Maier *et al.* (2003) pointed out that *Melampsora* either split from other rust genera a long time ago or its mutation rate has been considerably higher than that of the others. The life cycle of *Melampsora* is similar to that of other rusts having sessile teliospores, i.e. producing up to five spore stages, the majority forming aeciospores on different hosts and completing a single sexual life cycle annually. Therefore, it is unlikely that the rate of mutation has been much higher in *Melampsora* than in other rusts. It appears that *Melampsora* split from ancestors of other rust genera at early stages of evolution of rust fungi. Our results from the combined LSU and SSU data (Fig. 1.4) suggest that the split between the rusts having sessile teliospores and those having pedicelled teliospores had occurred before the split between *Melampsora* and other rusts having sessile teliospores. Phylogenetic studies carried out by Sjamsuridzal *et al.* (1999), using SSU sequences of rDNA, suggested that the fern rusts are not the most basal or 'primitive' rust fungi. Both the molecular evidence and the broad host specificity of the genus at the aecial stage indicate that the genus *Melampsora* arose from the earliest line of ancestral rusts.

Implications on the Timing of Divergence

The molecular clock concept is one of the most intriguing, yet hotly debated, topics in evolutionary biology. It was first proposed by Zuckerkandl and Pauling (1965), based on the approximate constancy of amino-acid substitution in the make-up of human haemoglobin. The molecular clock hypothesis assumes that the rate of amino-acid or nucleotide substitution is approximately constant over evolutionary time. Much of the controversy surrounding the molecular clock approach is whether the clock is accurately calibrated. Fossil evidence can be crucial in calibration of the molecular clock. Fungal fossils are rare in nature. Fortunately, some early fossils discovered so far provide precious clues to the evolutionary path of the fungi over the geological time span (see Berbee and Taylor, 2001). One such example is the amazingly well-preserved fossil ascomycetous fungus from 400 million-year-old Rhynie Chert fossil beds in Scotland (Taylor *et al.*, 1999).

Using SLU rDNA sequence data and a re-calibrated molecular clock, Berbee and Taylor (2001) estimated that *Ascomycetes* diverged from *Basidiomycetes* some 600 million years ago. According to their gene tree, the pine stem rust *Cronartium ribicola* and the mushroom fungus *Boletus satanas* split sometime around 500 million years ago (Ordovician Period). If such a time scale is applied to Fig. 1.3, we can roughly estimate that: (i) *Melampsora* may have split from other rusts some 250 million years ago; (ii) many rust genera differentiated between 100 and 50 million years ago; and (iii) speciation in *Melampsora* occurred over the past 50 million years.

Conclusions

Some 20 years ago, Hennen and Buriticá (1980) commented that there was 'little direct knowledge of the evolution of rusts and few ways of testing evolutionary hypotheses'. Since then, the rapid advances of DNA technology have revolutionized our understanding of the evolutionary paths of the living organisms on Earth. So far, rDNA sequence information has provided some unprecedented insights into the evolutionary relationships among the rust fungi. The rDNA sequence data suggest that the multi-spore form of life cycle is likely to have evolved at the very early stages of rust evolution. Molecular evidence also indicates that the genus *Melampsora* evolved early as a distinct group from ancestral rusts. Further molecular data from more and wider regions of the rust genome will provide answers to many intriguing questions on the evolution of the rust fungi.

Acknowledgements

This study was funded by the Department for the Environment, Food and Rural Affairs (DEFRA), UK and the European

Union. Rothamsted Research receives grant-aided support from the Biotechnology and Biological Sciences Research Council of the UK.

References

Ando, K. (1984) Phylogeny of the fern rusts (Uredinopsis, Milesina and Hyalopsora). *Transactions of the Mycological Society of Japan* 25, 295–304.

Arthur, J.C. (1929) *The Plant Rusts (Uredinales)*. John Wiley & Sons, New York.

Azbukina, Z.M. (1974) *Rust Fungi of the Soviet Far East* [in Russian]. Nauka Publishers, Moscow.

Begerow, D., Bauer, R. and Oberwinkler, F. (1997) Phylogenetic studies on nuclear large subunit ribosomal DNA sequences of smut fungi and related taxa. *Canadian Journal of Botany* 75, 2045–2056.

Berbee, M.L. and Taylor, J.W. (2001) Fungal molecular evolution: gene trees and geologic time. In: McLaughlin, D.J., McLaughlin, E.G. and Lemke, P.A. (eds) *The Mycota, VIIB, Systematics and Evolution*. Springer-Verlag, Berlin, pp. 229–245.

Berres, M.E., Szabo, L.J. and McLaughlin, D.J. (1995) Phylogenetic relationships in auriculariaceous basidiomycetes based on 25S ribosomal DNA sequences. *Mycologia* 87, 821–840.

Bessey, E.A. (1950) *Morphology and Taxonomy of Fungi*. The Blakiston Co., Philadelphia.

Castagne, L. (1843) Observations sur quelques plantes acotylédonées et dans les soustribus des Nemasporées et des Aecidinées, recueillies dans le Départment des Bouches-du-Rhône 2, 18. Aix.

Chaw, S.M., Parkinson, C.L., Cheng, Y.C., Vincent, T.M., and Palmer, J.D. (1997) Seed plant phylogeny inferred from all three plant genomes: monophyly of extant gymnosperms and origin of Gnetales from conifers. *Proceedings of the National Academy of Sciences* 97, 4086–4091.

Crane, P.R., Friis, E.M. and Pederson, K.R. (1995) The origin and early diversification of angiosperms. *Nature* 374, 27–33.

Cullings, K.W. and Vogler, D.R. (1998) A 5.8S nuclear ribosomal RNA gene sequence database: applications to ecology and evolution. *Molecular Ecology* 7, 919–923.

Cummins, G.B. and Hiratsuka, Y. (1983) *Illustrated Genera of Rust Fungi*, revised edn. APS Press, St Paul, Minnesota.

Cummins, G.B. and Hiratsuka, Y. (2003) *Illustrated Genera of Rust Fungi*, 3rd edn. APS Press, St Paul, Minnesota.

Dietel, P. (1900) Uredinales. In: Engler, A. and Prantl, K. (eds) *Die Natürlichen Pflanzenfamilien*, 1(1). Verlag von Wilhelm Engelmann, Leipzig, pp. 24–81.

Dietel, P. (1928) Hemibasidii (Ustilagiales und Uredinales). In: Engler, A. and Prantl, K. (eds) *Die Natürlichen Pflanzenfamilien*, 2(6). Verlag von Wilhelm Engelmann, Leipzig, pp. 24–98.

Doyle, J.A. (1998) Molecules, morphology, fossils, and the relationship of angiosperms and Gnetales. *Molecular Phylogenetics and Evolution* 9, 448–462.

Durrieu, G. (1980) Phylogeny of Uredinales on Pinaceae. *Report of Tottori Mycological Institute* 18, 283–290.

Fischer, E. (1898) Entwicklungsgeschichtliche Untersuchungen über Rostpilze. *Beiträge zur Kryptogamenflora Schweiz* 1, 1–120.

Gäumann, E. (1959) *Die Rostpilze Mitteleuropas. Beiträge zur Kryptogamenflora Schweiz* 12. Buchdruckerei Büchler and Co., Bern, Germany.

Hawksworth, D., Kirk, P., Sutton, B. and Pegler, D. (1995) *Ainsworth's and Bisby's Dictionary of the Fungi*, 8th edn. CAB International, Wallingford, UK.

Hennen, J.F. and Buriticá, P.C. (1980) A brief summary of modern rust taxonomic and evolutionary theory. *Reports of Tottori Mycological Institute* 18, 243–256.

Hiratsuka, Y. and Sato, S. (1982) Morphology and taxonomy of rust. In: Scott, K.J. and Charkravorty A.K. (eds) *The Rust Fungi*. Academic Press, London, pp. 1–36.

Jackson, H.S. (1935) The nuclear cycle in *Herpobasidium filicinum* with a discussion of the significance of homothallism in Basidiomycetes. *Mycologia* 27, 553–572.

Klebahn, H. (1904) *Die wirtswechselnde Rostpilze. Versuch einer Gesamtdarstellung ihrer biologischen Verhältnisse*. Borntraeger, Berlin.

Laundon, G.F. (1973) Uredinales. In: Ainsworth G.C., Sparrow, F.K. and Sussman, A.S. (eds) *The Fungi, an Advanced Treatise, IVB, A Taxonomic Review with Keys: Basidiomycetes and Lower Fungi*. Academic Press, London, pp. 247–279.

Leppik, E.E. (1965) Some viewpoints on the phylogeny of rust fungi. V. Evolution of biological specialization. *Mycologia* 57, 6–22.

Leppik, E.E. (1972) Evolutionary specialization of rust fungi (Uredinales) in the Leguminoaceae. *Annales Botanici Fennici* 9, 135–148.

Maier, W., Begerow, D., Weiß, M. and Oberwinkler, F. (2003) Phylogeny of the rust fungi: an approach using large subunit ribosomal DNA sequences. *Canadian Journal of Botany* 81, 12–23.

Mendgen, K. (1997) Uredinales. In: Carroll, G. and Tudzynski, P. (eds) *The Mycota, VB, Plant Relationships*. Springer, Heidelberg, pp. 79–94.

Savile, D.B.O. (1955) A phylogeny of Basidiomycetes. *Canadian Journal of Botany* 33, 60–104.

Savile, D.B.O. (1976) Evolution of the rust fungi (Uredinales) as reflected by their ecological problems. *Evolutionary Biology* 9, 137–207.

Sjamsuridzal, W., Nishida, H., Ogawa, H., Kakishima, M. and Sugiyama, J. (1999) Phylogenetic positions of rust fungi parasitic on ferns: evidence from 18S rDNA sequence analysis. *Mycoscience* 40, 21–27.

Sydow, P. and Sydow, H. (1915) *Monographia Uredinearum III*. Fratres Borntraeger, Leipzig.

Taylor, T.N., Hass, H. and Kerp, H. (1999) The oldest fossil ascomycetes. *Nature* 399, 648.

Tehler, A., Little, D.P. and Farris, J.S. (2003) The full-length phylogenetic tree from 1551 ribosomal sequences of chitinous fungi, *Fungi*. *Mycological Research* 107, 901– 916.

Wilson, M. and Henderson, D.M. (1966) *The British Rust Fungi*. Cambridge University Press, Cambridge.

Zuckerkandl, E. and Pauling, L. (1965) Molecules as determinants of evolutionary history. *Journal of Theoretical Biology* 8, 357–366.

2 A Brief Review of *Melampsora* Rusts on *Salix*

Ming Hao Pei
Rothamsted Research, Harpenden, Hertfordshire AL5 2JQ, UK

Melampsora on Willow

Rust fungi found on willows (and also on poplars) belong to the genus *Melampsora*. The main characteristic of *Melampsora* species is formation of crust-like telia which comprise sessile, laterally adherent single-celled teliospores. In fact, the name *Melampsora* originated from the Greek words *Melas* (black) and *psora* (scab). Of some 80–100 *Melampsora* spp. (Hiratsuka and Sato, 1982; Hawksworth *et al.*, 1995), more than half have been described on *Salicaceae*.

Host Genus *Salix*

The family of *Salicaceae* consists of two major genera: *Populus* L. (poplar) and *Salix* L. (willow). *Populus* is wind pollinated, has loose and pendulous catkins, bears numerous stamens (5–80) and produces thin-walled pollen grains. Its flowers also lack glands. *Salix*, on the other hand, is insect-pollinated, forms compact and erect catkins, usually has two stamens (1–7) and produces thick-walled pollen grains. Flowers of *Salix* usually have nectar-secreting glands. The genus *Chosenia* Nakai, which has only a single species *Ch. arbutifolia* (Pallas) A. Skvortsov (syn. *Salix arbutifolia* Pallas), is similar to *Salix* in many characteristics but resembles *Populus* in being wind-pollinated and having pendulous catkins.

Willow species are dioecious, tend to hybridize relatively easily, and often vary morphologically at different stages of growth. Probably for these reasons, *Salix* taxonomy is notoriously difficult. According to different authorities, there are between 300 and 500 *Salix* species worldwide. Within the genus, the species are grouped into 2–4 subgenera and, within a subgenus, they are further grouped into various sections (Skvortsov, 1968; Dorn, 1976; Argus, 1986). There is a general consensus among taxonomists on the recognition of subgenus *Salix*, which comprises the species typically forming trees and large shrubs. The rest of *Salix* species (more than two-thirds of the total) range from dwarf shrubs to small trees and are extremely variable in their morphology. Within these willows, Skvortsov (1968) recognized two subgenera: subgenus *Vetrix* and subgenus *Chamaetia*, but stated that boundaries can be drawn between them only with stipulations and conditionally, and that they are considerably more closely related to each other than to the subgenus *Salix*. Dorn (1976) recognized only one subgenus, *Vetrix*, to include all the species outside the subgenus *Salix*. In this chapter, we adopt

Dorn's (1976) treatment at subgenus level and Bean's (1980) treatment in classification of sections and species. The names of willow species, which were not included in Bean (1980), were adopted from Skvortsov (1968), Dorn (1976) and Wang and Fang (1984).

Geographic distribution of *Salix* species

Species diversity of willow is richest in China (*c*. 270 species) and then in the former Soviet Union (*c*. 120 species) (Argus, 1997). There are some 65 willow species in Europe and over 100 species in North America. Willows are not native to Australasia, and only a single species, *S. humboldtiana*, is naturally distributed in South America.

In general, the natural distribution of a *Salix* species is confined to a certain geographical region and the willow species occurring in Europe usually differ from those native to the Far East and to North America. For the willows occurring in arctic, subartic or alpine regions, there are often no distinct boundaries between continents.

Taxonomy of Willow Rusts

The taxonomy of *Melampsora* on *Salix* has long been in a state of confusion. Many *Melampsora* species on willows were described in the late 19th to early 20th century. By 1915, a total of 22 *Melampsora* species on *Salix* were complied in the *Monographia Uredinearum III* by Sydow and Sydow (1915). Willow *Melampsora* species were mainly established based on their morphology, alternate hosts and the telial host range. Some of the great difficulties in identification of the species on willow arise from the fact that the host ranges of different rust species often overlap and there is no clear morphological distinction between those that are separated by having different alternate hosts. Usually, only one or two spore stages can be found at the time of observation and there is no information on the alternate hosts.

Aecial hosts of willow *Melampsora* include conifers, dicotyledonous and monocotyledonous plants. Alternative hosts can be identified by inoculating suspected plants in the spring, using overwintered telia. However, such work can be very demanding because, even if the potential alternate hosts have been identified, maturity of the teliospores, conditions of overwintering and teliospore germination, and growing conditions of the plants may still affect successful outcome. With many willow rust species, only uredial and telial stages have been described to the present day, and it is not known whether, and which, aecial hosts are involved in their life cycle.

Facing the problems in identification of *Melampsora* on *Salix*, Hylander *et al.* (1953) recognized *M. epitea* Thüm. as a complex species to include those species (*sensu stricto*) similar in morphology. Wilson and Henderson (1966) adopted *M. epitea* as a collective species made up of various races and groups of races which are morphologically indistinguishable, but alternate on different hosts. They distinguished two varieties: var. *reticulatae* alternates on *Saxifraga* and has larger urediniospores and paraphyses than var. *epitea*, which includes all other forms alternating on a number of different hosts. Hiratsuka and Kaneko (1982) and Pei *et al.* (1993, 1996) also adopted Wilson and Henderson's treatment. However, a remaining problem is that, because *M. epitea* contains many widely different groups, the name means very little in terms of the biological characteristics and genetic identity for the group or, in a narrow sense, the species.

In this chapter, for the convenience of summarizing existing information, the species names are used in the narrow sense (*sensu stricto*) and the complex species *M. epitea* is specified as '*M. epitea* complex'.

Morphology of Willow Rusts

A summary of the morphological characteristics of willow rust is listed in Table 2.1.

Table 2.1. Morphological characteristics of *Melampsora* species/forms on *Salix*.

Species/form	Size of aecia (mm)	Size of aeciospores (μm)	Position and size of uredinia	Size and wall of urediniospores	Size and wall of paraphyses	Position and size of telia	Size and wall of teliospores
M. abieti-caprearum Tubeuf	0.25–0.75	19–21 × 12–14; wall thickened at the apex in N. American population	Hypophyllous; *c.* 0.5 mm	13–20 × 12–16 μm; evenly echinulate; wall 1.5 μm thick	38–410 × 16–24 μm; wall 1.5–3 μm thick	Hypophyllous; sub-epidermal	19–30 × 9–12; wall 1 μm thick
M. allii-fragilis Kleb.	2 × 0.5–1	18–25 × 12–19	Mainly hypophyllous; *c.* 0.5 mm	22–33 × 13–15; smooth at apex; wall up to 3 μm thick	50–70 × 15–20 μm; wall 3–5 μm thick	Mainly epiphyllous; sub-cuticular; 0.25–1.5 mm	30–48 × 7–14; wall 1 μm thick
M. amygdalinae Kleb.	Up to 10	18–23 × 14–19	Hypophyllous; *c.* 0.5 mm	19–32 × 11–15; smooth at apex; wall 1.5 μm	35–80 × 10–18 μm; wall 2–5 μm thick	Hypophyllous; sub-epidermal: 0.3–0.5 mm	18–42 × 7–14; wall 1 μm thick
M. arctica Rostr.	0.3–0.6	16–17 × 13–24	Mainly hypophyllous	17–25 × 15–22; evenly echinulate; wall 2.5–4 μm thick	55–60 × 17–24 μm; wall 3–7 μm thick	Epiphyllous; sub-epidermal	20–45 × 11–14; wall 1 μm thick
M. galanthi-fragilis Kleb.	1–2	17–22 × 14–19	Mainly hypophyllous; 0.5–1 mm	25–38 × 12–16; smooth at apex; wall up to 3 μm thick	50–70 × 17–23 μm; wall 2–5 μm thick	Mainly epiphyllous; sub-cuticular; 0.25 mm to, rarely, 1 mm across	25–45 × 8–15; wall 1 μm thick
M. caprearum Thüm.	0.25 × 1	15–25 × 12–17	Hypophyllous; 1–2 mm or up to 5 mm on juvenile leaves	14–21 × 13–15 μm; evenly echinulate; wall 2–2.5 μm thick	50–60 × 18–26 μm; thickened at apex, 5–6 μm	Epiphyllous; sub-cuticular; 1 mm or more across	30–45 × 7–14; wall porous and up to 10 μm thick at apex and 1 μm thick at sides
M. caprearum Thüm. (Japanese form)			Mainly hypophyllous	16–26 × 14–23 μm; evenly echinulate; wall 2.5–4 μm thick	50–70 × 18–26, thickened at apex, up to 6 μm	Epiphyllous; sub-cuticular;	25–45 × 10–17; wall porous and up to 10 μm thick at apex and 1 μm thick at sides
M. chelidonii-pierotii Matsumoto		13–20 × 10–16	Mainly hypophyllous	16–26 × 12–17 μm; evenly echinulate; wall 1.5–2 μm thick	Thickened at apex, up to 7 μm	Mainly epiphyllous; sub-cuticular	18–60 × 5–11; wall 0.8 μm thick

continued

Table 2.1. *Continued.*

Species/form	Size of aecia (mm)	Size of aeciospores (μm)	Position and size of uredinia	Size and wall of urediniospores	Size and wall of paraphyses	Position and size of telia	Size and wall of teliospores
M. choseniae Azb. et Kov.			Hypophyllous; 0.3–0.8 mm	14–25 × 14–18 μm; evenly echinulate; wall 1.2–2 μm thick	25–35 × 12–18 μm; wall 2–2.5 μm thick	Mainly epiphyllous;	28–64 × 10–14; wall 1.5–2.5 μm thick
M. coleosporioides Dietel			Mainly hypophyllous	18–29 × 13–19 μm; evenly echinulate or occasionally smooth at apex; wall 1.5–2 μm thick	35–450 × 14–20 μm	Hypophyllous; mainly sub-epidermal; 0.5–1 mm	20–45 × 6.5–15; wall 0.5–1 μm thick
M. dimorphospora Kaneko and Hiratsuka			Mainly hypophyllous; 0.2–1 mm	19–30 × 15–25 μm; evenly echinulate or verrucose; wall 2.5–3 μm thick	35–103 × 22–30 μm; wall 3–4 μm thick	Hypophyllous; sub-epidermal; 0.2–0.5 mm	20–50 × 7.5–13.5, occasionally septate; wall *c.* 1.5 μm thick
M. epiphylla Dietel		15–20 × 13–18	Mainly hypophyllous	12–20 × 10–15 μm; evenly echinulate; wall 1.5–2 μm thick		Mainly epiphyllous; sub-cuticular	22–50 × 7–12.5; porous and 3.5–5 μm thick at apex, 0.5–1 μm at sides
M. epitea Thüm. (*M. epitea* complex)		15–25 × 10–21	Amphigenous; 0.5–1.5 mm	12–25 × 10–18 μm; evenly echinulate; wall 1.5–3 μm thick	35–80 (occasionally 90) × 15–24 (occasionally 35) μm	Mainly hypophyllous; sub-epidermal	25–50 × 7–14; wall 2 μm or slightly thicker at apex
M. epitea f. sp. *tsugae*	2–4	14–21 × 15–24, wall thickened at apex	Hypophyllous; 0.05–0.15 mm	12–17 × 14–20 μm; evenly echinulate; wall 0.6–1 μm thick	10–18 × 20–45 μm; wall up to 3 μm thick	Hypophyllous; sub-epidermal; 0.15–0.5 mm	6–14 × 16–30; wall 1 μm thick
M. euonymi-caprearum Kleb.	Up to 1.5	18–23 × 14–19; wall up to 5	Hypophyllous; *c.* 0.5 mm	18–23 × 14–19 μm; evenly echinulate; wall up to 5 μm thick	50–70 × 18–25 μm; wall up to 8 μm thick at apex, 2 μm at sides	Hypophyllous; sub-epidermal; *c.* 0.5 mm	25–40 × 7–13; wall 1 μm thick
M. humilis Dietel			Mainly hypophyllous	16–20 × 13–18 μm; evenly echinulate; wall 2 μm thick	Up to 22 μm wide, wall 2–2.5 at sides, slightly thickened at apex	Mainly hypophyllous; sub-epidermal	13–33 × 6.5–15; wall 1 μm thick, sometimes slightly thicker at apex

M. kamikotica Kaneko et Hiratuska			Amphigenous; 0.3–0.5 mm	20–33 × 15–18 μm; evenly echinulate; wall 2–2.5 μm thick at sides, 5–6.5 μm thick at apex	40–75 × 15–25 μm; wall 5–6.5 μm thick at apex, 2–2.5 μm at sides	Amphigenous; sub-epidermal; 0.3–0.8 mm	25–53 × 10–16; wall 1 μm thick
M. kiusiana Hiratsuka			Hypophyllous	11–17 × 10–15 μm; evenly echinulate; wall 2.5 μm thick		Mostly epiphyllous; sub-epidermal	30–62 × 7–10; wall 1 μm thick, slightly thicker at apex
M. kupreviczii Zenk.			Hypophyllous; 0.25–0.5 mm	10–15 × 10–12 μm; evenly echinulate; wall up to 2 μm thick	17–20 μm long; wall 2–2.5 μm thick	Epiphyllous; sub-epidermal; 0.25–0.5 mm	30–37 × 10–15; wall 0.5–0.7 μm thick
M. lapponum Linfors			Hypophyllous; *c.* 0.25 mm	20–21 × 15–16 μm; evenly echinulate; wall	15–30 μm wide: wall 1–4 μm thick	Mainly epiphyllous; sub-epidermal; 0.2–0.5 mm	30–50 × 6–12; wall 1 μm thick
M. larici-pentandrae Kleb.		19–27 × 18–20, wall up to 3 μm thick	Mainly hypophyllous; *c.* 1 mm	26–44 × 12–16 μm; smooth at apex	40–60 × 12–22 μm	Hypophyllous; sub-epidermal; *c.* 0.5 mm	19–30 × 9–12; wall 1 μm thick
M. larici-epitea Kleb.	0.25 × 1	18–26 × 13–20	Hypophyllous; 0.25–1.5 mm	12–25 × 9–19 μm; evenly echinulate; wall 1.5–3.0 μm thick	35–80 × 15–24 μm; wall 3–5 μm, sometimes up to 10 μm thick	Mostly hypophyllous; sub-epidermal; 0.25–1 mm	20–50 × 7–14; wall 1 μm thick
M. larici-urbaniana Matsumoto	0.5–1.5	15–25 × 10–21	Hypophyllous	15–27 × 13–17 μm; evenly echinulate; wall 2–2.5 μm thick at sides, 4–7.5 μm at apex	Slightly thickened at apex	Hypophyllous; sub-epidermal	35–70 × 8–16; wall 1–2 μm thick at sides, 2.5–3.5 μm thick at base and apex
M. microsora Dietel		15–25 × 13–21	Hypophyllous; less than 0.2 mm	16–26 × 11–15 μm; evenly echinulate; wall 1.5–2 μm thick	Wall up to 8 μm thick at apex, 1.5–2 μm at sides	Hypophyllous, sub-epidermal	20–35 × 8.5–20; wall 1 μm thick
M. paradoxa Diet. & Holw.			Mainly hypophyllous; 0.3–0.5 mm	15–19 × 17–24 μm; evenly echinulate; wall 2.5–3.5 μm thick	50–70 × 22–26 μm; wall 3–5 μm thick	Amphigenous; sub-epidermal; *c.* 0.5 mm	11–14 × 29–43; wall 1 μm thick
M. repentis Plowr.	0.1–0.2	15–220 × 18–27	Mainly hypophyllous; 0.25–0.5 mm	13–17 × 12–14 μm; evenly echinulate; wall 1.5 μm thick	40–70 × 16–20 μm; wall 3–5 μm thick	Mainly hypophyllous; sub-epidermal; 0.2–0.5 mm	16–48 × 7–14; wall 1 μm thick

continued

Table 2.1. *Continued.*

Species/form	Size of aecia (mm)	Size of aeciospores (μm)	Position and size of uredinia	Size and wall of urediniospores	Size and wall of paraphyses	Position and size of telia	Size and wall of teliospores
M. reticulatae Blytt	In groups, 10–20	15–20 × 11–15	Hypophyllous; 0.5–1 mm	17–35 × 15–23 μm; evenly echinulate; wall 2–3.5 μm thick	60–90 × 18–41 μm; wall up to 10 μm thick	Mainly epiphyllous; sub-epidermal; 0.3–0.5 mm	35–44 × 10–13; wall 1 μm thick
M. ribesii-epitea Kleb.	0.5–1	16–25 × 14–20	Hypophyllous; 0.5–1 mm	16–20 × 14–18 μm; evenly echinulate; wall 3–3.5 μm thick	55–70 × 16–24 μm; wall 2.5–4 μm thick	Hypophyllous; sub-epidermal; up to 0.5 mm	20–30 × 7–11; wall 1 μm thick
M. ribesii-purpureae Kleb.	0.5–1.5	17–24 × 15–24	Mainly hypophyllous; up to 5 mm on juvenile leaves	15–23 × 14–19 μm; evenly echinulate; wall 2.5 μm thick	40–70 × 15–21 μm; wall 3–5 μm thick	Mainly hypophyllous; sub-epidermal; 0.25–0.5 mm	25–30 × 7–10; wall 1 μm thick
M. ribesii-viminalis Kleb.	0.5–1.5	15–23 × 12–19	Hypophyllous; *c.* 0.25 mm	15–19 × 14–16 μm; evenly echinulate; wall 2 μm thick	50–70 × 18–25 μm; wall 1–2 μm thick	Epiphyllous, sub-cuticular; 0.25–0.5 mm	25–40 × 7–10; wall 1 μm thick
M. salicis-albae Kleb.	1.5	18–23 × 14–17	Mainly hypophyllous; 0.5 mm; up to 5 mm on young twigs and up to 2 mm on young leaves	20–36 × 11–17 μm; smooth at apex; wall 2 μm thick	50–70 × 15–20 μm; wall 2–2.5 μm thick	Mainly hypophyllous; sub-epidermal	25–45 × 7–10; wall 1 μm thick
M. salicis-cavaleriei Tai			Epiphyllous; 0.2–0.5 mm	18–37 × 11–18; angular? wall 1.5–2 μm thick		Epiphyllous; sub-epidermal; 0.2–0.5 mm	36–69 × 7–11; wall 1–1.5 μm thick
M. salicis-cupularis Wang			Hypophyllous	21–23 × 17–18; wall 1.5–2 μm thick	c. 98 × 24–35 μm; wall 7 μm thick	Hypophyllous; sub-epidermal	28–39 × 11–14
M. salicis-warburgii Sawada	1	17–26 × 15–18	Amphigenous; 0.1–0.3 mm	15–18 × 13–28 μm; evenly echinulate; wall 2–2.5 μm thick	30–65 × 12–16 μm; wall up to 7 μm thick at apex	Mostly hypophyllous	34–42 × 8–15; wall 1–1.8 μm thick
M. yezoensis Miyabe et Matsumoto			Mainly hypophyllous	18–29 × 13–18 μm; evenly echinulate; wall 2–2.5 μm thick	Wall up to 8 μm thick at apex	Mostly hypophyllous; mostly sub-cuticular	20–35 × 6.5–12.5; wall 0.5–0.8 μm thick
Caeoma salicis-miyabeana Kaneko et Hiratsuka	0.2–1.5	17.5–29 × 13.5–18.5					

Uredinia and urediniospores

Broadly, there are two types of urediniospores in willow *Melampsora*, one type having relatively small urediniospores with an evenly echinulate surface and the other having relatively large urediniospores with a smooth apex. In Europe, all the *Melampsora* species occurring on tree willows (subgenus *Salix*), i.e. *M. allii-fragilis*, *M. amygdalinae*, *M. galanthi-fragilis*, *M. larici-pentandrae*, produce urediniospores having a smooth surface at the apex. In contrast to the European *Melampsora* species on subgenus *Salix*, the species occurring on tree willows in Japan and the Far East, *M. chelidonii-pierotii*, *M. larici-urbaniana*, *M. microsora*, *M. salicis-warbugii* and *M. yezoensis*, produce evenly echinulate urediniospores. The urediniospores of *M. coleosporioides* are mostly evenly echninulate, but the rust also occasionally produces uredinospores that are smooth at the apex. From the existing description, it is not clear whether urediniospores of *M. salicis-cavaleriei*, which was found on a tree willow in China, have an unevenly echinulate surface. In *M. dimorphospora* Kaneko and Hiratsuka, urediniospores are either echinulate or verrucose. In North America, all the rusts on willows have evenly echinulate urediniospores (all could be placed under the *M. epitea* complex).

The size and position of uredinia, thickness of spore walls and, to a lesser extent, the morphology of paraphyses also serve as criteria to distinguish the species of willow rusts. Some species, such as *M. caprearum* and *M. ribesii-purpureae* form uredinia up to 5 mm in diameter on juvenile leaves. In contrast, *M. microsora* produces uredinia less than 0.2 mm in diameter (Hiratsuka and Kaneko, 1982). The majority of the species produce uredinia mainly on the lower sides of leaves (hypophyllous). *Melampsora salicis-cupularis* is the only species described as forming uredinia on the upper sides of host leaves (epiphyllous). For most species, the walls of urediniospores are evenly thick, measuring 1.5–2 μm. However, urediniospore walls in *M. kamikotica* and *M. larici-urbaniana* are thickened at the apex. Several species, such as *M. arctica* and *M. euonymi-caprearum*, produce urediniospores having even, but slightly thick walls. Helfer (1992) described the spine density on the surface of urediniospores in *M. lapponum* as considerably higher than that of other willow rusts recorded in Britain.

Telia and teliospores

Historically, teliospore morphology formed the basis of classification at family and sub-family level under the order of *Uredinales*. A classic example is the separation of two major groups, *Pucciniaceae* and *Melampsoraceae*, based on the presence and absence of a pedicel (Arthur, 1934). With willow rusts, the position of telia on the host and the thickness of spore walls are important features in species identification. The species forming epiphyllous telia include *M. allii-fragilis*, *M. arctica*, *M. gallanti-fragilis*, *M. choseniae*, *M. caprearum*, *M. epiphylla*, *M. kiusiana*, *M. kupreviczii*, *M. lapponum*, *M. reticulatae*, *M. ribesii-viminalis* and *M. salicis-cavalariei*. The remainder are either hypophyllous or amphigeneous (on both sides of leaves). Some variation in the appearance of telia was observed within *M. larici-epitea*. For example, telia of f. sp. *larici-daphnoides* on *S. daphnoides* were button-shaped, round and confluent, usually with a sunken centre, compared with irregularly shaped telia of the other *M. larici-epitea* samples (Pei *et al.*, 1993). In some species, teliospores form above the epidermal cells of the hosts (subcuticular), while in others teliospores are produced beneath the epidemal cells (subepidermal).

One of the most prominent morphological features in willow rusts is the thickness of teliospore walls. The majority of willow rust species have teliospores with thin, evenly thickend walls. However, *M. caprearum* and *M. epiphylla* have teliospore walls thickened at the apex. Teliospore walls in *M. larici-urbaniana* are slightly thicker both at the apex and at the base.

Basidiospores and spermogonia

Basidia and basidiospores are produced from overwintered teliospores in spring. Each basidium becomes four-septate and produces four spherical basidiospores, which are vigorously discharged into the air. Basidia and basidiospores are thin walled and relatively short lived. Spermogonia are either subcuticular or subepidermal, and often form as groups, especially on dicotyledonous plants. Most willow *Melampsora* species lack descriptions of basidia, basidiospores and spermogonia, and therefore these features are not considered to be taxonomically significant.

Aeciospores

The surface of aeciospores of willow rusts is ornamented with numerous warts, each resembling stacked rings (verrucose). The aeciospore walls of most species are evenly thick. Exceptions can be found in *M. epitea* f. sp. *tsugae* and the North American populations of *M. abieti-caprearum*, both forming aeciospores having a thickened wall at the apex.

Life Cycle, Host Alternation and Overwintering

Life cycle

Typically, *Melampsora* species are macrocyclic, producing five different spore stages, i.e. basidiospores, spermatia, aeciospores, uredinospores and teliospores, during their life cycle. The rusts spread on the willow hosts through repeated cycling of urediniospores (usually within 2 weeks from one cycle to the next) during the growing season. They develop teliospores in late summer and autumn, and overwinter on fallen willow leaves. In spring, teliospores germinate to produce basidiospores that infect aecial hosts. Meiosis takes place in the early stages of teliospore germination. In *M. larici-epitea*, spermogonia start to appear on the alternate host larch 6–7 days after inoculation with basidiospores (Pei *et al.*, 1996, 1999). Spermogonial formation is associated with production of light-brown honeydew and a strong fragrance of incense which attracts insects and facilitates fertilization between spermogonia. Fertilization between spermogonia on aecial hosts results in formation of aeciospores (in the case of *M. larici-epitea*, aecia start to form 2–3 days after the appearance of spermogonia). Studies (Pei *et al.*, 1999) have shown that in *M. larici-epitea*, there are two mating types among the spermogonia, and only the combination between one and the other leads to formation of aecia. The aeciospores infect the telial host willow to produce urediniospores. The basidiospores and spermatia are monokaryotic, having a single haploid nucleus, while aeciospores, urediniospores and early phases of teliospores are dikaryotic, containing two nuclei.

For 11 species/forms of willow rusts, only uredinio- and teliospore stages are known. In recent years, the stem-infecting form (SIF) on *S. viminalis* has been studied in some detail, due to its importance in short-rotation coppice plantations for renewable energy. The SIF is highly specialized in its pathogenicity, being found only on willow genotypes belonging to *S. viminalis* L., which also serves as the telial host of *M. larici-epitea* (Pei *et al.*, 1995, 1996). *M. larici-epitea* infects only expanded willow leaves and readily completes a full sexual life cycle, during which five spore stages are produced (Pei *et al.*, 1993, 1996). Unlike *M. larici-epitea*, SIF infects shoot-tips and young stems, and causes stem cankers (Pei *et al.*, 1995). In the field, SIF occurs as the uredinial stage and overwinters in buds or stems of infected willows. Although SIF can produce teliospores morphologically similar to those of *M. larici-epitea* under laboratory conditions, field observations and molecular evidence showed that it is an asexual population and may have a clonal lineage (Pei *et al.*, 1995; Pei and Ruiz, 2000). Our recent result showed that the stem-infecting form shares the same sequences with *M. larici-epitea* and

M. caprearum in the large subunit (LSU) region of ribosomal DNA (rDNA) and only differs slightly in the internally transcribed spacer (ITS) region of rDNA (M.H. Pei and C. Bayon, unpublished results). This indicates that the SIF is likely to have evolved relatively recently from its heteroecious ancestor by shifting its niche from fully expanded leaves to very young leaves and shoots, a process known as 'sympatric speciation'. Its ability to overwinter as an asexual uredinial stage on the telial host may have facilitated the shortening of its life cycle.

Host alternation

In a review of the rust fungi occurring in central and western Europe, Gäumann (1959) summarized that most *Melampsora* species infecting willows are heteroecious, with *Abies, Allium, Euonymus, Larix, Ribes, Saxifraga, Viola* and some *Orchidaceae* as alternate hosts. Some willow *Melampsora* species were separated solely on the basis of having different alternate hosts, often identified in early inoculation experiments. For example, both *M. allii-fragilis* and *M. galanthi-fragilis* share the same morphology, telial host range (on *S. fragilis* and *S. pentandra*) and distribution, but they differ in their alternate hosts. The connection of the spore stages between the willow hosts and *Allium* spp. in *M. allii-fragilis* was first demonstrated by Klebahn at the beginning of the 20th century and later confirmed by Mayor (1933, 1936). The occurrence of the aecial stage of *M. galanthi-fragilis* on snowdrop *Galanthus nivalis* was first studied by Schröter (1893) and confirmed by Klebahn (1902).

M. amygdalinae Kleb., which was described on the tree willows *S. triandra* and *S. pentandra*, is the only known autoecious, macrocyclic species of willow *Melampsora* (Sydow and Sydow, 1915; Gäumann, 1959). In addition to this, a rust apparently belonging to *Melampsora* was found on *S. miyabeana* in Japan and named *Ceoma salicis-miyabeana* (Hiratsuka and Kaneko, 1982; Hiratsuka *et al.*, 1992). With *C. salicis-miyabeana*, only spermogonia and aecia were described on *S. miyabeana* but it was assumed that this rust completes its life cycle on the same willow host.

Of 14 species of willow *Melampsora* in Japan, four were shown by artificial inoculation to be capable of alternating on *Larix*, one on *Corydalis* and one on both *Corydalis* and *Chelidonium* (Hiratsuka and Kaneko, 1982). In North America, *Larix, Abies, Tsuga* and *Saxifraga* are involved as alternate hosts (Ziller, 1974). Pei *et al.* (1993) reported that *M. caprearum* and most of the telial collections of *M. larici-epitea* readily infected the European larch, *L. decidua*, but not the Japanese larch *L. kaempferi*. In contrast, Hiratsuka (1932) infected *L. kaempferi* without difficulty using four species of *Melampsora*, including several samples of *M. larici-epitea* and *M. caprearum*. It is possible that, within the same species, host–parasite coevolution in geographically isolated populations may also have resulted in the differences in selection of the alternate hosts.

For some willow rust species in which only uredinio- and teliospore stages have been described, their existence may depend on host alternation, but alternate hosts are yet to be found. For others, host alternation may not be vitally important for their survival or may no longer be required, as seen in the case of SIF. Among willow rusts, asexual populations with a shortened life cycle may not be unique to SIF. Pei *et al.* (1993) inoculated *Ribes* spp. with two UK teliospore collections of *M. ribesii-purpureae*. As a result, only one of the collections formed poorly developed spermogonia on *Ribes*. However, despite prolonged incubation, no aecia were produced. *M. ribesii-purpureae* has a doubtful record of the aecial stage in the UK (Wilson and Henderson, 1966) and the rust appears to be capable of overwintering as the uredinial stage (Pei *et al.*, 1993). Similar observations with *M. ribesii-purpureae* were made by Scaramella (1932) in Italy. A possible speculation is that some populations of *M. ribesii-purpureae* may have lost the ability to infect the alternate host *Ribes* to complete a full macrocyclic life cycle.

Overwintering

The rusts with a shortened life cycle, such as SIF on *S. viminalis*, must overwinter as the uredinial stage and re-infect the telial hosts in next growing season to continue their existence. In contrast, the typical life cycle of the heteroecious, macrocyclic rusts involves overwintering as teliospores and producing basidiospores in the next spring to infect their alternate hosts. However, overwintering as the uredinial stage on willow has been reported with many heteroecious species, including *M. caprearum*, *M. larici-pentandrae* (Hylander *et al.*, 1953), *M. arctica* (Savile, 1953; Wilson and Henderson, 1966), *M. lapponum* (Wilson and Henderson, 1966), *M. ribesii-purpureae* (Scaramella, 1932), *M. ribesii-viminalis* (Scaramella, 1932), *M. paradoxa* (Ziller, 1970) and *M. salicis-albae* (Klebahn, 1905). Thus, their continuation may not depend on the availability of alternate hosts. In Britain, the autoecious *M. amygdalinae* can also overwinter as the uredinial stage in buds and stem cankers of the host (Ogilvie, 1932; M.H. Pei, unpublished observations). For willow rusts, the ability to infect willow buds/stems is essential to overwintering as the uredinial stage.

Host Specificity in Willow Rusts

Historical studies

Melampsora rusts are specialized in their pathogenicity, and their telial host range can be useful in distinguishing different *Melampsora* species. At sub-species level, the host range provides the basis for the identification of *formae speciales* or pathotypes (races). The host range of willow rusts is determined by inoculations of willows with urediniospores or aeciospores. Since the late 19th century, extensive inoculation experiments conducted by various workers have contributed much to the present knowledge of the host range of *Melampsora* species on *Salix*. In Europe, the pioneering work on host specificity in willow rusts was conducted by Klebahn and others in the late 19th to early 20th centuries (see Gäumann, 1959). In the Far East, most studies on willow *Melampsora* were carried out by Japanese mycologists (Matsumoto, 1915, 1919, 1926; Hiratsuka, 1927, 1932; Hiratsuka and Kaneko, 1982). In North America, Ziller conducted a number of experiments to determine the alternate hosts and telial host range of North American willow rusts around the mid-20th century (see Ziller, 1974).

Rusts on tree willow

Some 11 species of *Melampsora* occur exclusively on the subgenus *Salix* (Table 2.2). Three species, *M. allii-fragilis*, *M. galanthi-fragilis* and *M. ari-salicina*, have been described on *S. fragilis* in central Europe. Five species, *M. larici-pentandrae*, *M. salicis-albae*, *M. allii-fragilis*, *M. galanthi-fragilis* and *M. amygdalinae* have been described on *S. pentandra*. Contrary to the experience of Mayor (1933), inoculation attempts with urediniospores of *M. amygdalinae* from England failed to produce symptoms on *S. pentandra* (M.H. Pei, unpublished results). Also, the record of *S. pentandra* as the host of *M. salicis-albae* (see Gäumann, 1959) is doubtful. Over the past 12 years of observation, neither *M. amygdalinae* nor *M. salicis-albae* have been found on *S. pentandra* in the national willow collection, south-west England, where both the rust species occur annually (M.H. Pei, unpublished observations). In Scandinavia, there have been no records of either *M. amygdalinae* or *M. salicis-albae* on *S. pentandra*.

M. coleosporioides occurring on tree willow is native to the Far East, from maritime provinces of Russia to Taiwan, but was spread to Australasia in the 1970s (Latch, 1980). In North America, *M. abieti-caprearum* has been recorded on *S. exigua* and *M. paradoxa* on *S. lasiandra*. These two rusts occur mainly on shrub willows.

Table 2.2. List of hosts and distribution of *Melampsora* species/forms on *Salix*.

Melampsora sp.	Aecial host	Telial host (subgenus, section)	Distribution	References
M. abieti-caprearum Tubeuf (= *M. humboldtiana* Speg.)	*Abies* spp.	*S. appendiculata, aurita, caprea, cinerea, scouleriana* (V., Vetrix), *incana* Shrank (V., Canae), *nigricans* (V., Nigricantes), *purpurea* (V., Helix), *repens* (V., Incubaceae), *commutata, mackenzieana* (V., Cordatae), *exigua* (S., Longifoliae)	Northern and central Europe, North and South America	Gäumann (1959), Ziller (1974)
M. allii-fragilis Kleb.	*Allium* spp.	*S. fragilis* (S., Salix), *pentandra* (S., Pentandrae)	Central Europe	Sydow and Sydow (1915), Gäumann (1959)
M. amygdalinae Kleb.	*S. triandra, S. pentandra*	*S. triandra* (S., Amygdalinae), *pentandra* (S., Pentandrea)	Eurasia	Gäumann (1959), Wilson and Henderson (1966), Azbukina (1974)
M. arctica Rostr. (= *M. alpina* Juel)	*Saxifraga* spp.	*S. herbacea* (V., Retusae), *reinii* (V., Glabrella), *auilonia, groenlandica, subreniformis, yezoalpina* (S., Glaucae)	Arctic, subarctic and alpine regions of the northern hemisphere	Sydow and Sydow (1915), Gäumann (1959), Hiratsuka and Kaneko (1982), Parmelee (1989)
M. caprearum Thüm. (= *M. larici-caprearum* Kleb.)	*Larix* spp.			Sydow and Sydow (1915), Gäumann (1959), Wilson and Henderson (1966)
f.sp. *typica* Kleb.		*S. caprea* (V., Vetrix)	Eurasia	Gäumann (1959)
f.sp. *grandifoliae* Schneid.		*S. aurita, caprea* (V., Vetrix)	Eurasia	Gäumann (1959)
'Japanese form'*		*S. bakko, leucopithecia* (V., Vetrix)	Japan	Hiratsuka (1932), Hiratsuka and Kaneko (1982)
'Chinese form'*		*S. caprea, caprea* var. *sinica* (V., Vetrix); *S. spathulifolia* (V., Psilostigmatae)	China	Tai (1979)
M. chelidonii-pierotii Matsumoto	*Corydalis incisa*	*S. pierotii* (S., Subalbaceae), *chaenomeloides* (S., Glandulosae)	Japan and maritime provinces of Russia	Hiratsuka and Kaneko (1982)
M. choseniae Azb. Et Kov.	?	*Chosenia arbutifolia* (Chosenia)	Maritime provinces of Russia	Azbukina (1974)
M. coleosporoides Dietel	?	*S. babylonica, matsudana* (S., Subalbaceae), *chilensis* (S., Humboldtianae)	Maritime provinces of Russia, China, Japan, Taiwan, Australasia	Sydow and Sydow (1915), Hiratsuka and Kaneko (1982), Spiers and Hopcroft (1996)

continued

Table 2.2. *Continued.*

Melampsora sp.	Aecial host	Telial host (subgenus, section)	Distribution	References
M. dimorphospora Kaneko et Hiratsuka	?	*S. koriyanagi* (V., Helix)	Japan	Hiratsuka and Kaneko (1982)
M. epiphylla Dietel (= *M. larici-opaca* Miyabe et Matsumoto)	*Larix kaempferi*	*S. kinuynagi*, *sachalinensis*, *pet-susu* (V., Vimen), *rorida* (V., Daphnella)	Japan, North-East Asia	Azbukina (1974), Tai (1979), Hiratsuka and Kaneko (1982)
M. epitea		*S. futura* (V., Vetrix), *gilgiana* (V., Helix)	Eurasia, North and South Americas, Australasia	Hylander *et al.* (1953), Wilson and Henderson (1966)
M. epitea f. sp. *tsugae*	*Tsuga* spp.	*S. scouleriana* (V., Vetrix), *sitchensis* (V., Sitchenses)	North America	Ziller (1974)
M. euonymi-caprearum Kleb.	*Euonymus europaeus*		Europe	Gäumann (1959)
f.sp. *typica* Kleb.		*S. aurita, caprea, cinerea* (V., Vetrix)		Gäumann (1959)
f.sp. *euonymi-incanae* Schneid.		S. *incana* Shrank. (V., Canae)		Gäumann (1959)
M. galanthi-fragilis Kleb.	*Galanthus nivalis*	*S. fragilis* (S., Salix), *pentandra* (S., Pentandrae)	Central Europe	Sydow and Sydow (1915), Gäumann (1959)
M. humilis Dietel	?	*S. integra, koriyanagi* (V., Helix)	Sakhalin, Taiwan, Japan	Hiratsuka and Kaneko (1982)
M. kamikotica Kaneko et Hiratuska	*Larix*?	*Chosenia arbutifolia* (Chosenia)	Japan	Hiratsuka and Kaneko (1982)
M. kiusiana Hiratsuka		*S. subopposita* (V., Incubaceae)	Japan	Hiratsuka and Kaneko (1982)
M. kupreviczii Zenk.	?	*S. rosmarinifolia* var. *brachypoda* (V., Incubaceae)	Amur River, Russia	Azbukina (1974)
M. lapponum Linfors	*Viola epipsila*	*S. lapponum* (V., Villosae)	Northern and central Europe	Gäumann (1959)
M. larici-pentandrae Kleb.	*Larix* spp.	*S. pentandra* (S., Pentandrae)	Northern and central Europe	Sydow and Sydow (1915), Gäumann (1959)
M. larici-epitea Kleb.	*Larix* spp.		Eurasia, North Africa	
f. sp. *larici-epitea typica* Kleb.		*S. aurita, caprea, cinerea* (V., Vetrix), *viminalis* (V., Vimen), *repens* var. *argentea* (V., Incubaceae)	Eurasia and North Africa	Sydow and Sydow (1915), Gäumann (1959), Pei *et al.* (1996)

f. sp. *larici-daphnoides* Kleb.		*S. acutifolia, daphnoides* (V., Daphnella)	Europe	Sydow and Sydow (1915), Gäumann (1959), Pei *et al.* (1996)
f. sp. *larici-nigricantis* Schnei.		*S. glabra, nigricans* (V., Nigricantes), *hegetschweileri* (V., Arbuscella)	Europe	Sydow and Sydow (1915), Gäumann (1959)
f. sp. *larici-purpureae* Schneid.		*S. aurita* (V., Vetris), *daphnoides* (V., Daphnella), *purpurea* (V., Helix)	Europe	Sydow and Sydow (1915), Gäumann (1959)
f. sp. *larici-retusae* Fischer		*S. alpicola, hibernica, nigricans,* (V., Nigricantes), *dasyclados* (V., Vimen), *S. disperma* (V., Vetrix), *foetida, waldsteiniana* (V., Arbuscella), *hastata* (V., Cordatae), *helvetica* (V., Villosae), *herbacea, retusa, serphyllifolia* (C., Retusae)	Europe	Gäumann (1959), Pei *et al.* (1996)
f. sp. *larici-reticulata*		*S. hastata* (V., Cordatae), *reticulata* (V., Chamaetia)	Europe	Sydow and Sydow (1915), Gäumann (1959)
'Japanese race 1'		*S. schwerinii* (V., Vimen)	Japan	Hiratsuka (1932), Hiratsuka and Kaneko (1982)
'Japanese race 2'		*S. miyabeana* (V., Helix)	Japan	Hiratsuka (1932), Hiratsuka and Kaneko (1982)
M. larici-urbaniana Matsumoto	*Larix kaempferi*	*S. urbaniana* (S., Urbanianae)	Japan and Kurile Islands	Azbukina (1974), Hiratsuka and Kaneko (1982)
M. microsora Dietel		*S. subfragilis* (S., Amygdalinae)	Japan	Hiratsuka and Kaneko (1982)
M. paradoxa Diet. & Holw. (= *M. bigelowii* Thüm.)	*Larix* spp.	*S. barrattiana* (V., Lanatae), *glaucops, petrophila, vestita* (V., Glaucae), *lasiandra* (S., Pentandrae), *mackenzieana* (V., Cordatae), *sitchensis* (V., Sitchensis)	North and South America	Sydow and Sydow (1915), Ziller (1974)
M. repentis Plowr. (= *M. orchidi-repentis* Kleb.)	*Orchidaceae*	*S. repens* (V., Incubaceae), *aurita* (V., Vetrix)	Europe	Gäumann (1959)
M. reticulatae Blytt	*Saxifraga* spp.	*S. reticulata* (V., Chamaetia)	Arctic, subarctic or alpine regins of the northern hemisphere	Gäumann (1959)

continued

Table 2.2. *Continued.*

Melampsora sp.	Aecial host	Telial host (subgenus, section)	Distribution	References
M. ribesii-epitea Kleb.			Europe	
f. sp. *ribesii-auritae* Kleb.	*Ribes* spp.	*S. aurita* (V., Vetrix), *arbuscula* (V., Arbuscella)		Gäumann (1959)
f. sp. *ribesii-grandifoliae* Schneid.	*Ribes alpina*	*S. appendiculata, aurita* (V., Vetrix)		Gäumann (1959)
M. ribesii-purpureae Kleb.	*Ribes* spp.	*S. purpurea* (V., Helix)	Eurasia and North America	Gäumann (1959), Tai (1979), Ziller (1974)
M. ribesii-viminalis Kleb.	*Ribes* spp.	*S. viminalis* (V., Vimen)	Europe	Gäumann (1959), Wilson and Henderson (1966)
M. salicis-albae Kleb. (= *M. allii-salicis-albae* Kleb.)	*Allium* spp.	*S. alba* (S., Salix)	Eurasia	Gäumann (1959), Tai (1979)
M. salicis-cavaleriei Tai	?	*S. cavaleriei* (S., Wilsonianae)	South-west China	Tai (1979)
M. salicis-cupularis Wang	?	*S. cupulariris* (V., Scleophylla)	North-west China	Tai (1979)
M. salicis-warburgii Sawada	?	*S. warburgii* (S., Glandulosae), *pierotii* (S., Subalbaceae), *subfragilis* (S., Amygdalinae)	Taiwan, Japan	Hiratsuka and Kaneko (1982)
M. yezoensis Miyabe et Matsumoto	*Corydalis ambigua*	*S. yessoensis (S., Subalbaceae)*	Japan	Hiratsuka and Kaneko (1982)
'Stem-infecting form'*	–	*S. viminalis*	British Isles	Pei *et al.* (1995)
Caeoma salicis-miyabeana Kaneko et Hiratsuka	*S. miyabeana*	?*S. miyabeana* (V., Helix)	Japan	Hiratsuka and Kaneko (1982)

*Tentative name

Rusts on shrub and dwarf willow

Of all the *Melampsora* species on willow, *M. larici-epitea* is the most widespread and most complex in its host range. Much of the knowledge on the pathogenic specialization in this rust came from inoculation experiments carried out in continental Europe (reviewed by Gäumann, 1959), Japan (Hiratsuka, 1932) and the UK (Pei *et al.*, 1996). Six *formae speciales* within *M. larici-epitea* were recognized in Europe (Sydow and Sydow, 1915; Gäumann, 1959). While there is no clear morphological distinction between the *formae speciales*, the host range of a *forma specialis* appears largely confined to certain sections of willow. For example, f. sp. *larici-epitea typica* has been recorded on willows belonging to Secs *Vimen*, *Vetrix* and *Incubaceae*, while f. sp. *larici-daphnoides* occurs on Sec. *Daphnella*. *M. larici-epitea* f. sp. *larici-retusae* has been recorded on about a dozen willow species belonging to several sections of the subgenus *Vetrix*. In Japan, two 'races' have been recognized, out of four reported originally (Hiratsuka, 1932; Hiratsuka and Kaneko, 1982). From their host range, these two races appear to be equivalent to different *formae speciales*, as treated by European mycologists. In the UK, three *formae speciales* – *larici-epitea typica*, *larici-retusae* and *larici-daphnoides* – were identified within *M. larici-epitea* (Pei *et al.*, 1996). In New Zealand, a new rust morphologically similar to *M. epitea* caused severe infections on *S. viminalis* and its hybrids in 1985 (Spiers and Hathaway, 1987; Spiers and Hopcroft, 1996). From the host lists given by these authors, the *M. epitea* on *S. viminalis* in New Zealand corresponds with the host range of f. sp. *larici-epitea typica* (Pei *et al.*, 1996).

M. caprearum is also one of the most widespread species in Eurasia. *Salix aurita*, *S. caprea* and *S. cinerea* (section *Vetrix*) have been recorded as the telial hosts of *M. caprearum* (Wilson and Henderson, 1966). *S. cinerea* was included in the host range of *M. caprearum* based on the inoculation experiment conducted by Schneider at the turn of the 20th century (see Gäumann, 1959). However, Schneider's inoculations produced only a trace infection (a single uredinium) on *S. cinerea* and, therefore, cannot be regarded as sufficient evidence to prove that *S. cinerea* is a natural host of *M. caprearum*. There have been no records of *M. caprearum* on *S. cinerea* in Scandinavia, where both *M. caprearum* and *S. cinerea* commonly occur (Hylander *et al.*, 1953). Also, in the UK, *M. caprearum* has not been observed on any *S. cinerea* clones in the past 12 years of observation (M.H. Pei, unpublished observations). Therefore, it seems unlikely that *S. cinerea* serves as telial host of *M. caprearum* in nature. In Europe, two *formae speciales* have been recognized in *M. caprearum*; f. sp. *typica*, infecting only *S. caprea*, and f. sp. *grandifoliae*, infecting both *S. aurita* and *S. caprea*. In Japan, *M. caprearum* occurs on two Japanese willow species of section *Vetrix*, *S. bakko* and *S. leucopithecia*. *M. caprearum* has also been recorded on *S. caprea* var. *sinica* Hao and *S. spathulifolia* Seem. (section *Psilostigmatae* of subgenus *Vetrix*) in central China (Tai, 1979).

M. abieti-caprearum, occurring in Europe and in North and South America, and the native American species *M. paradoxa* have also been recorded on a wide range of *Salix* species. *M. paradoxa*, the American counterpart of *M. larici-epitea*, can be distinguished from other willow rusts in the Americas by having relatively large aeciospores. The early account of the host range of *M. paradoxa* included some 37 willow species belonging to various sections of the subgenus *Vetrix* or subgenus *Salix* (Sydow and Sydow, 1915). Later, Ziller (1974) stated that it was futile to try to identify a North American willow rust without knowing its aecial state, and listed seven willow species belonging to subgenus *Vetrix* or *Salix* as the telial hosts of *M. paradoxa*.

Several species, such as *M. arctica*, *M. lapponum* and *M. reticulatae* mainly infect small shrub or dwarf willows occurring in arctic, subarctic or alpine regions in the northern hemisphere. Two morphologically distinct species, *M. choseniae* and *M. kamikotica*, occur on *Chosenia arbutifolia*, the unique species in *Salicaceae* by having characteristics resembling both *Salix* and *Populus*.

Historical Records and Host Range

Host records of willow rusts in the literature are based largely on the morphology at uredinial and telial stages. For many willow *Melampsora* species (*sensu stricto*) which are morphologically indistinguishable, host records without the backing of inoculation experiments are often doubtful. Misidentification of the host species may add further confusion. Therefore, for many willow rusts, historical records on the host range need to be treated with caution. It is not difficult to understand that such complexities led to the wide acceptance of *M. epitea* as complex species to encompass immensely diverse groups of willow rusts.

It needs to be mentioned that whether *M. epitea* is the legitimate name is questionable. *M. epitea* was originally described by Thümen (1879) based on the specimen on *S. alba*. However, to date, there have been no credible records of *M. epitea*, which has evenly echinulate urediniospores, occurring on *S. alba*. It is likely that either the host species was misidentified in the type specimen or the rust described was not the *M. epitea* that we recognize today, probably because microscopes available at the time may have been too primitive to reveal morphological details. In fact, Thümen (1879) proposed the name *M. vitellinae* Thüm. to include the rusts on *S. alba* var. *vitellina*, *S. fragilis* and *S. pentandra*. All the rust species on these willows are known to produce urediniospores having a smooth surface at the apex. However, this feature was not included in Thümen's description of *M. vitellinae*.

Occasionally, results obtained under experimental conditions do not reflect the situation in the field because, in a protected environment, the pathogenicity of a given rust isolate is likely to be wider than in its natural habitat (Samborski *et al.*, 1958; Anikster, 1984; Helfer, 1989). For example, Pei *et al.* (1996) reported that *S. discolor*, a North American willow species, became infected by several isolates of *M. larici-epitea* in laboratory inoculation experiments. However, no rust was found on this species in the field (national willow collection, south-west England) where all the pathotypes occurred. In leaf disc inoculation experiments, young leaves of *S. daphnoides* also supported weak pathogen development after inoculations with f. sp. *larici-epitea typica* and f. sp. *larici-retusae*. However, over a period of 12 years, *S. daphnoides* has never been infected by the two *formae speciales* in the field (M.H. Pei, unpublished observations).

Concluding Remarks

In the past decade, the application of molecular tools in mycology has revolutionized our understanding of the phylogeny and population biology of the fungi. Recently, several species of *Melampsora* on willows collected from central Europe and the British Isles were examined for variation in rDNA sequences (Chapter 1). It appears that *M. larici-epitea*, *M. caprearum*, SIF and *M. epiphylla* share a common ancestral lineage. In contrast, *M. larici-epitea* and *M. ribesii-purpureae*, both of which can be grouped under the *M. epitea* complex, are phylogenetically distinct.

Urgent revision of the taxonomy of *Melampsora* on *Salix* has long been called for (Ziller, 1974; Helfer, 1992; Spiers and Hopcroft, 1996). The purpose of this chapter was to give a brief review of some of the existing information on the *Melampsora* species on *Salix*. Because of the extent of complexity and the lack of sufficient understanding of many aspects, appropriate revision of the taxonomy and nomenclature of willow rusts cannot be achieved until further knowledge of the genetic and phylogenetic relationships among the rusts is available.

Acknowledgements

This study was funded by the Department for Environment, Food and Rural Affairs

(DEFRA), UK, and the European Union. Rothamsted Research receives grant-aided support from the Biotechnology and Biological Sciences Research Council of the UK.

References

Anikster, Y. (1984) The *formae speciales*. In: Bushnell, W.R. and Roelfs, A.P. (eds) *The Cereal Rusts*, Vol. I. *Origins, Specificity, Structure and Physiology*. Academic Press, Orlando, Florida, pp. 115–130.

Argus, G.W. (1986) The genus *Salix* (Salicaceae) in the southeastern United States. *Systematic Botany Monographs*, Vol. 9. The American Society of Plant Taxonomists, Michigan.

Argus, G.W. (1997) Infrageneric Classification of *Salix* (Salicaceae) in the New World. *Systematic Botany Monographs*, Vol. 52. The American Society of Plant Taxonomists, Michigan.

Arthur, J.C. (1934) *Manual of the Rusts in the United States and Canada*. Purdue Research Foundation, Lafayette, Indiana.

Azbukina, Z.M. (1974) *Rust Fungi of the Soviet Far East* [in Russian]. Nauka Publishers, Moscow, pp. 120–140.

Bean, W.J. (1980) *Trees and Shrubs Hardy in the British Isles*, Vol. IV, 8th edn. John Murray, London, UK, pp. 246–312.

Dorn, R.D. (1976) A synopsis of American *Salix*. *Canadian Journal of Botany* 54, 2769–2789.

Gäumann, E. (1959) *Die Rostpilze Mitteleuropas. Beiträge zur Kryptogamenflora Schweiz 12*. Buchdruckerei Büchler and Co., Bern, Germany.

Hawksworth, D., Kirk, P., Sutton, B. and Pegler, D. (1995) *Ainsworth's and Bisby's Dictionary of the Fungi*, 8th ed. CAB International, Wallingford, UK.

Helfer, S. (1989) The culture of cereal leaf rusts for physiological and taxonomical studies. *Notes from the Royal Botanic Garden, Edinburgh* 46, 131–139.

Helfer, S. (1992) The rust diseases of willows in Britain. *Proceedings of the Royal Society of Edinburgh* 98B, 119–134.

Hiratsuka, N. (1927) Japanese species of *Melampsora* (Studies on *Melampsoraceae* of Japan IV). *Journal of the Society of Agriculture and Forestry, Sapporo* 19, 180–195 [in Japanese].

Hiratsuka, N. (1932) Inoculation experiments with some heteroecious species of the *Melampsoraceae* in Japan. *Japanese Journal of Botany* 6, 1–33.

Hiratsuka, N. and Kaneko, S. (1982) A taxonomic revision of *Melampsora* on willows in Japan. *Reports of the Tottori Mycological Institute* 20, 1–32.

Hiratsuka, Y. and Sato, S. (1982) Morphology and taxonomy of rust. In: Scott, K. J. and Charkravorty A. K. (eds) *The Rust Fungi*. Academic Press, London, pp. 1–36.

Hiratsuka, N., Sato, T., Katsuya, K., Kakishima, M., Hiratsuka, Y., Kaneko, S., Ono, Y., Sato, S., Harada, Y., Hiratsuka, T. and Nakayama, K. (1992) *The Rust Flora of Japan*. Tsukuba Shuppankai, Ibaraki, Japan, pp. 271–300.

Hylander, N., Jøstad, I. and Nanfeldt, J.A. (1953) Enumeratio Uredinearum Scandinavicarum. *Opera Botanica* 1, 1–102.

Klebahn, H. (1902) Kulturversuche mit Rostpilzen. X. *Zeitschrift für Pflanzenkrankhreiten* 12, 17–44.

Klebahn, H. (1905) Kulturversuche mit Rostpilzen. XII. *Zeitschrift für Pflanzenkrankhreiten* 15, 65–108.

Latch, B.J. (1980) Weeping willow rust in New Zealand. *New Zealand Journal of Agricultural Research* 23, 535–538.

Matsumoto, T. (1915) Impfversuche mit *Melampsora* auf Japanischen Weiden. *Transactions of the Sapporo Natural History Society* 6, 22–37.

Matsumoto, T. (1919) Some experiments with *Melampsora* in Japan. *Annals of Missouri Botanical Garden* 6, 309–316.

Matsumoto, T. (1926) On the relationship between *Melampsora* on *Salix pierotii* Miq. and *Caeoma* on *Chelidonium majus* L. and *Corydalis incissa* Pers. *Botanical Magazine, Tokyo* 40, 43–47.

Mayor, E. (1933) Notes mycologiques VIII. *Bulletin de la Societe neuchâteloise des Sciences naturelles* 58, 7–31.

Mayor, E. (1936) Notes mycologiques IX. *Bulletin de la Societe neuchâteloise des Sciences naturelles* 61, 105–123.

Ogilvie, L (1932) Notes on the rusts of basket willows and their control. *Annual Report of the Agricultural and Horticultural Research Station, Long Ashton, Bristol*, for 1931, pp. 133–138.

Parmelee, J.A. (1989) The rust (Uredinales) of Arctic Canada. *Canadian Journal of Botany* 67, 3315–3365.

Pei, M.H. and Ruiz, C. (2000) AFLP evidence of distinct patterns of life-cycle in two forms of *Melampsora* rust on *Salix viminalis*. *Mycological Research* 104, 937–942.

Pei, M.H., Royle, D.J. and Hunter, T. (1993) Identity and host alternation of some willow rusts (*Melampsora* spp.) in England. *Mycological Research* 97, 845–851.

Pei, M.H., Royle, D.J. and Hunter, T. (1995) A comparative study of stem-infecting and leaf-infecting forms of *Melampsora* rust on *Salix viminalis* in the UK. *Mycological Research* 99, 357–363.

Pei, M.H, Royle, D.J. and Hunter, T. (1996) Pathogenic specialization of *Melampsora epitea* var. *epitea* on *Salix*. *Plant Pathology* 45, 679–690.

Pei, M.H., Royle, D.J. and Hunter, T. (1999) Hybridisation in larch-alternating *Melampsora epitea* (*M. larici-epitea*). *Mycological Research* 103, 1440–1446.

Samborski, D.J., Forsyth, F.R. and Person, C. (1958) Metabolic changes in detached wheat leaves floated on benzimidazole and the effect of these changes on rust reaction. *Canadian Journal of Botany* 36, 591–601.

Savile, D.B.O. (1953) Short-season adaptations in rust fungi. *Mycologia* 45, 75–87.

Scaramella, P. (1932) Sullo svernamento delle Melampsorae dei salici in alta montagna. *Chanousia* 2, 72–76 [in Italian].

Schröter, J. (1893) Zur Entwicklungsgeschichete der Uredineen. 71. Jahresbericht der Schlesischen Gesellschaft für vaterländische Cultur, II, Abt., Botanische Sektion, 31–32.

Skvortsov, A.K. (1968) *Willows of the USSR* [in Russian with English summary]. Nauka, Moscow.

Spiers, A.G. and Hathaway, R.L. (1987) *Melampsora* rusts of *Salix* in New Zealand. In: *IEA/CBP Joint Report, Proceedings IEA/BA Task II Workshop Biotechnology Development. Sep. 1987, Uppsala, Sweden*. Faculty of Forestry, University of Toronto, Toronto, Canada, pp. 65–69.

Spiers, A.G. and Hopcroft, D.H. (1996) Morphological and host range studies of *Melampsora* rusts attacking *Salix* species in New Zealand. *Mycological Research* 100, 1163–1175.

Sydow, P. and Sydow, H. (1915) *Monographia Uredinearum III*. Fratres Borntraeger, Leipzig, pp. 334–350.

Tai, F.L. (1979) *Sylloge Fungorum Sinicorum* [in Chinese]. Science Press, Beijing, p. 536.

Thümen, F.V. (1879) *Melampsora salicina*, der Weidenrost. Eine monographische Stuie. *Mitteil Forstl. Versuchswesen Österreichs* 2, 25–46.

Wang, C. and Fang, C.F. (1984) *Salicaceae*. *Flora of the People's Republic of China* Vol. 20 (2) [in Chinese]. Science Press, Beijing, pp. 2–78.

Wilson, M. and Henderson, D.M. (1966) *The British Rust Fungi*. Cambridge University Press, Cambridge, UK.

Ziller, W.G. (1970) Studies of western tree rusts. VII. Inoculation experiments with conifer needle rusts (*Melampsoraceae*). *Canadian Journal of Botany* 48, 1313–1319.

Ziller, W.G. (1974) *The Tree Rusts of Western Canada*. Publication No. 1329, Canadian Forestry Service, Environmental Canada, Ottawa, Canada.

3 The Species of *Melampsora* on *Salix* (*Salicaceae*)

Gaddam Bagyanarayana
Department of Botany, Osmania University, Hyderabad 500 007 (A.P), India

Introduction

This chapter describes a morphotoxonomic study of the species of *Melampsora* infecting the host genera *Salix* and *Chosenia* (*Salicaceae*). A total of 19 species and ten f. spp. are recognized and described. Out of the ten f. spp., six are proposed as new combinations. *M. allii-fragilis, M. amygdalinae, M. capraearum, M. chelidonii-pierotii, M. choseniae, M. dimorphospora, M. epiphylla, M. epitea, M. epitea* f. sp. *euonymi, M. epitea* f. sp. *repentis, M. epitea* f. sp. *reticulatae, M. epitea* f. sp. *tsugae, M. kamikotica, M. kupreviczii, M. larici-pentandrae, M. larici-urbaniana, M. paradoxa, M. ribesii-viminalis, M. salicis-albae, M. salicis-cavaleriei, M. salicis-cupularis, M. salicis-warburgii* and *M. yezoensis* are recognized as valid. *M. allii-fragilis* f. sp. *galanthi-fragilis, M. epitea* f. sp. *abieti-capraearum, M. epitea* f. sp. *arctica, M. epitea* f. sp. *lapponum, M. epitea* f. sp. *laricis-epitea* and *M. epitea* f. sp. *ribesii-purpureae* are treated as new combinations. A key to the described species, detailed morphological characters of the spore forms, host range and geographical distribution are provided. This study is based on examination of hundreds of *Melampsora*-infected specimens including types, original collections and other authentic specimens obtained from several International Herbaria (Table 3.1).

Telia on *Chosenia*

Urediniospores evenly thickened — ***M. choseniae*** **Azb. et Kov.**
Urediniospores: 14–25 × 14–18 μm
Teliospores: 28–64 × 10–14 μm
Urediniospores apically thickened — ***M. kamikotica*** **Kaneko et Hirats.**
Urediniospores: 20–33 × 15–18 μm (apically 5–6.5 μm)
Teliospores: 25–53 × 64–10 μm

Telia on *Salix*

Autoecious macrocyclic — ***M. amygdalinae*** **Kleb.**
Urediniospores: 20–40 × 11–18 μm
Teliospores: 18–50 × 7–14 μm

 Rust Diseases of Willow and Poplar (eds M.H. Pei and A.R. McCracken)

Table 3.1. Abbreviations, names and addresses of herbaria.

Abbreviation	Name and address of the herbarium
B	Botanischer Garten and Museum, Berlin-Dahlem, 1000 Berlin 33, Konigin Luise Strasse 6-8, Berlin
C	The Botanical Museum and Herbarium, University of Copenhagen, 130 Gothersgade, DK-1123, Copenhagen K, Denmark
DAOM	National Mycological Herbarium, Research Branch, Biosystematics Research Institute, Wm. Saunders Building, C. E. F., Ottawa, Canada K1A OC6
DAVFP	Canadian Forestry Service, Forest Pathology, Forest Research Laboratory, Victoria, BC, Canada
FH	The Forlow Herbarium of Harvard University, 20 Divinity Avenue, Cambridge, Massachusetts, USA
G	Conservatoire et Jardin Botanique, Institut de Botanique Systematique de l'Universite, Case Postale 60, CH-1292 Chambésy Suisse, Genève, Switzerland
HCIO	Herbarium Cryptogamiae Indiae Orientalis, Division of Mycology and Plant Pathology, Indian Agricultural Research Institute, New Delhi-12, India
K	Royal Botanic Gardens, Kew, Richmond, Surrey, England
LEV	New Zealand Department of Agriculture Horticultural Research Centre, Levin Mycological Herbarium, New Zealand
PAV	Instituto Di Botanica Orto Botanico, Laboratoria Cryttogamico, via. S. Epifanio, Case Postale 99, 27100-Pavia, Italy
S	Swedish Museum of Natural History, Section for Botany, 104 05 Stockholm 50, Sweden
WRSL	Instytut Botaniczy, Universytetu Wroclawskiego, ul Kanonia 6/8, Wroclaw, Poland
IMI	IMI Herbarium CABI Bioscience UK Centre, Bakeham Lane, Egham, Surrey TW20 9TY, UK

Heteroecious or the aecial stages not known
Telia sub-cuticular
Telia sub-epidermal

Telia sub-cuticular

1. Teliospores apically thickened 2
1. Teliospores apically not thickened 3

2. Apical thickening 3–10 μm **_M. capraearum_ Thum.**
Urediniospores: 18–22 × 13–16 μm
Teliospores: 30–45 × 7–18 μm
2. Apical thickening 2.5–4.8 μm **_M. epiphylla_ Diet.**
Urediniospores: 12.8–16 × 11.2–15 μm
Teliospores: 24–45 × 8–16 μm

3. Urediniospore wall not thickened 4
3. Urediniospore wall thickened **_M. yezoensis_ Miyabe et Mats.**
Urediniospores: 16–28 × 12.5–20 μm
Teliospores: 16–33.6 × 6.4–13 μm

4. Teliospores short 5
4. Teliospores long, 64 μm in length **_M. chelidonii-pierotii_ Mats.**
Urediniospores: 16–24 × 13–16 μm
Teliospores: 20–64 × 6–8 μm

5. Urediniospores small, up to 22 µm — ***M. ribesii-viminalis*** **Kleb.**
   Urediniospores: 15–22 × 14–18 µm
   Teliospores: 25–40 × 7–15 µm

5. Urediniospores lengthy, more than 32 µm
   Aecia on *Allium* — ***M. allii-fragilis*** **Kleb.**
      Urediniospores: 22–33 × 12–15 µm
      Teliospores: 30–48 × 7–14 µm
   Aecia on *Galanthus* — ***M. allii-fragilis*** **f. sp. *galanthi-fragilis* (Kleb.) comb. nov.**
      Urediniospores: 25–38 × 12–16 µm
      Teliospores: 25–46 × 8–16 µm

Telia sub-epidermal

1. Urediniospores one type — 2
1. Urediniospores two types — ***M. dimorphospora*** **Kaneko & Hirat.**

2. Urediniospores apically smooth — 3
2. Urediniospores evenly echinulate — 4

3. Urediniospores apically thick — ***M. larici-pentandrae*** **Kleb.**
   Urediniospores: 26–44 × 12–16 µm
   Teliospores: 19–30 × 9–12 µm
3. Urediniospores apically not thickened — ***M. salicis-albae*** **Kleb.**
   Urediniospores: 20–36 × 11–17 µm
   Teliospores: 25–45 × 7–10 µm

4. Urediniospores apically thickened — ***M. larici-urbaniana*** **Mats.**
   Urediniospores: 15–27 × 13–17 µm
   Teliospores: 35–70 × 8–16 µm
4. Urediniospores apically not thickened — 5

5. Urediniospores more than 15 µm — 6
5. Urediniospores up to 15 µm — ***M. kupreviczii*** **Zenk.**
   Urediniospores: 10–15 × 10–12 µm
   Teliospores: 30–37 × 10–15 µm

6. Urediniospores length up to 37 µm — ***M. salicis-cavaleriei*** **Tai.**
   Urediniospores: 18–37 × 11–18 µm
   Teliospores: 36–69 × 7–11 µm
6. Urediniospores less than 30 µm — 7

7. Urediniospore wall thickness up to 3 µm — 8
7. Urediniospore wall thickness less than 2.5 µm — ***M. salicis-warburgii*** **Sawada.**
   Urediniospores: 13–28 × 15–18 µm
   Teliospores: 34–42 × 8–15 µm

8. Teliospores length up to 40 µm — ***M. paradoxa*** **Diet & Holw.**
   8A Urediniospores: 15–26 × 15–21 µm — ***M. paradoxa*** **Diet & Holw.**
      Teliospores: 29–43 × 11–14 µm
   8B Urediniospores: 21–23 × 17–18 µm[wall 1.5–2 µm] — ***M. salicis-cupularis***
      Teliospores: 28–39 × 11–14 µm
8. Teliospores length up to 50 µm — ***M. epitea*** **(Kunze & Schm.) Thum.**

A collective species having nine *formae speciales* on different aecial hosts

Urediniospores: 2–25 × 10–19 μm
Teliospores: 20–50 × 7–15 μm

Aecia on *Saxifraga*
Urediniospore wall 2 μm thick — ***M. epi.* f. sp. *arctica* (Rostr.) comb. nov**
Urediniospore wall more than 2 μm thick (2.5–3.5) — ***M. epi.* f. sp. *reticulatae* (A. Blytt.) Jorst.**

Aecia on *Abies* — ***M. epi.* f. sp. *abieti-capraearum* (Tub.) comb. nov.**

Aecia on *Euonymus* — ***M. epi.* f. sp. *euonymi* (Kleb.) Boerema & Verhoev.**

Aecia on *Larix* — ***M. epi.* f. sp. *larici-epitea* (Kleb.) comb. nov.**

Aecia on *Orchidaceae* — ***M. epi.* f. sp. *repentis* (Plowr.) Boerema & Verhoeven**

Aecia on *Ribes* — ***M. epi.* f. sp. *ribesii-purpureae* (Kleb.) comb. nov.**

Aecia on *Tsuga* — ***M. epi.* f. sp. *tsugae* Ziller**

Aecia on *Viola* — ***M. epi.* f. sp. *lapponum* (Lindf.) comb. nov.**

Melampsora allii-fragilis Kleb., *Pringsh. Jahrab. Wiss. Bot.* 35, 674; 1901

= *Uredo allii-fragilis* Arth., *Result, Scient. Congr. Intern. Bot. Vienne* p. 338; 1905, 1906.

Spermogonia foliicolous, lenticular, sub-epidermal, 200 μm wide. Aecia caeomoid, amphigenous and caulicolous, aggregated, 0.5–1.0 mm wide, sub-epidermal, erumpent, pulverulent, orange; aeciospores 18–26 × 12–20 μm, irregularly angular or ellipsoid, rarely globoid, wall 1–2 μm thick, verrucose. Uredinia minute, amphigenous, mainly hypophyllous, 0.5 mm, sub-epidermal, erumpent, pulverulent, orange; urediniospores 22–33 × 12–15 μm, ellipsoid, ovate or oblong, wall 3 μm thick, echinulate; paraphyses capitate or clavate, up to 70 μm long, 15–20 μm wide, wall 2–5 μm thick. Telia minute, amphigenous, mostly epiphyllous, scattered or aggregated, 0.2–0.5 mm wide, sub-cuticular, not erumpent, dark brown; teliospores 30–48 × 7–14 μm, rounded at both the ends, wall 1 μm thick, pale brown.

Type: On *Salix fragilis* L., Triglitz in der Prignitz, Brandenburg, Germany.

Hosts: Spermogonia and aecia on *Allium ampeloprasum* L., *A. ascalonicum* L., *A. carinatum* L., *A. cepa* L., *A. fistulosum* L., *A. moly* L., *A. neapolitanum* Cyr., *A. obligum* L., *A. ochroleucum* W. et K., *A. odorum* Ten., *A. oleraceum* L., *A. pedemontanum* Willd., *A. porrum* L., *A. pulchellum* Don., *A. sativum* L., *A. schoenoprasum* L., *A. scenescens* L., *A. sphaerocephalum* L., *A. ursinum* L., *A. victorialis* L., *A. vineale* L. Uredinia and telia on *Salix fragilis* L., *S. pentandra*.

Distribution: Europe.

Specimens examined: FH & S.

Morphologically, *Melampsora allii-fragilis* Kleb. is strikingly similar to *M. salicis-albae* Kleb. and both have *Allium* as their aecial host. The only distinguishing feature is that the telia are sub-cuticular in the former species whereas they are sub-epidermal in the latter species. Host-alternation and host selection of *M. allii-fragilis* was studied by Klebahn (1902, 1903) and Mayor (1934, 1936, 1958).

Melampsora allii-fragilis Kleb., f. sp. *galanthi-fragilis* (Kleb.) Bagyanarayana comb. nov.

= *Uredo galanthi* Kirchner, *Lotos* 6, 179; 1856.

= *Caeoma galanthi* Schroet., *Abhandl. Schles. Ges. Vaterl. Kult.* 1869/72. 72; 1869. = *Melampsora galanthi-fragilis* Kleb., *Zeitschr. F. Pl. Krankh.* 12, 27; 1902.

Spermogonia flat, slightly elevated, lens shaped, sub-epidermal, 80–100 µm high, 120–155 µm wide. Aecia caeomoid, amphigenous, often hypophyllous, in small circular groups, often fusing, 1–2 mm long, sub-epidermal, erumpent, pulverulent, orange; aeciospores 16–25 × 12–20 µm, globose, sub-globose or angular, wall 1–2 µm thick, verrucose. Uredinia mostly hypophyllous, rarely epiphyllous, scattered or aggregated, 0.5–1 mm in diameter, sub-epidermal, erumpent, pulverulent, orange; urediniospores 25–38 × 12–16 µm, ovoid, ellipsoid, sometimes pyriform, wall 3 µm thick, echinulate; paraphyses mostly capitate, up to 70 µm long, 16–24 µm wide, wall 2–5 µm thick. Telia mostly epiphyllous, rarely hypophyllous, scattered or aggregated, 0.25–1 mm in diameter, sub-cuticular, not erumpent, dark brown; teliospores 24–46 × 8–16 µm, more or less round at both the ends, wall 1.5 µm thick, pale brown.

Hosts: Spermogonia and aecia on *Galanthus nivalis* L., *G. plicatus* M.B. Uredinia and telia on *Salix fragilis* L., *S. pentandra* L.

Distribution: Europe.

Specimens examined: S.

Melamspora allii-fragilis f. sp. *galanthi-fragilis* Bagyanarayana very much resembles *M. allii-fragilis* Kleb., except it differs in having *Galanthus* as its aecial host. Indeed, some of the earlier workers have treated this as a synonym of *M. allii-fragilis*. Schroeter (1893) noticed the host alternation of *M. allii-fragilis* f. sp. *galanthi-fragilis*, which was later proven by Klebahn (1902, 1903).

Melampsora amygdalinae Kleb., *Pringsh. Jahrb. Wiss. Bot.* 34, 352; 1900

Spermogonia amphigenous, lenticular, slightly elevated, orange, 100 µm wide, 50 µm high. Aecia caeomoid, on young leaves, usually hypophyllous, rarely epiphyllous, sometimes on branches up to 1 mm, but when in groups up to 3–10 mm, fusing, sub-epidermal, erumpent; aeciospores 16–24 × 14–20 µm, globose, angulato-globose or ovoid, wall 2 µm thick, minutely verrucose. Uredinia minute, hypophyllous, rarely epiphyllous and culmicolous, scattered, 0.5 mm, orange, sub-epidermal, erumpent, pulverulent; urediniospores 20–40 × 12–20 µm, ovate to oblong, wall 1–5 µm thick, echinulate; paraphyses capitate, up to 80 µm in length, 10–15 µm in width, wall 2–5 µm thick. Telia minute, 0.3–0.5 mm, usually hypophyllous, rarely epiphyllous, aggregated in small groups, sub-epidermal, non-erumpent, in the beginning reddish-brown, later blackish-brown; teliospores 18–40 × 7–14 µm, cylindrical, often irregular, round at both ends, wall 1 µm thick, yellowish-brown.

Type: On *Salix amygdalina* L., Elsfleth in Oldenburg, Germany.

Hosts: On *Salix amygdalina* L., *S. laevigata* Bebb., *S. lucida*, *S. triandra* L., *S. pentandra* L., *S. pentamira*.

Distribution: Austria, Canada, Europe and UK.

Specimens examined: FH, G, S.

This is the only autoecious macrocyclic species of the genus *Melampsora* on *Salix*. Morphologically *M. amygdalinae* is very close to *M. larici-pentandrae* Kleb., but the urediniospores are not apically thickened as in the case of *M. larici-pentandrae*. Raabe (1939) found that the mycelium can overwinter in the sprouts and buds of *Salix triandra* and can produce urediniospores in the following spring without requiring a fresh infection.

Melampsora capraearum (D.C.) Thüm., *Mith. Forstl. Versuch SW, Oestree* 2, 34, 36; 1879

= *Uredo farinosa* var. *Uredo salicis-capreae* Pers., *Syn. Meth. Fung.* p. 217; 1801.
= *U. capraearum* D.C., Lam & D.C. *Syn. Pl. Gall* p. 48; 1806.
= *Sclerotium salicinum* Pers., Moug & Nestl. *Strip Crypt. Veg. Rehn.* 386; 1810. Pers. Ex DC *Fl. Fr.* 5, 114; 1815.

= *Lecythea salicinia* Lev., *Ann. Sc. Nat.* p. 375; 1847. Tul. *Ann Science Natt.* VII, 6–7; 1854.
= *Melampsora salicinia* (D.C.) Lev., *Ann. Sci. Nat. Bot. III* 8, 375; 1847.
= *Caeoma laricis* Hartg, *Wichtige Krankh. d. Waldaume* p. 93; 1874.
= *M. farinosa* (Pers.) Schroet., *Cohn. Krypt. Fl. Schles* 1, 360; 1887.
= *M. larici-caprearum* Kleb., *Frostl. Naturw. Zeitschr.* 6, 469–470; 1897. *Zeistchr. f. Pfl. Krankh* 7, 326; 1897.
= *M. capraearum* (D.C.) Thüm f. sp. *typica* Klebh.
= *Uredo larici-caprearum* Arth., *Result Scient. Congr. Intern. Bot. Vienne* p. 305; 1906.

Spermogonia amphigenous, scattered or in small groups, sub-cuticular, up to 80 µm wide, 45 µm high. Aecia caeomoid, minute, 0.25–1 mm, scattered or aggregated, pale orange, sub-epidermal, erumpent, pulverulent; aeciospores 12–25 × 12–17 µm, globose, angulato-globose or ellipsoid, wall 2 µm thick, finely verrucose. Uredinia hypophyllous, scattered or aggregated, 1–2 mm diameter. Up to 5 mm on juvenile leaves, orange; urediniospores 14–24 × 12–16 µm, ovoid or ellipsoid, wall 2–2.5 µm thick, echinulate; paraphyses numerous, capitate, up to 80 µm in length, 19–25 µm in width, wall 5–6 µm thick. Telia epiphyllous, dense, aggregated, often fusing, sub-cuticular, non-erumpent, 0.5–3 mm, dark orange-brown to dark reddish-brown or blackish-brown; teliospores (21) 25–45 × 8–16 µm, wall 1–2 µm, thick at sides and apically up to 10 µm thick, dark cinnamon brown.

Type: On *Salix caprea* L., Germany, Bavaria, De Jhumen, March 1896.

Hosts: Spermogonia and aecia on *Larix decidua* Mill., *L. leptolepis* Murr., *L. occidentalis* Nutt., *L. sibirica* Led., *Pseudolarix kaempferi* (Fort.) Gond. Uredinia and telia on *Salix aurita* L., *S. bakka* Kimura, *S. caprea* L., *S. cinerea* L., *S. daphnoides* Vill, *S. dasyclados* Wimm., *S. elegans* Wall., *S. grandifolia* Ser., *S. hastate* L., *S. imprimis*, *S. phlomoidis* Bieb., *S. phylicifolia*, *S. rosmarinifolia* L., *S. smithiana* Willd., *S. tetrasperma* Roxb.

Distribution: Eurasia.

Specimens examined: G, S.

According to Klebahn (1914) *M. capraearum* is one of the most commonly seen rusts of central Europe. This rust is easily distinguishable from other melampsoras on *Salix* by having teliospores with heavy apical thickening and a conspicuous germ pore. On the basis of experiments carried out by Jacky (1899), Klebahn (1897, 1899, 1900, 1905, 1908), Mayor (1918) and O. Schneider (1906), Gäumann recognized two *formae speciales*. (i) f. sp. *typica* Kleb.; and (ii) f. sp. *grandifoliae* O. Schneid. But this treatment does not look justified as both the f. spp. have the same aecial hosts, the only difference being in the infection types produced on the telial hosts. Gäumann (1959) himself further commented that 'on the other hand the results of Schneider (1905) allow one to presume the existence of small local races'.

Melampsora chelidonii-pierotii Mats., *Bot. Mag. Tokyo* 40, 46; 1926

Spermogonia epiphyllous, deposited in circular groups. Aecia caeomoid, hypophyllous, occasionally caulicolous, scattered or aggregated 1–2 mm wide, often fusing, erumpent, pulverulent, orange-red, aeciospores 13–20 × 12–15 µm, globoid or ovoid, wall 2 µm thick, verrucose. Uredinia hypophyllous, scattered, erumpent, pulverulent, orange–yellow; urediniospores 16–24 × 13–16 µm, ovoid or ellipsoid, wall 2–3 µm, thick, echinulate; paraphyses capitate, up to 65 µm long, 17–22 µm wide, wall thickened apically up to 12 µm. Telia amphigenous, mostly epiphyllous, scattered, sub-cuticular, non-erumpent, reddish-brown; teliospores 20–64 × 6–8 µm, cylindrical or wedge-shaped, wall 1 µm thick.

Type: On *Salix pierotii* Miq., Morioka, Japan.

Hosts: Spermogonia and aecia on *Chelidonium majus* L., *Corydalis incisa* Pers. Uredinia and telia on *Salix pierotii* Miq. (*S. pseudo-korensis* Koidz.) and *S. chaenomeloides* Kimura (*S. glandulosa* Seem.).

Distribution: Japan and maritime provinces of Russia.

Matsumoto (1926) discovered the relationship between the *Melampsora* on *Salix pierotii*, *Chelidonium majus* and *Corydalis incissa*. To this rust he gave the name *M. chelidonii-pierotii*. The other *Salix* rust having sub-cuticular telia, with which *M. chelidonii-pierotii* is comparable, is *M. ribesii-viminalis* Kleb. However, morphologically *M. chelidonii-pierotii* has lengthy teliospores; in addition it differs in its aecial host range.

Melampsora choseniae Azb. et Koval, *Novit. Syst. Pl. Vasc.* 247–248; 1967

Spermogonia and aecia not found. Uredinia hypophyllous, 0.25–0.8 mm in diameter, round or oval; urediniospores round to oval, 14–25 × 14–17.8 µm, wall densely verrucose, 1.5–2 µm thick, hyaline; paraphyses capitate rarely clavate, 11.6–17.6 µm wide, wall 2–2.5 µm thick. Telia epiphyllous, subepidermal, blackish orange-brown, 0.1–0.5 mm; teliospores prismatic, 28.5–63.9 × 10–14 µm, apically not thickened, wall 1.5–2.5 µm, thick.
Type: On *Chosenia arbutifolia* (Pall.) A. Skv., 3–09–1955, coll. By E.Z. Koval, 15–08–1964, Leg M. Azbukina, Inst. Bot. Acad. Sci. URSS (Leningrad) Conservatur.
Hosts: *Chosenia arbutifolia*.
Distribution: Maritime provinces of Russia.

Melampsora dimorphospora Kaneko et Hiratsuka, f., *Trans. Mycol. Soc. Japan* 23, 201–210; 1982

Spermogonia and aceia unknown. Uredinia hypophyllous, occasionally epiphyllous, scattered, rounded, erumpent, 0.2–1 mm in diameter, brownish-yellow in dry state, sometimes containing one- or two-celled free teliospores; urediniospores are of two types: (1) borne singly on pedicels, echinulate, subglobose, broadly ellipsoid or globose, 19–30 × 17–25 µm, wall 2.5–3 µm thick, colourless, germ pores scattered; (2) catenulate in short chains, finely and densely verrucose, subglobose or broadly ellipsoid, 20–24 × 15–22 µm, wall 2–2.5 µm thick, colourless, germ pores scattered; paraphyses numerous, intermixed, capitate to clavate, 35–103 × 22–30 µm, wall 3–4 µm thick, slightly thickened at apex. Telia amphigenous, subepidermal, usually surrounding uredinia, rounded, 0.2–0.5 mm in diameter, reddish brown in dry state; teliospores adhering laterally, prismatic in surface view of telium, oblong and rounded at both ends in side view, rarely acute above, arranged in a single layer, one-celled, occasionally two-celled with a horizontal or oblique septum, 20–50 × 7.5–13.5 µm, wall *c.* 1.5 µm thick, yellowish-brown, not thickened at apex.
Type: On *Salix koriyanagi* Kimura, Pref. Aichi, Anjyo-cho, 5 Dec. 1938, K. Kuwazuka, HH. 77875.
Hosts: *Salix koriyanagi* Kimura.
Distribution: Japan.

M. dimorphospora is a unique species which produces two kinds of spores, e.g. caeomid uredinia producing catenulate, verrucose uredio spores and uredinoid uredinia producing catenulate urediniospores. No other species of *Melampsora* is known to produce this kind of dimorphic urediniospore.

Melampsora epiphylla Diet., *Englers Bot. Jahrb.* 32, 50; 1902

= *M. larici-opaca* Miyabe et Matsumoto, *Trans. Sapporo Nat. Hist. Soc.* 6, 22–37; 1915.

Spermogonia amphigenous; aecia hypophyllous; aeciospores globose, 15–20 × 12.5–18 µm; uredinia minute, mostly hypophyllous, very rarely epiphyllous, scattered, sub-epidermal, erumpent, pulverulent, circular, pale yellowish-orange, 0.6 mm, urediniospores 12.8–16 × 11.2–15 µm, usually globoid, rarely ovate, wall 1–2 µm thick, finely and prominently echinulate; paraphyses clavate to capitate, sometimes curved, 35 µm long, 10–15 µm wide, hyaline.

Telia epiphyllous, in dense clusters, very closely aggregated sub-cuticular, not erumpent up to 1 mm, dark cinnamon-brown; teliospores 24–45 × 8–16 µm, cylindrical, sessile, single-celled, wall 1 µm thick, brown, apically the corners are thickened, up to 2.5–4.8 µm, dark cinnamon-brown.
Type: On *Salix sachalinensis* Fr. Schm. (= *Salix shikokiana* Makino) Japan: S. Kusanoi, November 22, 1899.
Hosts: Spermogonia and aecia on *Larix decidua* Mill. and *L. leptolepis*; uredinia and telia on *Salix sachalinensis* Fr. Schm. (=*Salix shikokiana* Makino).
Distribution: China, Japan, former USSR.
Specimens examined: Type only.

Melampsora epiphylla Diet., resembles *M. capraearum* Thuem., in having epiphyllous, sub-cuticular telia, but differs from it in having small urediniospores and the apical walls of the teliospores are not thickened to the same extent as those of *M. capraearum*.

Melampsora epitea (Kunze. & Schm.) Thüm., *Mith. Forstl. Versuchsw. Oesterr* 2, 38, 40; 1879

= *Uredo epitea* Kunze. & Schm. *Mycol. Hefte* 1, 68; 1817.
= *M. hartigii* Thüm., *Mith. Forstl, Versuchsw. Oesterr* 2, 41; 1879.
= *M. coleosporioides* Diet., *Englers Bot. Jahrb.* 32, 50; 1902.
= *M. humilis* Diet., *Englers Bot. Jahrb.* 32, 50; 1902.
= *M. microsora* Diet., *Englers Bot. Jahrb.* 32, 50; 1902.
= *M. humboldtiana* Speg., *Ann. Mus. Mat. Buenos-Aires* 23, 28; 1912.
= *M. larici-miyabeana* Miyabe et Mats., *Trans. Sapporo Nat. Hist. Soc.* 6, 22–35; 1915.
= *M. kiusiana* Hirat. f., *Bot. Mag. Tokyo*, 57, 281; 1943.

Spermogonia amphigenous, sub-cuticular, variable in shape and size, 60–120 µm high, 150–200 µm wide. Aecia caeomoid, amphigenous, scattered or aggregated, sub-epidermal, erumpent, pulverulent, rarely having a rudimentary peridium, orange; aeciospores 15–25 × 10–19 µm, globoid, angulately globoid or ovoid, wall 1.5–2.5 µm, finely verrucose. Uredinia amphigenous, sub-epidermal, in the beginning covered by the rudimentary peridium, later on erumpent, pulverulent, 0.5–1.5 mm; urediniospores 12–25 × 10–19 µm, wall 1.5–3 µm thick, echinulate; paraphyses numerous, capitate, up to 80 µm long, 15–25 µm wide. Telia minute, amphigenous, but mostly hypophyllous, scattered or aggregated, often fusing, sub-epidermal, non-erumpent, 0.3–1 mm in diameter, blackish-brown; teliospores 20–50 × 7–15 µm, rounded at both the ends, wall 2 µm or slightly thicker at the apex.
Type: On *Salix alba* L., Bayreuth, Bavaria, Germany (Thüm, Hert, Mye, Oec. 388. sub-nom. *M. salicina*).
Hosts: Spermogonia and aecia on species of *Abies*, *Euonymus*, *Larix*, *Ribes*, *Saxifraga* and several genera of *Orchidaceae*. Uredinia and telia on species of *Salix*.
Distribution: Canada, China, Europe, India, USA.
Specimens examined: DAVFP, LEV, PAV, S, G, WRSL, HCIO, B.

The willow rust, *M. epitea* has various races which do not differ greatly in their morphological characters but differ only in their aecial hosts; all these should be considered as *formae speciales*. Most of the willow rusts are included under this collective species. This follows Hylander *et al.* (1953), Wilson and Henderson (1966), Boerema and Verhoeven (1972) and Ziller (1974).

Melampsora epitea f. sp. *abieti-capraearum* (Tub.) Bagyanarayana comb. nov. (Fig. 3.1)

= *Caeoma abietis-capraearum* Tub. *Centralbl. Bak.* II, 241; 1902.
= *M. abieti–capraearum* Tub. *Centralbl. Bak.* II, 241; 1902.
= *M. americana* Arth., *Bull. Torrey Club* 47, 465; 1920.

Spermogonia amphigenous, pale yellow, 70–115 µm wide, 60–70 µm high. Aecia hypophyllous, solitary, parallel to the

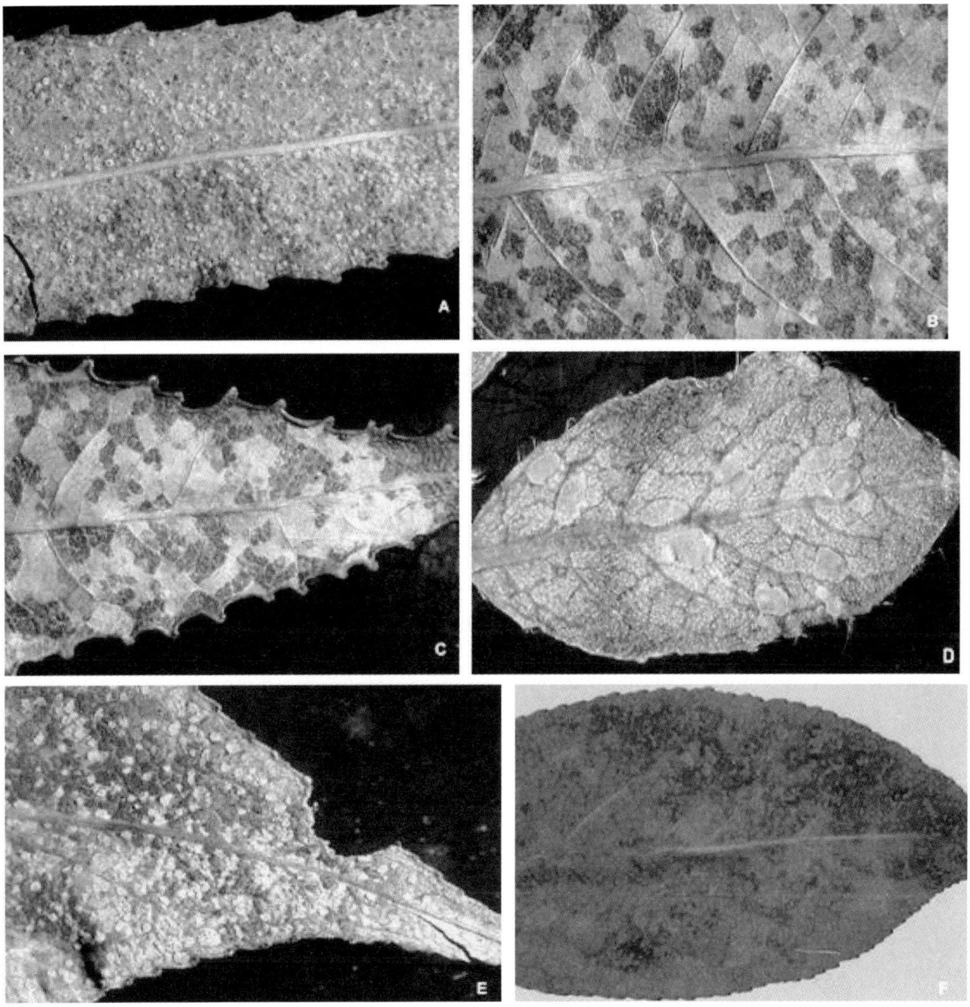

Fig. 3.1. (A) *M. epitea* f. sp. *abieti-capraearum* on *Salix nigra* (Uredinia & Telia); (B) *M. allii-fragilis* f. sp. *galanthi-fragilis* on *Salix fragilis* (Telia); (C) *M. amygdalinae* on *Salix triandra* (Telia); (D) *M. epitea* f. sp. *arctica* (Uredinia); (E) *M. capraearum* on *Salix* sp. (Uredinia & Telia); (F) *M. larici-pentandrae* on *Salix pentandrae* (Uredinia & Telia).

midrib, oblong, 0.4–0.6 mm long, often fusing, sub-epidermal, erumpent, pulverulent, orange yellow; aeciospores 15–24 × 12–18 µm, globoid or ellipsoid, sometimes angular, wall 1.5–2.5 µm thick, verrucose. Uredinia minute, hypophyllous, scattered, 0.5 mm, sub-epidermal, erumpent, pulverulent, orange; urediniospores 15–20 × 10–16 µm, globoid to broadly ellipsoid, wall 1.5–2 µm thick, echinulate; paraphyses capitate, up to 50 µm long, 15–25 µm wide, wall 1.5–3.5 µm thick. Telia hypophyllous, scattered, 2 mm wide, more or less fusing, sub-epidermal, non-erumpent, blackish-brown; teliospores 20–30 × 7–12 µm, oblong, wall 1 µm thick, light brown.

Type: On *Salix caprea* L., Barnau, near Chiemsee, Bavaria, Germany.

Hosts: Spermogonia and aecia on *Abies amabilis* (Dougl.) Forb., *A. balsamea* (L.) Mill., *A. cephalonica* Loud., *A. concolor* Lindl. & Gord., *A. grandis* (Dougl.) Lindl., *A. lasiocarpa* (Hook) Nutt., *A. nordmanniana* (Stev.) Spach., *A. pectinata* D.C., *A. pinaspo* Boiss. Uredinia and telia on *Salix amygdalina* L., *S. appendiculata* Vill.,

S. argyrocarpa Anderss., *S. aurita* L., *S. bebbiana* Sarg., *S. caprea* L., *S. cinerea* L., *S. commutata* Bebb., *S. cordata* Muhl., *S. exigua* Nutt., *S. humilis* Marsh., *S. humboldtiana* Willd., *S. incana* Schrank, *S. interior* Rowlee, *S. longifolia* Wall ex Anderss, *S. mackenzieana* Barr. Apud Anderss, *S. nigra* Marsh, *S. nigricans* Sm., *S. pedicellaris* Pursh., *S. petiolaris* Sm., *S. purpurea* L., *S. repens* L., *S. scouleriana* Barr., *S. sericea* Muhl.

Distribution: Argentina, Canada, Europe, USA.

Specimens examined: DAVFP, FH, HCIO, S.

Tubeuf (1902) created the species *Melampsora abieti-capraearum* to accommodate a rust on *Salix* having *Abies* as its aecial host. Arthur (1920) termed this *Salix–Abies* rust as *M. americana*. Later, Arthur (1934) identified this rust as equivalent to *M. abieti-capraearum*. However, morphologically *M. abieti-capraearum* is similar to *M. epitea*, a collective species having several *formae speciales* on different aecial hosts. As such *M. abieti-capraearum* is reduced to the rank of a *forma specialis* of *M. epitea*.

Host alternation was established by Tubeuf (1902, 1905). Fraser (1912, 1913) and Weir and Hubert (1918) made cultural studies of this rust fungus.

Melampsora epitea (Kunze & Sch.) Thüm., f. sp. *arctica* (Rostr.) Bagyanarayana, comb. nov.

= *M. alpina* Juel., *Ofvens K. Vetensk. Akad. Forh.* 51, 417; 1894.
= *Uredo alpina* Arth., *Res. Sci. Congr. Bot. Vienne* 338, 1906.
= *U. rostrupiana* Arth., *N. Am. Flora* 7, 100; 1907.
= *M. arctica* Rostr., *Medd on Gronl.* 3, 535; 1888.

Spermogonia amphigenous, scattered or in groups, sub-epidermal, 150–160 μm wide, 90–130 μm high; aecia minute, caeomoid, amphigenous, chiefly hypophyllous, 0.3–0.5 mm, sub-epidermal, erumpent, pulverulent; aeciospores 15–26 × 15–21 μm, globoid, ovoid, rarely angular, wall 2–3 μm thick, finely verrucose.

Uredinia minute, amphigenous, chiefly hypophyllous, scattered or aggregated, 0.5 mm in diameter, sub-epidermal, erumpent, pulverulent, bright orange-yellow; urediniospores 15–25 × 12–20 μm, globoid, broadly ellipsoid, wall 2 μm thick, echinulate; paraphyses numerous, capitate, up to 65 μm long, 15–24 μm wide, wall 4–6 μm thick. Telia minute, amphigenous, scattered, 0.3–0.5 mm in diameter, often fusing, reddish-brown, sub-epidermal, non-erumpent; teliospores 25–45(50) × 8–17 μm, oblong, wall 1.5 μm thick, pale golden-brown.

Lectotype: On *Salix groenlandica* (Anderss.) Lundst., Christianshaab, Greenland.

Hosts: Spermogonia and aecia on *Antiphylla oppositifolia* (L.) Forr., *Leptasea flagellaris* (Willd.) Small, *Saxifraga adscendens* L., *S. bracteata* D. Don., *S. cernus* L., *S. caespitosa* L., *S. exilis* Steph., *S. groenlandica* (Anderss.) Lundst., *S. hypnoides* L., *S. oppositifolia* L. Uredinia and telia on *Salix anglorum* Cham., *S. aquilonica* Kimura, *S. arctica* Pall., *S. balsamifera* (Hook.) Barratt, *S. chamissoni* Anderss, *S. fuscescens* Anderss, *S. glauca* L., *S. glaucops* Anderss., *S. groenlandica* (Anderss.) Lundst., *S. herbacea* L., *S. lanata* L., *S. myrtillifolia* Anderss., *S. myrsinites* L., *S. ovalifolia* Trautr., *S. petrophila* Rydb., *S. poloris* Wah., *S. pyrenaica* Gou., *S. pyrifolia* Anderss., *S. reptans* Rupr., *S. rotundifolia* Trautr., *S. sitchensis* Sanson, *S. stolonifera* Cav., *S. subreniformis* Kimura, *S. yezoalpina* Koidz.

Distribution: Canada, Europe, Greenland, Japan, USA.

Specimens examined: DAOM, FH, S.

Melampsora arctica Rostrup, is morphologically very much similar to *M. epitea*, a collective species having *Salix* as its telial host. As such it has been reduced to the level of a *forma specialis*. The other *Salix Melampsora* having *Saxifraga* as its aecial host is *M. epitea* f. sp. *reticulatae* (Blytt) Jorst., from which *M. epitea* f. sp. *arctica* differs in having smaller urediniospores with thinner walls.

Gäumann (1959) recognized two biological races under the species *M. alpina* Juel. They are *M. alpina* Juel and *M. reticulatae* Blytt. However, Arthur (1934) considered *M. alpina* Juel as identical with *M. arctica* and treated *M. alpina* as a synonym of *M. arctica*. But Gäumann (1959) preferred to retain *M. alpina* and considered *M. arctica* as different. The present study reveals that there is no justification in the retention of the species *M. alpina*, accordingly the author has followed Jørstad (1923) and Arthur (1934).

Melampsora epitea Thüm., f. sp. *euonymi* (Kleb.) Boerema et Verhoeven, *Neth. J. Pl. Path.* 78, 24; 1972

= *M. euonymi-capraearum* Kleb., *Forstl. Naturw. Zeischr.* 6, 469, 1897.
= *M. euonymi-capraearum* Kleb. ex Kleb., *Pringsh. Jahrb. Wiss.* 34, 358, 1900.
= *M. euonymi-incarnae* O. Schneid., Central Blatt. f.
= *Uredo euonymi-capraearum* Arth., *Result. Scient.Congr. Intern. Bot. Vienne* 338, 1905, 1906.

Spermogonia prominent, 200 µm wide, 80 µm high. Aecia caeomoid, mostly hypophyllous, occasionally epiphyllous, often aggregated in groups, up to 2 mm, sub-epidermal, erumpent, pulverulent, bright orange; aeciospores 18–26 × 12–20 µm, sub-globose, ovoid or ellipsoid, slightly angular, wall up to 5 µm thick, verrucose. Uredinia minute, hypophyllous, scattered or aggregated, 0.5 mm, sub-epidermal, erumpent, pulverulent, orange; urediniospores 14–20 × 12–17 µm, globose, sub-globose or ovoid, walls unevenly thickened, 1.5–4 µm, echinulate; paraphyses capitate, up to 70 µm long 18–25 µm wide, wall 2.7 µm thick. Telia minute, hypophyllous, deposited in groups, 0.5 mm in diameter, sub-epidermal, non-erumpent, dark brown; teliospores 24–40 × 7–14 µm, rounded at both the ends, wall 1 µm, slightly thickened apically, light brown.

Type: On *Salix cinerea* L., Steinbek, Hamburg, Germany.

Hosts: Spermogonia and aecia on *Euonymus europaeus* L., *E. latifolia* Scop., *E. verrucosa* Scop.; uredinia and telia on *Salix aurita* L., *S. caprea* L., *S. cinerea* L., *S. incana* Schrank., *S. viminalis* L.

Distribution: Europe.

Specimens examined: C, G, S.

Nielsen and Rostrup (1884, see Gäumann, 1959) studied the genetic relationship between the caeoma developed on *Euonymus europaeus* L. and a *Melampsora* grown on *Salix caprea* L. and *S. cinerea* L. Klebahn (1899) extended these studies and created the species *M. euonymi-capraearum*. However, Boerema and Verhoeven (1972) reduced it to the level of a *forma specialis*. The author agrees with the treatment of Boerema and Verhoeven.

Melampsora epitea Thüm., f. sp. *lapponum* (Lindf.) Bagyanarayana comb. nov.

= *Caeoma violae* Lindf., *Sv. Bot. Tidskr.* 4, 193, 1910.
= *M. lapponum* Lindf., *Sv. Bot. Tidskr.* 7, 48, 1913.

Spermogonia not known. Aecia hypophyllous, aggregated in groups, circularly deposited, 0.3–1 mm in diameter, sub-epidermal, erumpent, pulverulent, orange; aeciospores 14–27 × 12–20 µm, globose to ellipsoid, wall 2–2.5 µm thick, verrucose. Uredinia minute, hypophllous, 0.2–0.4 mm in diameter, sub-epidermal, erumpent, pulverulent, orange; urediniospores 14–24 × 10–20 µm, globose, sub-globose, ovate or ellipsoid, sometimes even angular, echinulate; paraphyses clavate to capitate, up to 55 µm in length, 15–30 µm in width, wall 1–4 µm thick. Telia minute, usually epiphyllous, 0.2–0.5 mm in diameter, sub-epidermal, non-erumpent, orange-brown; teliospores 30–50 × 6–12 µm, rounded at both ends, wall 1 µm thick, brown.

Type: On *Salix lapponum* L., Njuonjes, Lale lappomark, Sweden.

Hosts: Aecia on *Viola epipsila* Ledeb., *V. palustris* L., uredinia and telia on *Salix lapponum* L.

Distribution: Northern and central Europe.
Specimens examined: S.

According to Lindfors, the supposed difference between *M. lapponum* and typical *M. epitea* is that *M. lapponum* has wider uredo paraphyses. In addition, as the aecial host is *Viola*, *M. lapponum* is treated here as a *forma specialis* of *M. epitea*.

Melampsora epitea f. sp. *larici-epitea* Bagyanarayana comb. nov.

= *M. larici-epitea* Kleb., *Ztschr. Pflanzenkr.* IX, 1899.
= *Caeoma laricis* Hartig., *Wichtige Krankh. d. Waldbaume*, 1874, S. 93, pr. P.
= *Uredo larici-epitea* Arth., *Result. Scient. Congr. Intern. Bot. Vienne* 1905, 1906, p. 338.

Spermogonia subcuticular, round–conoid, 70–100 μm wide, 30–40 μm high. Aecia hypophyllous, single or in rows, with corresponding yellow spots on the upper side of the leaf, round or elongate, 0.5–1.5 mm long, pale orange-coloured; aecia of the fungus on *Salix retusa* are surrounded by a crown of capitate, thin-walled paraphyses. Aeciospores globoid, ellipsoid or slightly angular, 15–25 × 10–18 μm; wall about 1.5–3.0 μm thick, with a delicate, sparse, very short-ridged structure about 0.5 μm thick, at intervals of approximately 1 μm.

Uredinia either amphigenous or confined to the upper or lower side of the leaves, on yellow spots, 0.25–1.5 mm, orange-yellow. Urediospores usually ovoid, globoid or angular, 12–25 × 9–19 μm, slightly varying on different host plants; walls rather thick, 1.5–3.5 μm, echinulate, paraphyses capitate, 35–80 μm long, width 15–24 μm; wall 3–5 μm thick, apically thickened up to 10 μm.

Telia generally hypophyllous, on *Salix retusa*, *S. miyabeana*, *S. sachalinensis*, also epiphyllous; sub-epidermal, dark brown, sometimes tinged with violet, small, 0.25–1.0 mm, often closely packed or coalescing in small groups covering portions of the blade delimited by the small veins. Teliospores prismatic, wedge-shaped or irregular, usually rounded at both ends, 20–50 × 7–14 μm, walls light brown, uniformly thin, about 1 μm; pore imperceptible.
Hosts: Spermogonia and aecia on *Larix*. Uredinia and telia on *Salix aurita*, *S. caprea*, *S. cinerea*, *S. hastate*, *S. oxycarpa*, *S. repens* var. *argentea*, *S. viminalis*.
Distribution: Eurasia, North Africa, India.

M. larici-epitea Kleb. is morphologically indistinguishable from *M. epitea*, therefore reduced to the level of *forma specialis*. Uredinio- and teliospores on numerous species of *Salix* (comprising ten sections). Heterogeneous species break up into closely related biological forms adapted to different hosts.

Melampsora epitea f. sp. *repentis* (Plowr.) Boerema and Verhoeven *Neth. J. Pl. Path.* 78, 24; 1972

= *Uredo confluens* var. *orchidis* Arth. and Schw., *Consp. Fung. Nisk.* 122; 1805.
= *Caeoma orchidis* (Alb. and Schw.) Wint., *Rabh. Krypt. Fl.* 1, 256; 1882.
= *Melampsora repentis* Plowr., *Zeitschr. f. Pfl. Krankh.* 1, 131; 1891.
= *M. orchidi-repentis* Kleb., *Jahrb. Wiss. Bot.*, 34, 369; 1900.

Spermogonia lenticular, sub-epidermal, with flat hymenium, 170 μm wide and 80 μm high. Aecia caeomoid, hypophyllous, scattered or aggregated, often circularly deposited, 1–2 mm in diameter, often fusing, sub-epidermal, erumpent, pulverulent, orange-yellow; aeciospores 15–23 × 14–27 μm, globoid, angularly globose, ovoid, wall 1–1.5 μm thick, verrucose. Uredinia minute, hypophyllous, scattered or aggregated, 0.3–0.8 mm in diameter, sometimes fusing, sub-epidermal, erumpent, pulverulent, orange; urediniospores 12–20 × 12–16 μm, globose, ovate or ellipsoid, wall 1.5 μm thick, echinulate; paraphyses capitate, up to 75 μm long, 15–20 μm wide, wall 2.5 μm thick. Telia minute, hypophyllous, rarely epiphyllous, sub-epidermal, not erumpent, dark brown; teliospores 17–50 × 7–15 μm, both the ends

round, rarely irregular, wall 1 μm thick, light brown.
Type: On *Orchis latifolia* L., Germany, Otto Jaap, Hamburg, June 1, 1905.
Hosts: Spermogonia and aecia on *Gymnadenia comopsea* (L.) R. Br., *Leucoorchis albida* (L.) E. Mey, *Listera ovata* (L.) R. Br., *Orchis angustifolia* Lois., *O. incarnata* L., *O. insectifera* Crantz., *O. latifolia* L., *O. majalis* Rchb., *O. mascula* L., *O. militaris* L., *O. moris* L., *O. muscifera* Huds., *O. praetermissa*, *O. sambucina* L., *O. strictifolia* Opiz., *Platanthera bifolia* (L.) L.C. Rich. Uredinia and telia on *Salix aurita* L. and *S. repens* L.
Distribution: Europe and USA.
Specimens examined: FH, K, S.

Plowright (1891), transferred aeciospores from *Orchis maculata* to *Salix repens* and obtained uredinial and telial stages. Similarly, he also succeeded in infecting *Orchis maculata* using teliospores from *Salix repens*. To this fungus Plowright gave the name *M. repentis*. For the sake of better clarity, Klebahn (1900) named this fungus as *M. orchidi-repentis* (Plowr.). Kleb. Gäumann (1959) was not in agreement with the views of Klebahn. However, Boerema and Verhoeven reduced *M. repentis* Plowr., from the status of a species to that of a *forma specialis*. The author has followed Boerema and Verhoeven in treating *M. repentis* as a *forma specialis*.

Melampsora epitea f. sp. *reticulatae* (A. Blytt.) Jorst., *Skr. Norske. N. vidensk. Akad. Osla* 6, 31; 1940

= *Uredo saxifragarum* D.C., *Fl. Franc.* 6, 87; 1815.
= *Caeoma saxifragae* Wint., *Rabh. Krypt. Fl. Ed.* 1, 258; 1882.
= *M. reticulatae* A. Blytt., *Chria. Vidensk. Selsk. Forh. 1886* 6, 65; 1896.

Spermogonia scattered or aggregated, 90–125 μm high, 150 μm wide, light orange. Aecia caeomoid, amphigenous, mostly epiphyllous, solitary, 0.5–1 mm in diameter, sub-epidermal, erumpent, pulverulent, orange; aeciospores 16–25 × 14–20 μm, globoid, slightly angular, ovoid, wall 2–3 μm thick, verrucose. Uredinia minute, hypophyllous, scattered or aggregated, 0.5 mm in diameter, orange, sub-epidermal, erumpent, pulverulent; urediniospores 17–30 (35) × 15–20 (25) μm, globoid, broadly ellipsoid, ovoid, wall 2–3.5 (5) μm thick, echinulate; paraphyses numerous, capitate, up to 95 μm long, 20–30 μm wide, wall thickened up to 10 μm. Telia minute, amphigenous, dense, scattered, 0.4–0.8 mm in diameter, sub-epidermal, not erumpent, dark brown; teliospores 30–45 (50) × 7–10 (14) μm, rounded at both the ends, wall 1.5 μm thick, slightly thickened above, pale brown.
Type: On *Salix reticulata* L., Finland, Finmark, Norland, Alps, Dovre.
Hosts: Spermogonia and aecia on *Saxifraga aizoides* L., uredenia and telia on *Salix amygdalina* L., *S. bebbiana* Sarg., *S. brachycarpa* Nutt., *S. breweri* Bebb., *S. caprea* L., *S. coulteri* Anderss., *S. piperi* Bebb., *S. purpurea* L., *S. reticulata* L., *S. uva-ursi* Pursh.
Distribution: Europe.
Specimens examined: FH, HCIO, S, WRSL.

Melampsora epitea var. *reticulatae* (Blytt.) Jorst., differs from *M. epitea* var. *arctica* (Rostr.) Bagyanarayana in having bigger urediniospores with thick walls and longer uredinial paraphyses.

From their field observations, Jaap (1908) and Schneider (1910) found *Saxifraga aizoides* as an alternate host. Henderson (1953) determined the genetic relationship. Scaramella (1932) found the perennation of the mycelium in *Salix*, whereas Henderson (1953) found the perennation of the mycelium in the buds of *Saxifraga*. Dietel (1930) reported the presence of two-celled spores in the telia.

Melampsora epitea f. sp. *ribesii-purpureae* (Kleb.) Bagyanarayana comb. nov. (Fig. 3.2B)

= *M. ribesii-purpureae* Kleb., *Pringsh. Jahrb. Wiss. Bot.* 35, 667; 1901.
= *M. ribesii-salicum* Bub., *Arch. Naturw. Landesdurchf. Bohn.* 5, 200; 1908.

= *M. ribesii-auritae* Kleb., *Pringsh. Jahrb. Wiss. Bot.* 35, 670; 1901.
= *M. ribesii-grandifoliae* O. Schneid., *Central blatt f. Bakteriol.* 15, 238; 1905.
= *M. ribesii-epitea* Kleb., *Krypt. fl. Mark. Brandenb.* Va (Pilze III), 708; 1913.
= *M. confluens* (Pers.) Jackson, *Brookl. Bot. Gard. Mem.* I, 210; 1918.
= *M. ribesii-epitea* Kleb. f. sp. *ribesii-auritae* Kupr., Kuprevich & Transchel, *Fl. Pl. Crypt. URSS* 4, Ured. 1, 374; 1957.
= *M. ribesii-epitea* Kleb. f. sp. *ribesii-purpureae* Kupr., Kuprevich & Transhel, *Fl. Pl. Crypt. URSS* 4, Ured. 1, 374; 1957.

Spermogonia chiefly epiphyllous, aggregated, sub-epidermal 60–80 μm high, 150–200 μm wide. Aecia often hypophyllous, scattered or circularly aggregated, 0.5–1.5 mm in diameter, often fusing, sub-epidermal, erumpent, pulverulent, orange; aeciospores 15–26 × 12–20 μm, globose, angularly globose or ellipsoid, wall 1.5–3 μm thick, minutely verrucose. Uredinia chiefly hypophyllous, scattered or aggregated, 1–3 mm, orange, sub-epidermal, erumpent, pulverulent; urediniospores 15–25 × 14–20 μm, globose, sub-globose or ovate, wall 2–3.5 μm, echinulate; paraphyses clavate to capitate, up to 80 μm long, 15–25 μm wide. Telia minute, often hypophyllous, sub-epidermal, non-erumpent, scattered or aggregated, 0.2–0.5 mm in diameter, dark brown; teliospores 20–32 × 7–12 μm, round at both ends, wall 1 μm, light brown.

Type: On *Salix purpurea* L., Trignitz inder Prignitz, Brandenburg, Germany.

Host: Spermogonia and aecia on *Ribes alpinum* L., *R. atropurpureum* C.A.M., *R. aureum* Pursh., *R. grossularia* L., *R. nigrum* L., *R. reclinatum* (L.) Mill., *R. rubrum* L., *R. sanguineum* Pursh., *R. spicatum* Robs., *R. uva-crispa* L., *R. vallicola*. Uredinia and telia on *Salix arbuscula* L., *S. aurita* L., *S. bebbiana* Sarg., *S. bicolor* Fries., *S. caprea* L., *S. daphnoides* Vill., *S. glauca* L., *S. grandifolia* Seringe., *S. nigricantis*, *S. purpurea* L., *S. rubra* Huds., *S. scouleriana* Berr.

Distribution: Europe and USA.

Specimens examined: DAOM, S.

Klebahn's *Melampsora ribesii-purpureae* is morphologically very much similar to that of *M. epitea* the collective species. The other *Melampsora* on *Salix* having *Ribes* as its aecial host is *M. ribesii-viminalis*, but the telia in *M. ribesii-viminalis* are sub-cuticular whereas in *M. epitea* f. sp. *ribesii-purpureae* they are sub-epidermal.

The development and specialization of *M. epitea* f. sp. *ribesii-purpureae* was studied by Klebahn (1900, 1902, 1903), Schneider (1906) and Mayor (1924, 1958). According to Scaramella (1932), the dikaryotic mycelium can overwinter in the branches of *Salix purpurea* and some other *Salix* species, so that new infection need not take place in every spring.

Melampsora epitea f. sp. *tsugae* Ziller, *Can. J. Bot.* 37, 115; 1959

= *Caeoma dubium* C.A. Ludwig, *Phytopath.* 5, 281; 1915.

Spermogonia on current year's needles, occasionally on cones; minute, lenticular, originating under the epidermis. Aecia caeomoid, on current year's needles, on cones, scattered, sub-epidermal, erumpent, pulverulent, pale yellow, up to 1 mm; aeciospores 15–24 × 14–21 μm, globose, sub-globose, ovoid, wall apically thickened, verrucose. Uredinia minute, hypophyllous, scattered or aggregated, sub-epidermal, erumpent, pulverulent, pale yellow, up to 0.6 mm; urediniospores 14–20 × 12–17 μm, globoid to ellipsoid, wall 1–2 μm, uniformly thick, echinulate; paraphyses numerous, prismatic, wall 1 μm thick, brown. Telia hypophyllus, teliospores 6–14 × 16–30 μm.

Host: Spermogonia and aecia on *Tsuga heterophylla* (Raf.) Sarg., and *T. mertensiana* (Bong.) Carr. Uredinia and telia on *Salix scouleriana* Barr.

Distribution: Canada and USA.

Specimen examined: DAVFP.

Ziller (1959) created *M. epitea* f. sp. *tsugae* as a race of the *M. epitea* complex. Ziller (1974) is of the opinion that *M. epitea* f. sp. *tsugae* is morphologically indistinguishable from *M. abieti-carpraearum*. However, these two rusts differ in their aecial host range, as shown by Ziller (1959).

The basidiospores of *M. epitea* f. sp. *tsugae* do not infect *Abies* sp., but infect *Tsuga* sp., whereas the basidiospores of *M. epitea* f. sp. *abieti-capraearum* do not infect *Tsuga* sp. but infect the *Abies* species only.

Melampsora kamikotica Kaneko et Hiratsuka, f., *Trans. Mycol. Soc. Japan* 23, 201–210; 1982

Spermogonia and aecia unknown. Uredinia amphigenous, scattered, rounded, erumpent, 0.3–0.5 mm in diameter, pulverulent, yellow; urediniospores obovoid, ellipsoid or broadly ellipsoid, 20–33 × 15–18 μm, echinulate, wall 2–2.5 μm thick at sides, 2.5–7.5 μm thick at apex, colourless, germ pores 4–6, scattered, cytoplasm yellow; paraphyses numerous, intermixed, capitate to clavate, 40–75 × 15–25 μm, wall 2–2.5 μm thick at sides, 5–6.5 μm thick at apex. Telia amphigenous, sub-epidermal, scattered, rounded, 0.3–0.8 mm in diameter, reddish-brown; teliospores adhering laterally, prismatic in surface view of telium, one-celled, oblong and rounded at both ends in side view, arranged in a single layer, 25–53 × 10–16 μm, wall 1 μm thick, yellowish brown, not thickened at apex.
Hosts: *Chosenia arbutifolia.*
Distribution: Japan.
Type: On *Chosenia arbutifolia* (Pall.) A. Skvortz. Central Honshu, Pref. Nagano, Kamikochi, 10 September, 1979, Dr Shigeru Kaneko HH – 73060; TMI – 7105.

M. kamikotica differs from *M. choseniae*i in having longer, apically thickened walls.

Melampsora kuperviczii Zenk., *Novit. Syst. Pl.* Vasc. 2 Non Vasc., 197–199; 1964.

Spermogonia and aecia not found. Uredinia hypophyllus, 0.25–0.5 mm in diameter, round or oval, paraphyses hyaline; urediniospores globose, ovate, elliptical, angular, 10–15 × 10–12 μm, wall densely verrucose; 2 μm thick, hyaline; paraphyses coronate–capitate; 17.5–20 μm, wall 2.5 μm thick. Telia epiphyllous, sub-epidermal, prismatic, 30–37 × 10–15 μm, wall very thin 0.5–0.7 μm, apically not thickened, germ pore distinct.
Type: *Salix brachypoda* (Trautv. et May.) Kom.
Hosts: *Salix brachypoda.*
Distribution: Amur River, Svobodnensk, 1959, B.A. Tomolin, Russia.

Melampsora larici-pentandrae Kleb., *Forstl. Naturw. Zeitscher* 6, 470; 1897. (nomen provisorium) *Zeitschr f. Pfl. Krankh.* 7, 330; 1897 (Fig. 3.2A)

= *Uredo minutissima* Opiz., *Sezman Rostlin Kvitney Ceske* 152, 1852.
= *U. larici-pentandrae* Arth., *Result. Scient. Congr. Intern. Bot. Vienne* 338, 1; 1905, 1906.
= *M. minutissima* (Opiz.) Bub., *Arch., Naturw. Landesdurchf. Bohn* 5, 194; 1908.

Spermogonia amphigenous, subcuticular, 60–100 μm wide, 30–50 μm high. Aecia caeomoid, minute, mostly hypophyllous, scattered, orange; aeciospores 15–26 × 13–20 μm, ovoid, globoid or slightly angular, wall 2 μm thick, finely verrucose. Uredinia mostly hypophyllous, very rarely epiphyllous, scattered or aggregated, often fusing, up to 1 mm in diameter, subepidermal, erumpent, pulverulent, orange; urediniospores 25–45 × 17–17 μm, ovoid or broadly ellipsoid, wall 2 μm, apically thickened up to 5 μm, echinulate; paraphyses clavate to capitate, up to 70 μm long, 12–17 μm wide. Telia hypophyllous, minute, scattered or aggregated, subepidermal, not erumpent, often fusing, 0.5 mm, dark brown; teliospores 23–38 × 6–12 μm, round at both ends, wall 1.5 μm thick, light brown.
Type: On *Salix pentandra* L., Borsteler Jager, near Hamburg, Germany.
Hosts: Spermogonia and aecia on *Larix decidua* L., *L. sibirica* Ldb. Uredinia and telia on *Salix fragilis* L. and *S. pentandra* L.
Distribution: Europe.
Specimens examined: S.

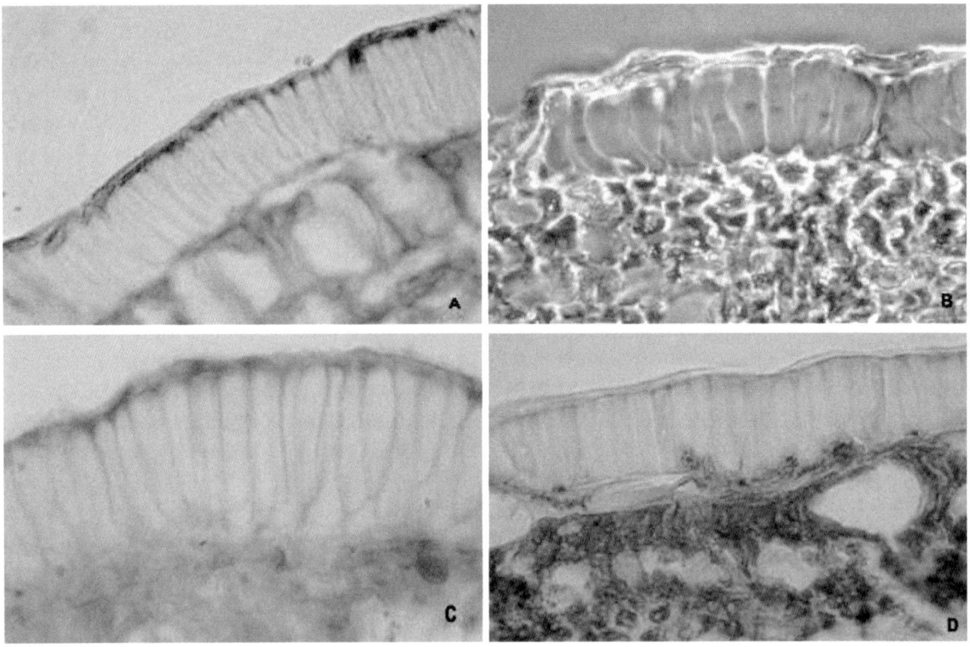

Fig. 3.2. V. S. of Telium (450×). (A) *M. larici-pentandrae*; (B) *M. epitea* f. sp. *ribesii-purpureae*: (C) *M. ribesii-viminalis*; (D) *M. yezoensis*.

Klebahn (1897) created the species *M. larici-pentandrae*, a *Salix Melampsora* having *Larix* as its aecial host. Later, Bubák (1908) named the *Salix–Larix* rust as *M. minutissima*. However, the name *M. larici-pentandrae* Kleb., is preferred over *M. minutissima* as it is the earlier, validly published name.

The characteristic feature of this species is the urediniospore wall, which is conspicuously thickened apically, a feature not commonly seen in other species of *Melampsora*. Klebahn (1897, 1899, 1902, 1903) made a detailed study of the development and host specialization of this species.

Melampsora larici-urbaniana Mats., *Ann. Missouri Bot. Gard.* 6, 311; 1919

Spermogonia not known. Aecia caeomoid, hypophyllous, scattered, pale, orange-yellow; aeciospores 15–26 × 13–20 µm, ovoid, globoid or ellipsoid, wall 3–4 µm thick, echinulate. Uredinia hypophyllous, dense, scattered, sub-epidermal, erumpent, pulverulent, orange-yellow; urediniospores 15–26 × 10–17 µm, ovoid, oblong or ellipsoid, wall 2–3 µm thick, echinulate; paraphyses capitate, up to 70 µm in length, 16–25 µm in width, wall 3–4 µm thick. Telia hypophyllous, scattered often fusing, sub-epidermal, not erumpent, dark reddish-brown; teliospores 35–70 × 8–15 µm, prismatic, cylindrical, round at both the ends, wall 1.5 µm, light brown.

Type: *Toisusu urbaniana* (Seem.) Kimura (= *Salix urbaniana* Seem.), Hokkaido, Sapporo-shi, 4 October 1901, J. Hanzawa SAPA: HH78307.

Host: Aecia on *Larix decidua*, *L. kaempferi* Sarg. Uredinia and telia on *Salix arctica* × *S. cuneata* and *Toisusu urbaniana* (Seem.) Kimura (= *S. urbaniana* Seem.)

Distribution: Japan.

Specimens examined: Herb. Hiratsuka.

Matsumoto, in 1916, carried out extensive inoculation experiments using teliospores from *Salix urbaniana* to inoculate *Larix decidua*, and aeciospores from *Larix decidua* to inoculate *Salix urbaniana*. In both ways he got positive results. To this rust, Matsumoto (1919) gave the name *M. larici-urbaniana.*

Another *Melampsora* species on *Salix*, having *Larix* as its aecial host and resembling *M. larici-urbaniana*, is *M. larici-pentandrae* Kleb. However, *M. larici-urbaniana* differs from *M. larici-pentandrae* both in its urediniospore and teliospore morphology. Hiratsuka *et al.* (1992) reported 4–6 scattered germ pores in the urediniospores and an apical thickening in the urediniospore walls. However, the present author failed to notice these features.

Melampsora paradoxa Diet. et Holw., *Hedw. Beibl.* 40, 32; 1901

= *M. bigelowii* Thüm., *Mitth. Forstl. Vers. Oest.* 2, 37; 1879.
= *Lecythea macrosora* Perck., *Bot. Gaz.* 5, 35; 1880.
= *M. larici-retusae* Ed. Fisch., *Ured. Schweiz.* 487, 1904.
= *M. larici-nigricantis* O. Schneid., *Centrbl. Bakteriol.* II, Abt. 13, 233; 1904.
= *M. larici-purpureae* O. Schneid., *Centrbl. Bakteriol.* II, Abt. 13, 233; 1904.
= *M. larici-reticulatae* O. Schneid., *Centrbl. Bakteriol.* II, Abt. 15, 233; 1905.
= *Uredo bigelowii* Arth., *Result. Scenti. Congr. Intern. Bot. Vienne*, 338, 1906.

Spermogonia minute, amphigenous, scattered or in small groups, pale yellow, inconspicuous, conical, sub-cuticular, 60–80 μm in diameter, and 50 μm height. Aecia caeomoid, mostly hypophyllous, scattered or aggregated in small groups, minute, 0.1–0.3 mm, pale orange, sub-epidermal, erumpent, pulverulent, aeciospores 17–28 × 15–23 μm, globose or sub-globose, wall 2–3 μm thick, finely verrucose. Uredinia mainly hypophyllous, scattered or aggregated, minute, 0.3–0.6 mm in diameter, orange, turning to pale yellow, sub-epidermal, erumpent, pulverulent; urediniospores 15–26 × 15–21 μm, globose or sub-globose, wall 2.5–3.5 μm thick, paraphyses capitate, up to 70 μm long and 20–25 μm in width, 3.5 μm thick. Telia amphigenous, scattered or aggregated, often fusing, 0.5 mm in diameter, sub-epidermal, non-erumpent, orange-brown or reddish-brown; teliospores 20–50 × 8–15 μm, prismatic or oblong, wall 2.5–3.5 μm thick, uniform, cinnamon-brown.

Type: On *Salix* sp., in high mountains, Gunnison County, Colorado, North America, Bartholomew.

Hosts: Spermogonia and aecia on *Larix americana* Michx., *L. decidua* Mill., *L. laricina* (Du Roi.) K. Koch, *L. lyallii* Parl., and *L. occidentalis* Nutt. Uredinia and telia on *Salix alba* L., *S. amygdaloides* L., *S. arctica* Pall., *S. babylonica* L., *S. barclayi*, *S. bigelowii* Torr., *S. bonplondiana* H.B. & K., *S. brachycarpa* Nutt., *S. candida* Flugge., *S. caprea* L., *S. cinerea* L., *S. cordata* Muhl., *S. delnortensis* C.K. Schneid., *S. discolor* Muhl., *S. exigua* Nutt., *S. fendleriana* Anderss., S. *flavescens* Nutt., *S. fluviatilis* Nutt., *S. geyriana* Anderss., *S. glaucops* Anderss., *S. gracilistyla* Miq., *S. herbacea* L., *S. humilis* Marsh., *S. irrorata* Anderss., *S. jepsoni* Benth., *S. lutea* Nutt., *S. mackenzieana* Barr., *S. missourensis* Bebb., *S. monticola*, *S. nigricans* Sm., *S. nigra* Marsh., *S. parksiana* Ball., *S. petrophila* Rydb., *S. phylicifolia* L., *S. planifolia* Bebb., *S. pseudocoradata* (Anders.) Rydb., *S. purpurea* L., *S. repens* L., *S. reticulata* L., *S. rostrata* Richards, *S. scouleriana* Barr., *S. silesiaca* Willd., *S. sitchensis* Sanson, *S. subcaerulea*, *S. yestita* Pursh., *S. vitellina* L., and *S. wardi* Bebb.

Distribution: North and South America.

Specimens examined: DAOM, DAVFP, HCIO, FH, S, G, WRSL.

Gäumann (1959) considers *M. paradoxa* (= *M. bigelowii)* as an American form of a *Salix–Larix Melampsora* complex, having many biological races. The three races he classified as belonging to *M. paradoxa* from Europe are *M. larici-capraearum*, *M. larici-epitea* and *M. larici-pentandrae.* However, in the present study *M. capraearum* (= *M. larici-capraearum*) and *M. larici-pentandrae* are treated as distinct species, whereas *M. larici-epitea* is treated as a part of the *M. epitea* complex.

Through his field observations and inoculation experiments, Ziller (1970) found that *M. paradoxa* overwinters on willow as uredinial mycelia. Weir and Hubert (1916), Overholts (1926) and Shope (1940) have studied host alternation of this species.

Melampsora ribesii-viminalis Kleb., *Pringsh. Jahrb. Wiss. Bot.* 34, 363; 1900 (Fig. 3.2C)

= *Uredo ribesii-viminalis* Arth., *Result. Scient. Congr. Intern. Bot. Vienne* 338; 1905, 1906.

Spermogonia generally epiphyllous, sometimes hypophyllous, sub-epidermal, aggregated, 60–80 µm high, 150–190 µm wide. Aecia caeomoid, hyphophyllous, scattered or mostly aggregated in circular groups, often fusing, up to 1.5 mm, orange, sub-epidermal, erumpent, pulverulent; aeciospores 16–24 × 14–17 µm, globoid to ovoid, rarely angular, wall 2–3.5 µm thick, verrucose. Uredinia minute, hypophyllous, scattered or aggregated, 0.2–0.4 mm, pale orange-yellow, sub-epidermal, erumpent, pulverulent; urediniospores 15–22 × 14–18 µm, globoid to ovoid, wall 2 µm long, 18–25 µm wide. Telia minute, epiphyllous, scattered or aggregated, 0.2–0.5 mm, sub-cuticular, non-erumpent, dark brown; teliospores 25–40 × 7–15 µm, prismatic or irregular, rounded at both ends, wall 1 µm, light brown.

Type: On *Salix viminalis* L., Bremen, Germany.

Hosts: Spermogonia and aecia on *Ribes alpinum* L., *R. aureum* Pursh., *R. grossularia* L., *R. nigrum* L., *R. reclinatum* (L.) Mill., *R. rubrum* L., and *R. uva-crispa* L. Uredinia and telia on *Salix triandra* L., and *S. viminalis* L.

Distribution: Europe.

Specimens examined: S.

Melampsora ribesii-viminalis Kleb., differs from *M. epitea* f. sp. *ribesii-purpureae* (Kleb.) Bagyanarayana, the other *Ribes–Salix Melampsora*, in having sub-cuticular telia.

Klebahn (1900, 1902, 1903) and Mayor (1927, 1929) studied the biological relationship of *M. ribesii-viminalis*. According to Scaramella (1932) the dikaryotic mycelium overwinters in the branches of *Salix viminalis*.

Melampsora salicis-albae Kleb., *Pringsh. Jahrb. Wiss. Bot.* 35, 679; 1901

= *Caeoma allii-ursini* (D.C.) Wint., *Rabh. Krypt. Fl.* 1, 255; 1822.
= *Melampsora allii-salicis-albae* Kleb., *Zeitschr. f. Pfl. Krankh* 12, 19; 1902.
= *Uredo allii-salicis-albae* Arth., *Result. Scient. Cong. Inter. Bot. Vienne* 338; 1905, 1906.

Spermogonia prominent, lenticular 120 µm high, 150–200 µm wide. Aecia caeomoid, hypophyllous and caulicolous, aggregated in small groups, 1 mm in diameter, sub-epidermal, erumpent, pulverulent, orange; aeciospores 17–26 × 15–20 µm, globoid or broadly ellipsoid, sometimes irregular or angular, wall 1.5–2 µm thick, verrucose. Uredinia foliicolous, on young shoots and on the bark of branches, on the leaves usually hypophyllous, rarely epiphyllous, densely aggregated, up to 2 mm long, on the bark usually solitary, projecting from the cracked periderm, up to 5 mm long, sub-epidermal, erumpent, pulverulent; uredinispores 20–36 × 11–17 µm, oblong, sometimes clavate or pyriform, wall 2–2.5 µm thick, echinulate; paraphyses often capitate, up to 70 µm long, 12–20 µm in width, wall 2–3 µm thick. Telia hypophyllous, rarely epiphyllous, scattered or aggregated, sub-epidermal, not erumpent, dark-brown; teliospores 24–45 × 7–10 µm, irregularly prismatic, rounded at both the ends, wall 1 µm thick, light brown.

Type: On *Salix alba* L., Bremen, Germany.

Hosts: Spermogonia and aecia on *Allium ampeloprasum* L., *A. ascalonicum* L., *A. carinatum* L., *A. cepa* L., *A. cernum* Roth., *A. coerulescens* Boiss., *A. gaditanum* Per. Lar., *A. glaucum* Hort., *A. kansuense* Regel, *A. kartaviense* Regel, *A. longicuspis* Regel, *A. moly* L., *A. nutans* L., *A. obliqum* L., *A.*

ochroleucum W. et K., *A. odorum* Ten., *A. oleraceum* L., *A. pedemontanum* Willd., *A. platyspathum* Schrenk, *A. polyanthum* Schult., *A. pulchellum* Don, *A. pyrenaicum* Costa, *A. sativum* L., *A. schoenoprasum* L., *A. scenescens* L., *A. sphaerocephalum* L., *A. strictum* Schrad., *A. suaveolens* Jacq., *A. tartaricum* L., *A. ursinum* L., *A. victorialis* L., *A. vineale* L., *A. wallichianum* Steud., *A. yunanense* Diels.

Uredinia and telia on *Salix alba* L., *S. babylonica* L., *S. fendleriana* Anderss., *S. geyriana* Anderss., *S. glandulosa* V. Seem., *S. hindsiana* Benth., *S. humboldtiana* Willd., *S. lasiandra* Benth., *S. longifolia* Wall ex Anderss., *S. melanopsis* Nutt., *S. pentandra* L., *S. pyrifolia* Anderss., *S. retusa* L., *S. scouleriana* Barr., *S. sitchensis* San., *S. tetrasperma* Roxb., *S. vitellina* L.

Distribution: Europe, India.

Specimens examined: G, HCIO, S, FH.

The development and host selection of *M. salicis-albae* was studied by Klebahn (1900, 1902, 1903, 1905), Schneider (1910) and Mayor (1929, 1934, 1936, 1958). The aecial host range is exactly the same as that of *M. allii-fragilis*. According to Klebahn (1905) the dikaryotic mycelium can overwinter in the bark of the susceptible varieties of *Salix*.

M. salicis-cavaleriei Tai., *Farlowia* 3(1), 102; 1947

Uredinia epiphyllous, scattered or slightly aggregated, round, 0.2–0.5 mm diameter, golden yellow; urediniospores angular–subglobose, oblong or clavate, verrucose, 18–37 × 11–18 µm, wall 1.5–2 µm thick, paraphysis capitate, hyaline. Telia epiphyllous, scattered or densely aggregated, round, 0.2–0.5 mm diameter, pale cinnamon-brown, sub-epidermal, teliospores cylindrical, prismatic or clavato-cylindric, yellow, apex flat or round, 36–69 × 7–11 µm, wall 1–1.5 µm thick.

Type: On *Salix cavaleriei* Yunnan, Kunming, 10 November 1942, W.F. Chiu 7607, Type 8486.

Host: *Salix cavaleriei.*

Distribution: Yunnan, China.

Epiphyllous uredinia and telia, slender teliospores, angular urediniospores are the distinguishing characters of this species.

Melampsora salicis-cupularis Wang *Nat. Acad of Peiping* 6(40), 225; 1949

Uredinia hypophyllous, minute, scattered, brown; urediniospores sub-globose, ovate, oblong to ellipsoid, yellow, 21–23 × 17–18 µm, echinulate, hyaline, 1.5–2 µm, germ pores obscure; paraphyses numerous, capitate, hyaline, up to 98 µm long, 24–35 µm wide, wall 7 µm, thick. Telia hypophyllous, sub-epidermal, minute, scattered or aggregated, confluent, blackish. Teliospores prismatic, oblong-cylindric, apically not thickened, orange-brown, 28–39 × 11–14 µm.

Type: On *Salix cupularis* Shanxi: Tai-Pai-Shan, from Pao-Ma-Liang down to under Wen-Kung Miao, alt. 3300 m. 14 August 1942. Y.C. Wang 4073.

Host: *Salix cupularis.*

Distribution: Shanxi, China.

Melampsora salicis-warburgii Sawada, *Trans. Nat. Hist. Soc. Formosa* 21, 227–235; 1931

Spermogonia and aceia unknown. Uredinia amphigenous, rounded, 0.1–0.3 mm in diameter, but often confluent; urediniospores mostly obovoid, 15–18 × 13–28 µm, walls 2–2.5 µm thick, echinulate, but smooth at apex, germ pores scattered; paraphysis wall thickened at apex, up to 7 µm thick. Telia mostly hypophyllous, rounded; teliospores 34–42 × 8–13 µm, wall not thickened at apex.

Type: On *Salix warburgii* Seem, Taiwan.

Hosts: *Salix warburgii*, *S. pierotii*, *S. subfragilis.*

Distribution: Taiwan and Japan.

Melampsora yezoensis Miyabe et Mats., *Matsumoto Trans., Sapporo Nat. Hist. Soc.* 6, 29; 1915 (Fig. 3.2D)

Spermogonia minute, epiphyllous, orange-brown, up to 0.2 mm. Aecia caeomoid, hypophyllous, just below the spermogonia, in circular rings, often fusing, 2–3 mm wide, sub-epidermal, erumpent, pale yellow; aeciospores 25–20 × 13–17 μm, ovoid or angular, wall 1.5–2.5 μm thick, echinulate. Uredinia minute, amphigenous, mostly hypophyllous, sub-epidermal, erumpent, pulverulent 0.4 mm, pale yellow; urediniospores 16–28 × 12.5–20 μm, ovoid or ellipsoid, rarely globoid, wall 2–5 μm thick, echinulate; paraphyses capitate to clavate, up to 50 μm long, 12–20 μm wide. Telia minute, amphigenous, scattered or aggregated in small groups, up to 0.4 mm, often fusing, sub-cuticular, non-erumpent, dark brown; teliospores 16–33.6 × 6.4–13(16) μm, cylindrical, wedge-shaped, rarely irregular, wall 1 μm thick, light brown.

Type: On *Salix jessoensis* V. Seem, Sapporo, Japan.

Hosts: Spermogonia and aecia on *Corydalis ambigua* Cham. et Schl. Uredinia and telia on *Salix breweri* Bebb., *S. hondoensis* Koidz., *S. jessoensis* V. Seem, *S. lasiandra* Benth., *S. longiflora* Wall ex Anderss., *S. scouleriana* Barr.

Distribution: China, Japan.

Specimens examined: Herb. Hiratsuka.

Matsumoto (1915) created the species *M. yezoensis* to accommodate a *Melampsora* having its spermogonial and aecial stage on *Corydalis* and the telial stage on *Salix*. The other *Salix* rust having *Corydalis* as its aecial host, with sub-cuticular telia, is *M. chelidonii-pierotii* Mats., from which *M. yezoensis* differs in its teliospore morphology and in the possession of thick-walled urediniospores.

Acknowledgements

The author expresses his deep sense of gratitude to the late Professor P. Ramachar, under whose guidance this work was carried out. The kind help of the Curators/Directors of the following international herbaria – B, C, DAOM, DAVFP, FH, G, HCIO, H H, IMI, K, LEV, PAV, WRSL, etc. – is gratefully acknowledged. I sincerely thank the following scientists and researchers for their help and co-operation: the late Dr D. Krishnamurthy, Dr Sesikaran, Mr Ramachandran, National Institute for Nutrition, Hyderabad, India; Dr Uwe Braun, Martin Luther University, Germany; Dr Halvor B. Gjaerum, Plant Protection Research Institute, Norway; the late Professor Nohide Hiratsuka, Tottori Mycological Research Institute, Japan; the late Dr D.B.O. Savile and Dr Mary Esther Elliott, Biosystematics Research Institute, Canada; the late Dr Wolf G. Ziller and Dr Y. Hiratsuka, Canadian Forestry, Canada; Miss Ann Ansell, International Mycological Institute, UK; Dr Ming H. Pei, Rothamsted Experimental Station, UK. I also thank the University Grants Commission and Council of Scientific and Industrial Research, New Delhi for financial assistance, and officials of Osmania University for the physical facilities.

References

Arthur, J.C. (1920) New species of Uredineae. *Bulletin of the Torrey Botanical Club* 47, 467.

Arthur, J.C. (1934) *Manual of the Rusts in United States and Canada*. Purdue Research Foundation, Lafayette, Indiana.

Boerema, G.H. and Verhoeven, A.A. (1972) Check-list for scientific names of common parasitic fungi. Series 1a: Fungi on trees and shrubs. *Netherlands Journal of Plant Pathology* 78 (suppl. 1).

Bubák, F.R. (1908) Die Pilze Böhmens. I. Uredinales. *Archiv der naturwissenschaflichen Landesdurchforschung von Böhmen* 13(5).

Dietel, P. (1930) Kleine Beitrage zur Uredineenkunde. *Jahresbericht des Vereins für Naturkunde Zwickauin Sachsen 1928–30*, p. 1–9.

Fraser, W.P. (1912) Cultures of heteroecious rusts. *Mycologia* 4, 188.

Fraser, W.P. (1913) Further cultures of heteroecious rusts. *Mycologia* 5, 238.

Gäumann, E. (1959) *Die Rostpilze Mitteleuropas. Beiträge zur Kryptogamenflora Schweiz 12.*

Buchdruckerei Büchler and Co., Bern, Germany.
Henderson, D.M. (1953) Some Scottish mountain rust fungi. *Transactions of the British Mycological Society* 36, 315–319.
Hiratsuka, N., Sato, S., Katsuya, K., Kakishima, M., Hiratsuka, Y., Kaneko, S., Ono, Y., Sato, T., Harada, Y., Hiratsuka, T., and Nakayama, K. (1992) *The Rust Flora of Japan*. Tsukuba Shuppankai, Tsukuba, Ibaraki, Japan.
Hylander, N., Jørstad, I. and Nannefeldt, J.A. (1953) Enumeratio Uredinearum Scandinavicarum. *Opera Botanica* 1, 18–26.
Jaap, O. (1908) Zweites verzeichnis zu meinem exsiccate nwerk Fungiselecti exsiccate Ser. V–VIII, nebst Beschreibungen never Arten und Bumerkungen. *Verhhandlungen des Botanischen Vereins fur die Provinz, Brandenburgund die angrenzenden Lander* 49, 1907, 7–29.
Jacky, E. (1899) Untersuchungen über einige Schweigerische Rostpilze. Ber. Schweiz. *Botanical Gazette* 9, 49–78.
Jørstad, A. (1923) *Chytridineae, Ustilagineae and Uredineae from Novaya Zemlya*. Report of the scientific results of the Norwegian Expedition of Novaya Zemlya 1921, No. 18, pp. 3–12.
Klebahn, H. (1897) Vorläufiger Bericht über Kulturversuche mit heteroecischen Rostpilizen. *Zeitschrift für Pflanzen Krankhieten 7*, 129–130.
Klebahn, H. (1899) Kulturversuche mit heteroecischen Rost pilzen. VII. *Zeitschrift für Pflanzen Krankhieten* 9, 14–26, 88–99, 137–160.
Klebahn, H. (1900) Kulturversuche mit heteroecischen Rost pilzen. VIII. *Jahrbucher für Wissenschaftlichen Botanik* 34, 660–710.
Klebahn, H. (1902) Kulturversuche mit Rost pilzen. X. *Zeischrift für Pflanzen Krankhieten* 12, 17–44, 132–151.
Klebahn, H. (1903) Kulturversuche mit heteroecischen Rost pilzen. XI. *Jahrbuch der Hamburgischen Wissenschaftlichen Anstalten* 20 (1902), 3, 1–56.
Klebahn, H. (1905) Kulturversuche mit heteroecischen Rost pilzen. XII. *Zeischrift für Pflanzen Krankhieten* 15, 65–108.
Klebahn, H. (1908) Kulturversuche mit heteroecischen Rost pilzen. XIII. *Zeischrift für Pflanzen Krankhieten* 17, 1907, 129–157.
Klebahn, H. (1914) Kulturversuche mit heteroecischen Rost pilzen. XV. Bericht (1912 and 1913). *Zeischrift fur Pflanzen Krankhieten* 24, 1–32.
Matsumoto, T. (1915) Impfversuche mit *Melampsora* auf Japani-schen weiden. *Sapporo Natural History Society Transactions* 6, 22–35.
Matsumoto, T. (1919) Culture experiments with *Melampsora* in Japan. *Annales of the Missouri Botanical Garden* 6, 309–316.
Matsumoto, T. (1926) On the relationship between *Melampsora* on *Salix pierotii* Miq. and *Caeoma* on *Chelidonium majus* L. and *Corydalis incissa* Pers. *Botanical Magazine, Tokyo* 40, 43–47.
Mayor, E. (1918) Contribution a l'etude de la flore mycologique des environs de leysin. *Bulletin de la Murthienne Societe Valaisanne des Sciences Naturelles* 52, 113–149.
Mayor, E. (1924) Notes mycologiques. *Bulletin de la Societe Neuchateloise des Sciences Naturelles* 48, 367–396.
Mayor, E. (1927) Notes mycologiques VI. *Bulletin de la Societe Neuchateloise des Sciences Naturelles* 51, 53–76.
Mayor, E. (1929) Notes mycologiques VII. *Bulletin de la Societe Neuchateloise des Sciences Naturelles* 54, 45–59.
Mayor, E. (1934) Notes mycologiques VIII. *Bulletin de la Societe Neuchateloise des Sciences Naturelles* 58, 1933, 7–31.
Mayor, E. (1936) Notes mycologiques X. *Bulletin de la Societe Neuchateloise des Sciences Naturelles* 61, 105–123.
Mayor, E. (1958) Etude experimentale de quelques Uredinees heteroiques. *Uredineana* 5, 149–324.
Overholts, L.O. (1926) Mycological notes for 1925. *Mycologia* 18, 179–185.
Plowright, W.B. (1891) Einige Impfversuche mit Rostpilzen. *Zeitschrift für Pflanzen Krankhieten* 1, 130–131.
Raabe, A. (1939) Parasitische Pilze der Umgebung von Tubingen. Ein Beitrag Zur Kryptogamen flora Subwestdeutschlands. *Hedwigia* 78, 1–106.
Scaramella, P. (1932) Contribute aila flora micdogica del piccolo S. Bernardo (Val d' aostal). *Malphigia* 31, 178–224.
Schneider, O. (1905) Weitere Versuche mit Schweizerischen Weidenmel ampsoren. Vorlaufige Mitteilung. *Zentralblatt für Bakteriologie, Parasitenkunde, Infektionskrankheiten und Hygiene. 2. Naturwissenschaftliche Abteilung* 15, 232–234.
Schneider, O. (1906) Experimentelle Untersuchungen über Schweizerischen Weidenrostpilze. *Zentralblatt für Bakteriologie, Parasitenkunde, Infektionskrankheiten und Hygiene. 2. Naturwissenschaftliche Abteilung* 16, 74–93, 159–176, 192.
Schneider, O. (1910) Beitrag zur Kenntnis der Schweizerischen Weidenmelampsoren. *Zentralblatt für Bakteriologie, Parasitenkunde, Infektionskrankheiten und Hygiene. 2. Naturwissenschaftliche Abteilung* 25, 436–439.

Schroeter, J. (1893) Zur Entwicklungsgechichete der Uredineen. *Jahresbericht Schlesischen Gesellschaft fur Vaterlandische Cultur. II. Botanische Sektion* 71, 31–32.

Shope, P.F. (1940) Colorado rusts of woody plants. *University of Colorado Stud. D.* 1, 105–127.

Tubeuf, C. Von. (1902) Infections versuche mit uredineen der Weibtanne. *Zentralblatt für Bakteriologie, Parasitenkunde, Infektionskrankheiten und Hygiene. 2. Naturwissenschaftliche Abteilung* 9, 241.

Tubeuf, C. Von. (1905) Infections versuche mit uredineen. *Naturwissenschaftliche Zeitschrift fur Forst.u Landwirtschaft (Stuttgart)* 3, 41–46.

Weir, J.R. and Hubert, E.E. (1916) Successful inoculations of *Larix occidentalis* and *Larix europea* with *Melampsora bigelowii. Phytopathology* 6, 37.

Weir, J.R. and Hubert, E.E. (1918) Notes on forest tree rusts. *Phytopathology* 8, 118.

Wilson, M. and Henderson, D.M. (1966) *British Rust Fungi*. Cambridge University Press, Cambridge, UK.

Ziller, W.G. (1959) Studies of Western trees rusts. V. The rusts of hemlock and fir caused by *Melampsora epitea. Canadian Journal of Botany* 37, 109–119.

Ziller, W.G. (1970) Studies of Western trees rusts. VIII. Inoculation experiments with conifer needle rusts (Melampsoraceae). *Canadian Journal of Botany* 48, 1471–1476.

Ziller, W.G. (1974) *The Tree Rusts of Western Canada*. Publication No. 1329. Canadian Forestry Service, Environmental Canada, Ottawa, Canada.

4 A Brief Summary of *Melampsora* Species on *Populus*

Ming Hao Pei[1] and Yan Zhong Shang[2]
[1]*Rothamsted Research, Harpenden, Hertfordshire, AL5 2JQ, UK;* [2]*College of Forestry, Inner Mongolia Agricultural University, Huhehot, China*

Background

Rust caused by *Melampsora* is one of the most important leaf diseases of poplars (*Populus*). Some 13 species and two hybrids of *Melampsora* have been described on *Populus*. A recent revision of the taxonomy of the poplar rusts was presented by Bagyanarayana (1998). This chapter will give a brief summary of the species of *Melampsora* described on *Populus*. No attempt has been made to revise the poplar rust taxonomy because more meaningful classification can be achieved when further information on evolutionary and genetic relationships among the poplar rusts becomes available.

The Genus *Populus*

Poplars occur naturally across the northern hemisphere between the latitudes 30° and 72°N (Wang and Fang, 1984). Poplars are fast growing, easily propagated and can be cultivated under various growing conditions. They are now widely grown in various parts of the world for the purpose of urban amenity, farmland sheltering and for timber, pulp and biomass production. The genus *Populus* L. is closely related to *Salix* (willow) and they are grouped together in the family *Salicaceae*. Diversity of poplars is centred in East Asia (China and Russia) and Pacific and Atlantic regions of North America. There is no consensus on how many species should be recognized in *Populus*. Taxonomists from the East tend to recognize many species by adopting the narrow species concept, while Western workers favour the broad species concept and recognize fewer species. Thus the number of species in *Populus* can be from 30 to 100, according to different authorities.

The genus is divided into the five sections: *Aigeiros* (European and American black poplars), *Tacamahaca* (balsam poplars) and *Leuce* (aspens and true white poplars), *Leucoides* and *Turanga*.

Section *Aigeiros* (black poplars)

The majority of cultivated poplars belong to the section *Aigeiros*. There are two genetic centres: the Mediterranean region and North America. The Eurasian species is *P. nigra* (black poplar) with several subspecies and with natural stands from the Mediterranean region extending northward to central Europe and eastward to central Asia. The most important endemic species in North America is *P. deltoides* (eastern cottonwood; eastern parts of North

America). *Populus* × *euramericana* (syn. *P.* × *canadensis* Moench.) is the name given to cover the vast numbers of hybrid clones derived from *P. deltoides* and *P. nigra*.

Section *Tacamahaca* (balsam poplars)

This section is the largest in the genus *Populus*. Balsam poplars are naturally distributed in Asia and in North America. In general, they occur naturally further north than black poplars. Wang and Fang (1984) included 36 species (*sensu stricto*), native or introduced, in the Chinese flora of *Salicaceae*. Some examples for balsam poplars from Asia are *P. laurifolia*, *P. maximowiczii*, *P. koreana*, *P. simonii* and *P. szechuanica*. Examples for the *Tacamahaca* section from North America are *P. trichocarpa* (black cottonwood), *P. balsamifera*, (balsam or southern poplar) and *P. angustifolia* (willow-leafed poplar).

Section *Leuce* (aspens and true white poplars)

This section is further divided into aspens (sub-section *Trepidae*) and white poplars (sub-section *Albidae*). Aspens occur in northern and mountainous regions of Eurasia and in North America. In the subsection *Trepidae*, the most important species are *P. tremula* (European or trembling aspen; distributed in Europe, western Asia and North Africa), *P. tremuloides* (American or quaking aspen, North America) and *P. tremula* var. *davidiana* (David's aspen, the Far East). In the sub-section *Albidae*, western taxonomists recognize only one species, *P. alba* L. (white poplar). This species has many sub-species and varieties (treated as species by eastern taxonomists) which occur naturally in the Mediterranean, eastern Europe and central Asia. The grey poplar, *P.* × *canescens* (Aiton) Sm., which is widely distributed in Europe, is considered to be a natural hybrid of *P. tremula* and *P. alba* (Bean, 1976).

Sections *Turanga* and *Leucoides*

Populus euphratica is the main species of the section *Turanga* and many other names placed in this section are often considered as local forms. *Populus euphratica* can tolerate poor soils, extreme heat and soil salinity. Its distribution extends from the Altai mountains (45°N) southward to the equator and westward to the Middle East (FAO, 1979). Except for one species in North America, the several species grouped in the section *Leucoides* are confined to China and the Himalayas.

Melampsora Species on Poplars

Most *Melampsora* species on poplars were described in the late 19th to early 20th centuries. By 1915, ten *Melampsora* species on *Populus* were complied in the *Monographia Uredinearum III* by Sydow and Sydow (1915). As in willow rusts, the rust species on poplars were established mainly based on their morphology, alternate hosts and telial host range. The aecial hosts of poplar *Melampsora* include conifers, dicotyledonous and monocotyledonous plants (Table 4.1). None of the *Melampsora* species described on poplars is known to be autoecious (capable of completing a full life cycle on the same host). Several species, *M. larici-tremulae*, *M. pinitorqua*, *M. magnusiana* and *M. rostrupii* are indistinguishable in their morphology but have different aecial hosts. Hylander *et al.* (1953) and Wilson and Henderson (1966) adopted *M. populnea* (Pers.) Karst. to include these morphologically indistinguishable species (*sensu stricto*, to be used in this chapter).

Morphology of Poplar Rusts

Typically, poplar rusts are macrocyclic, producing five different spore stages, i.e. basidiospores, spermatia, aeciospores, urediniospores and teliospores, during their life cycle. With *M. ciliata*, *M. pruinosae*, *M. microspora*, *M. osmaniensis*

Table 4.1. List of hosts and distribution of *Melampsora* species/forms on *Populus*.

Melampsora sp.	Aecial host	Telial host	Distribution	References
M. abietis-populi Imai	*Abies* spp.	**Sec. *Aigeiros*** *P. nigra* L., *P.* × *euramericana* (Dode) Guinier **Sec. *Tacamahaca*** *P. koreana* Rehd., *P. maximowiczii* Henry, *P. simonii* Carr.	Japan	Hiratsuka *et al.* (1992)
M. allii-populina Kleb.	*Allium* spp. and *Arum* spp.	**Sec. *Aigeiros*** *P. nigra*, *P. deltoides* Marsh, *P.* × *euramericana* **Sec. *Tacamahaca*** *P. balsamifera* L.	Europe, Russia northern Africa southern Africa	Sydow and Sydow (1915); Gäumann (1959)
M. ciliata Barcl.	?	**Sec. *Tacamahaca*** *P. ciliata* Wall.	India, Kashmir	Sydow and Sydow (1915); Bagyanarayana (1998)
M. × *columbiana*		**Sec. *Aigeiros*** × **Sec. *Tacamahaca*** *P. deltoides* × *P. trichocarpa* Hook.	USA	Newcombe *et al.* (2000)
M. cumminsii Bagyan. et Ram.	?	*Populus* sp.	USA	
M. larici-populina Kleb.	*Larix* spp.	**Sec. *Aigeiros*** *P. nigra*, *P. deltoides*, *P. fremontii* S. Wats., *P.* × *euramericana* **Sec. *Tacamahaca*** *P. angustifolia* James, *P. balsamifera*, *P. ciliata*, *P. koreana*, *P. laurifolia* Ledebour, *P. aximowiczii*, *P. suaveolens* Fischer, *P. simonii*, *P. ussuriensis* Kom., *P. trichocarpa* **Sec. *Leucoides*** *P. lasiocarpa* Oliv., *P. wilsonii* Schneider **Sect. *Turanga*** *P. euphratica* Oliv.	Eurasia, Australia, New Zealand	Sydow and Sydow (1915); Gäumann (1959) Hiratsuka *et al.* (1992) Shang and Pei (1988)
M. larici-tremulae Kleb. (= *M. laricis* Hartig.)	*Larix* spp.	***Sec. Leuce*** *P. tremula* L., *P. alba* L., *P.* × *canescens* (Ait.) Sm.	Eurasia, N. Africa	Gäumann (1959); Shang and Pei (1984)
M. magnusiana Wagner (= *M. pulcherrima* (Bub.) Maire)	*Chelidonium majus*, *Corydalis* spp.	***Sec. Leuce*** *P. tremula*, *P. sieboldii* Miq, *P. alba*, *P.* × *canescens*	Eurasia	Sydow and Sydow (1915); Gäumann (1959); Bagyanarayana (1998)

continued

Table 4.1. *Continued.*

Melampsora sp.	Aecial host	Telial host	Distribution	References
M. medusae f. sp. *deltoidae* Thüm. (= *M. albertensis* Arth. = *M. abietis-canadensis* (Farl.) C.A. Ludwig) f.sp. *tremuloidae*	*Abies* spp. *Larix* spp. *Picea sitkensis* *Pseudotsuga* spp. *P. taxifolia* *Tsuga* spp. *Pinus* spp.	**Sec. *Aigeiros*** *P. nigra, P. deltoides, P.* × *euramericana* **Sec. *Tacamahaca*** *P. maximowiczii, P. simonii, P. trichocarpa* **Sec. *Leuce*** *P. tremuloides, P. grandidentata* Mixchx.	North America southern Europe South Africa India Australia New Zealand	Ziller (1955, 1974); Shain (1988); Spiers and Hopcroft (1994)
M. medusae-populina Spiers		**Sec. *Aigeiros*** *P.* × *euramericana* **Sec. *Aigeiros*** × **Sec. *Tacamahaca*** *P. deltoides* × *P. yunnanensis*	New Zealand Australia?	Spiers and Hopcroft (1994)
M. microspora Tranz. et Eremeeva	?	**Sec. *Aigeiros*** *P. nigra, P. usbekistanica* Kom.	ex-USSR Iran	Azbukina (1974) Bagyanarayana (1998)
M. multa Shang *et al.*	?	**Sec. *Aigeiros*** *P.* × *euramericana* 'Robusta', 'Serotina' and 'I-214'	N.E. China	Shang *et al.* (1986)
M. occidentalis Jacks.	*Abies concolor* *Larix* spp. *Picea sitkensis* *Pseudotsuga menziesii* *Pinus* spp.	**Sec. *Aigeiros*** *P. nigra, P. fremontii* **Sec. *Tacamahaca*** *P. trichocarpa, P. angustifolia, P. balsamifera*	USA Canada	Ziller (1974)
M. osmaniensis Bagyan. et Ram.	?	***Sec. Leuce*** *P. sieboldii* Miq.	Germany	Bagyanarayana (1998)
M. pinitorqua Rostr.	*Pinus* spp.	***Sec. Leuce*** *P. tremula, P. alba, P.* × *canescens*	Europe	Sydow and Sydow (1915); Gäumann (1959)
M. pruinosae Tranzsch.	?	***Sec. Turanga*** *P. pruinosa* Schrenk.	West and Central Asia	Sydow and Sydow (1915); Gäumann (1959)
M. rostrupii Wagner	*Mercurialis perennis*	***Sec. Leuce*** *P. tremula, P. alba, P.* × *canescens*	Europe	Sydow and Sydow (1915); Gäumann (1959)

and *M. multa*, only uredinio- and teliospore stages have been described. Table 4.2 lists the main morphological characteristics of poplar rusts.

Uredinia and urediniospores

The urediniospores are capable of producing many cycles of the same form of spores during the season, causing disease epidemics on poplars. The majority of poplar rusts form uredinia mainly on the lower side of leaves (hypophyllous), except for *M. pruinosae*, which produces uredinia on both sides of leaves (amphigenous). The surface of urediniospores is covered with spines (echinulae). Broadly, there are two types of urediniospores, one type having relatively small urediniospores with an evenly echinulate surface and the other having relatively large urediniospores often with a smooth patch on the surface. Urediniospores of the *Melampsora* occurring on Sect. *Leuce* (*M. populnea* complex), *M. ciliata* and *M. pruinosae* are relatively small (less than 30 μm long) and evenly echinulate. Others on sections *Aigeiros* and *Tacamahaka* are relatively large (maximum length 35–50 μm) and often have a smooth area on the surface of spores. As an exception, *M. microspora* has small urediniospores and its urediniospore walls are mostly smooth and only faintly verrucose.

The location of the smooth area on the surface of urediniospores can be characteristic of certain species. For example, urediniospores of *M. larici-populina* and *M. allii-populina* have a smooth surface at the apex, while *M. medusae* has a smooth surface on the equatorial zone. *Melampsora* × *medusae-populina*, a presumed hybrid between *M. larici-populina* and *M. medusae*, has a smooth surface both at the apex and on the equatorial zone. In *M. larici-tremulae* (*M. laricis*), urediniospores are sometimes found to have a smooth area on the equator (Chapter 8). In some species, such as *M. larici-epitea*, *M. medusae* and *M. occidentalis*, urediniospore walls are often thickened at the equator.

Telia and teliospores

Teliospores overwinter on fallen leaves of hosts and, in spring, produce basidiospores which infect aecial hosts. The majority of the *Melampsora* species on *Populus* form telia mainly on the lower sides of leaves. In contrast, *M. larici-populina* produces telia on the upper sides of leaves (epiphyllous), and *M. pruinosae* and *M. multa* produce them on both sides of leaves (amphiogenous). All the *Melampsora* species on poplars are known to form telia beneath the epidermal cells of the hosts (sub-epidemal). The teliospore walls in most poplar rusts are uniformly thick (< 2.5 μm). In contrast, teliospores walls in *M. occidentalis* are noticeably thickened (up to 6 μm) at the apex. Teliospores of *M. cumminsii* are described as having thickened (3–5 μm) upper walls, but with prominent apical pores.

Basidiospores and spermagonia

As in willow rusts, basidia and basidiospores are produced from overwintered teliospores in spring. Each basidium produces four spherical basidiospores which infect alternate hosts and form spermagonia. Spermagonia are either sub-cuticular or sub-epidermal, and often form as groups, especially on dicotyledonous plants.

Aecia and aeciospores

Aeciospores of poplar rusts are produced on the alternate hosts as the consequence of fertilization between spermatia and spermogonia. The majority of macrocyclic poplar rusts produce small aecia (a few millimetres or less). However, *M. rostrupii* forms large aecia (up to 3 cm on stems) on the alternate host *Mercurialis perennis* (Wilson and Henderson, 1966). *M. pulcherrima*, which is confined to the Mediterranean, is usually treated as a synonym of *M. rostrupii* (Bagyanarayana, 1998; Cellerino, 1999). This rust is similar to *M. rostrupii* in

Table 4.2. Morphological characteristics of *Melampsora* species/forms described on *Populus*.

Species/form	Length of aecia	Size of aeciospores	Position and size of uredinia	Size and wall of urediniospores	Size and wall of paraphyses	Position and size of telia	Size and wall of teliospores
M. abietis-populi Imai		20–33 × 17–27 μm	Hypophyllous	21–35 × 15–20 μm; evenly echinulate; wall 1.5–2 μm thick	Wall up to 10 μm thick at the apex	Mostly hypophyllous, sub-epidermal	25–48 × 7–13 μm; wall 1 μm thick
M. alli-populina Kleb.	1 mm	17–23 × 14–19 μm, wall 2 μm thick	Mostly hypophyllous, usually less than 1 mm	24–38 × 11–18 μm; smooth at the apex; wall evenly thick, 2–4 μm	50–60 × 14–22 μm; wall evenly thick, 2–3 μm	Hypophyllous, sub-epidermal; 0.25–1 mm	35–60 × 6–10 μm; wall 1–1.5 μm thick
M. ciliata Barcl.			Hypophyllous	18–26 × 13–18 μm; wall 1.5–2 μm thick	35–50 × 12–18 μm; wall 3–6 μm thick	Hypophyllous; minute	30–40 × 8–9 μm
M. × *columbiana* Newc.			Mainly hypophyllous	19–38 μm long; often smooth at the equator			40–53 μm long
M. larici-populina Kleb.	0.5–1 mm	17–22 × 14–19 μm	Mainly hypophyllous	30–50 × 14–22 μm; wall 2 μm thick, smooth at the apex; often up to 7 μm thick at the equator	45–85 × 12–18 μm; wall up to 10 μm thick at the apex	Epiphyllous, sub-epidermal	25–45 × 10–17 μm; wall 1 μm thick, slightly thickened at the apex (2.5–3 μm)
M. larici-tremulae Kleb. (European)	0.75 mm	14–17 × 12–16 μm	Hypophyllous, *c.* 0.5 mm	15–22 × 10–15 μm; evenly echinulate; wall 1.5–2 μm thick	40–45 × 8–17 μm; wall 3–5 μm thick	Hypophyllous, sub-epidermal; *c.* 1 mm	40–60 × 7–12 μm; wall 1–2 μm thick
(Far Eastern)		14–27 × 13–22 μm	Hypophyllous	16–29 × 13–20 μm; evenly echinulate	34–65 × 7–15 μm		23–51 × 3–11 μm
M. magnusiana Wagner	Up to 4 mm	14–23 × 12–20 μm	Hypophyllous; 0.3–0.5 mm	17–26 × 12–19 μm; evenly echinulate; wall 2.5–3 μm thick	40–54 × 13–22 μm; wall 3–5 μm thick	Hypophyllous, sub-epidermal; 1 mm	40–55 × 7–10 μm; wall 1–2 μm thick
M. medusae Thüm.	0.3–1.5 mm	19–26 × 16–21 μm; wall thickened at the equator	Hypophyllous; 0.3–0.5 mm	23–35 × 15–23 μm; thickened up to 10 μm at the equator	Up to 60 × 13–25 μm; wall 3–6 μm thick	Mostly hypophyllous, sub-epidermal; up to 0.5 mm	20–45 × 10–15 μm; wall 1–1.5 μm thick

M. medusae-populina Spiers			Mainly hypophyllous; sub-epidermal	25–50 × 12–25 μm; smooth at the apex and the equator, thickened up to 10 μm at the equator	18–50 × 8–16 μm; thickened up to 12 μm at the apex	Mainly epiphyllous	20–50 × 6–16 μm; wall 1–2.5 μm thick
M. microspora Tranz. et Eremeeva			Mostly hypophyllous	11–17 × 11–13 μm; faintly verrucose, almost smooth; up to 4 μm thick	60 × up to 15 μm; wall 5 μm thick	Mostly hypophyllous, sub-epidermal	30–68 × 10–13 μm
M. multa Shang, Pei et Yuan			Hypophyllous; 0.2–0.4 mm	26–44 × 16–25 μm; smooth at the apex; often thickened at the equator	42–62 × 16–25 μm; wall up to 16 μm thick at the apex	Amphigenous, sub-epidermal; 0.3–0.8 mm	Teliospores in 1 or 2, rarely 3, layers, 30–45 × 6.5–13 μm; wall 1 μm thick
M. occidentalis Jacks.	0.3–1 mm	26–35 × 22–27 μm; wall thickened at the equator	Mainly hypophyllous; 0.5–1.5 mm	30–50 × 17–13 μm; evenly echinulate, wall 2–3 μm thick, thickened up to 8 μm at the equator	Up to 75 × 15–25 μm; up to 6.5 μm thick at the apex	Hypophyllous, sub-epidermal, up to 0.6 mm	40–66 × 10–20 μm; wall 1–2 μm thick at sides, thickened up to 6 μm at the apex
M. pinitorqua Rostr.	Up to 20 mm	15–22 × 11–17 μm	Hypophyllous, 0.3–0.5 mm	14–23 × 12–16 μm; evenly echinulate; wall 2 μm or thick or thickened up to 6 μm in the equator	40–60 × 20–25 μm; wall 3–7 μm thick	Hypophyllous, sub-epidermal; *c.* 0.5 mm	22–45 × 7–12 μm; wall 1 μm thick
M. pruinosae Tranzsch.			Amphigenous	20–28 × 16–18 μm; wall 3–4 μm thick	45–60 × 16–20 μm; wall 1.5–4 μm thick	Amphigenous, sub-epidermal	40–50 × 10–13 μm
M. rostrupii Wagner	Up to 5 mm on leaves, up to 30 mm on stems	13–24 × 11–17 μm	Hypophyllous, *c.* 1 mm	18–25 × 14–18 μm; wall 3 μm thick	46–55 × 15–23 μm; wall 3–6 μm thick	Hypophyllous; 0.5–1 mm	40–52 × 7–11 μm; wall 1–1.5 μm thick

its host range (*M. pulcherrima* occurs on *P. alba* and alternates on *Mecurialis annosum*). According to the literature (Gäumann, 1959), aeciospores of *M. rostrupii* are ovoid to oblong, rarely spherical and almost always polygonal, while aeciospores of *M. pulcherrima* are spherical to ellipsoid, rarely oblong, and polygonal only when immature. *M. pinitorqua* also forms aecia extended (up to 2 cm) along the pine needles. The size of aeciospores in European populations of *M. larici-tremulae* appears to be smaller than those in the Far East (Table 4.2).

Host Alternation and Overwintering

The two rust species native to North America, *M. occidentalis* and *M. medusae*, have a wide range of conifers as their alternate hosts. *M. occidentalis* infects *Abies*, *Larix*, *Picea*, *Pseudotsuga* and *Pinus* at the aecial stage. In addition to these conifers, *M. medusae* is also capable of infecting *Tsuga*. Such wide aecial host range is rather unusual among the rust fungi, considering that the majority of rust genera alternating on conifers have only one host genus as the aecial host.

Some poplar rust species are separated solely on the basis of having different alternate hosts. For example, *M. larici-tremulae*, *M. pinitorqua*, *M. rostrupii* and *M. magnusiana*, all occur on aspens or white poplars and share a similar morphology. However, *M. larici-tremulae* alternates on *Larix*, *M. pinitorqua* on two-needled species of *Pinus*, *M. rostrupii* on *Mercurialis* and *M. magnusiana* on *Chelidonium* and *Corydalis*. Of the poplar rusts native to Eurasia, *M. abietis-populi* is the only species having *Abies* as the alternate host. In *M. allii-populina*, two *formae speciales*, f. sp. *allii-populina* having *Allium* and *Arum* as aecial hosts and f. sp. *muscaridis-populina* having *Muscari* as aecial host, were recognized (Gäumann, 1959).

With *M. multa*, several inoculation attempts using basidiospores failed to produce aecia on the tested *Larix principis-rupprechtii* Mayer, *Picea wilsonii* Mast, *Picea meyeri* Rehd. et Wils., *Pinus tabuliformis* Carr and *Sabina chinensis* (L.) Antoine (Shang *et al.*, 1986). However, the possibility of *M. multa* having alternate host(s) may not be ruled out as teliospores of the rust germinated well and produced abundant basidiospores in the inoculation experiments.

For both *M. occidentalis* and *M. medusae*, host alternation appears to be obligatory for their existence in western Canada (Ziller, 1974). It appears that *M. larici-populina* and *M. larici-tremulae* also require the alternate host larch to continue their existence in north-eastern China (M.H. Pei and Y.Z. Shang, north-eastern China, 1981, unpublished observations). On the other hand, *M. pinitorqua* was found to produce uredinia on young shoots of aspen before signs of infection appeared on the aecial host, *Pinus*, in costal Italy (Longo *et al.*, 1975). *M. magnusiana* and *M. rostrupii* have also been reported to overwinter as mycelia in infected buds of poplars in China (Xu, 1998).

Distribution and Spread

Current distribution maps of poplar rusts provide some classic examples of the spread of plant pathogens between continents. *M. larici-populina*, the most widespread of all poplar rust species, is native to Eurasia but has now spread to Australasia, North and South Americas. Both *M. occidentalis* and *M. medusae* are native to North America. While *M. occidentalis* is still confined to North America, *M. medusae* now occurs in Europe, Australasia and southern Africa. In Europe, *M. medusae* was first found near Barcelona, Spain (Fragoso, 1925) and then in southern France (Dupias, 1943). Later, it was found in Portugal and further French locations (J. Pinon, Nancy, 2003, personal communications).

Members of the *Salicaceae* are not native to Australasia, and all the poplars grown in Australia and New Zealand have been introduced. *M. medusae* was first detected in Australia in January 1972 on poplars near Sydney (Walker and Hartigan, 1972). Within 2 months this rust had spread over a large area from Melbourne to south Queensland. The Eurasian *M. larici-populina* was found near Sydney 13 months later, and within 2 months had spread to most of New South Wales. It is not yet clear how these pathogens were introduced into Australia, but infected propagating material that had been imported illegally was the most likely avenue (Wilkinson and Spiers, 1976). Both rusts were found in New Zealand in 1973. The introduction of these rusts to New Zealand most likely occurred via high-trajectory wind currents from Australia (Wilkinson and Spiers 1976; Close *et al.*, 1978).

M. larici-populina is now well established in New Zealand, unlike *M. medusae*, which occurs rarely. The widespread occurrence of *M. larici-populina* is mainly due to the strong wind currents generally experienced within New Zealand, and successful overwintering on semi-evergreen Lombardy poplars widely planted in the North Island and Nelson (Viljanson-Rollinson and Cromey, 2002). Alternate hosts, such as larch, may also play an important role in providing inoculum throughout the year (Spiers, 1990).

In North America, Newcombe and Chastagner (1993) reported that in 1991 two poplar leaf rusts (*M. larici-populina* and *M. medusae* f. sp. *deltoidae*) were detected for the first time in the Pacific Northwest. Pinon *et al.* (1994) reported that *M. larici-populina* was found in 18 counties of California. They collected uredinia from two separate locations in California and one location in Washington, and compared their virulence patterns with European races (coded as E1, E2, . . .) using a set of poplar differentials (see Chapter 12, this volume). It was shown that that race E1 was predominant in both California and Washington, but that three other un-named races existed at a low frequency. Races E2 and E3 were not found in any of the locations.

M. microspora, which differs from other poplar rusts by having small and faintly verrucose urediniospores, was recorded in the former USSR and Iran (Bagyanarayana, 1998). In the former USSR, it was found on *P. nigra* in Uzbekistan and on *P. usbekistanica* (= *P. nigra* var. *afghanica* Ait. et Heml.) in Tadjikistan (Z. Azbukina, Vladivostok, 2003, personal communications). *M. multa*, which forms teliospores in 1–3 layers, was found on *P.* × *euramericana* clones in Liaoning Province, north-eastern China (Shang *et al.*, 1986).

Interspecific Hybridization

Two interspecific hybrids, *M. medusae-populina* and *M.* × *columbiana* have been described on poplars.

M. medusae-populina caused an outbreak of rust on some previously resistant clones of *P.* × *euramericana* and *P. deltoides* × *P. yunanensis* in New Zealand in March 1991 (Spiers and Hopcroft, 1994). This rust exhibited various morphological and physiological characteristics of both *M. larici-populina* and *M. medusae*. For example, urediniospores of *M. medusae-populina* had smooth areas both at the apex and the equator. This rust appeared only in 1991 and died out (Spiers and Hopcroft, 1994; Spiers, 1998). It is probable that *M. medusae-populina* may not be fit enough to complete a full sexual life cycle. There are many examples in nature where hybrids between species or forms are often unfit to produce the next generation. In the willow rust *M. larici-epitea*, it was found that F_1 hybrids between f. sp. *larici-epitea typica* and f. sp. *larici-retusae* were predominantly sterile (Pei *et al.*, 1999).

Newcombe *et al.* (2000) reported the occurrence of natural hybrids between *M. medusae* and *M. occidentalis* in the Pacific Northwest, USA, and named them as *M.* × *columbiana*. This hybrid taxon showed intermediate characters between *M. medusae* and *M. occidentalis* in uredinial and telial morphology. The hybrids frequently contained ribosomal ITS sequences from both *M. medusae* and *M. occidentalis*. Some also contained homozygous ITS sequences of either *M. medusae* or *M. occidentalis*, indicating that the primary F_1 hybrids may have produced further generations through the sexual life cycle. Examination of the herbarium specimens, which were nearly a century old, suggested that this hybrid group was widely distributed in North America.

Concluding Remarks

As there are fewer species in *Populus* (30–100 species) than in *Salix* (300–500 species), the taxonomy of poplar rust is expected to be less complicated than that of willow rusts. Yet, current knowledge does not allow satisfactory revision of the taxonomy of *Melamspora* on poplars. It should be pointed out that existing records of the host range of some poplar rusts need to be treated with caution. For example, according to the records (Bagyanarayana, 1998), *P. alba* has been included in the host range of *M. latici-populina*. However, mycologists with experience of working on this rust would feel that such records may be incorrect, since there is no confirmed evidence indicating that *M. larici-populina* causes infections on aspens or white poplars. A similar case may be found with *M. occidentalis*, which has also been recorded on *P. alba*.

Several species of poplar rusts, *M. microspora*, *M. cumminsii*, *M. osmaniensis* and *M. multa*, were described on the basis that they are morphologically different from other species of poplar rusts. Apart from the descriptions of the uredinial and telial morphology, little is known of their occurrence, life cycle and genetic relationships with other poplar rusts. It is not yet certain whether they are genetically distinct species or occasional variants which may occur in nature. Further surveys of rusts from their original localities will help to understand whether they exist as established populations.

Recent studies of ribosomal DNA sequences of poplar rusts (Tian *et al.*, 2004; Chapter 8) have given revealing insights into the genetic identity and evolutionary relationships among poplar rusts. It is expected that applications of other molecular techniques, such as amplified fragment length polymorphism (AFLP) and microsatellites, will benefit research on population genetics and disease epidemiology of poplar rusts.

Acknowledgements

This study was funded by the Department for Environment, Food and Rural Affairs (DEFRA), UK, and the European Union. Rothamsted Research receives grant-aided support from the Biotechnology and Biological Sciences Research Council of the UK.

References

Azbukina, Z.M. (1974) *Rust Fungi of the Soviet Far East* [in Russian]. Nauka Publishers, Moscow, pp. 120–140.

Bagyanarayana, G. (1998) The species of *Melampsora* on *Populus* (*Salicaceae*). In: Jalkanen, R., Crane, P.E., Walla, J.A. and Aalto, T. (eds) *Proceedings of the First IUFRO Rusts of Forest Trees W.P. Conference*, 2–7 August, Saariselkä. Finnish Forest Research Institute, Rovaniemi, pp. 37–51.

Bean, W.J. (1976) *Trees and Shrubs Hardy in the British Isles*, Vol. IV, 8th edn. John Murray, London, pp. 293–328.

Cellerino, G.P. (1999) Review of poplar diseases: 3. Diseases caused by fungi. Available at website http://www.efor.ucl.ac.be/ipc/pub/celle01/celle01.htm (accessed 27 September 2004).

Close, R.C., Moar, N.T., Tomlinson, A.I. and Lowe, A.D. (1978) Aerial dispersal of biological material from Australia to New Zealand. *International Journal of Biometeorology* 22, 1–19.

Dupias, G. (1943) Contribution à l'étude des Urédinées de la Haute-Garonne. *Bulletin de la Societe d'Histoire Naturelle de Toulouse* 78, 32–52.

FAO (1979) *Poplars and Willows in Wood Production and Land Use*. FAO Forestry series 10. Food and Agriculture Organisation of the United Nations, Rome.

Fragoso, R.G. (1925) *Flora Iberica. II. Uredinales*. Museo Nacional de Ciencias Naturales, Madrid, p. 424.

Gäumann, E. (1959) *Die Rostpilze Mitteleuropas. Beiträge zur Kryptogamenflora Schweiz 12.*

Buchdruckerei Büchler and Co., Bern, Germany, pp. 130–144.

Hiratsuka, N., Sato, T., Katsuya, K., Kakishima, M., Hiratsuka, Y., Kaneko, S., Ono, Y., Sato, S., Harada, Y., Hiratsuka, T. and Nakayama, K. (1992) *The Rust Flora of Japan*. Tsukuba Shuppankai, Ibaraki, Japan, pp. 271–300.

Hylander, N., Jørstad, I. and Nanfeldt, J.A. (1953) Enumeratio Uredinearum Scandinavicarum. *Opera Botanica* 1, 1–102.

Longo, N., Moriondo, F. and Naldini Longo, B. (1975) The status of *Melampsora pinitorqua* Rostr. in Italy. *European Journal of Forest Pathology* 5, 147–152.

Newcombe, G. and Chastagner, G.A. (1993) First report of the Eurasian poplar leaf rust fungus, *Melampsora larici-populina*, in North America. *Plant Disease* 77, 532–535.

Newcombe, G., Stirling, B., McDonald, S. and Bradshaw, H.D. (2000) *Melampsora* × *columbiana*, a natural hybrid of *M. medusae* and *M. occidentalis*. *Mycological Research* 104, 261–274.

Pei, M.H., Royle, D.J. and Hunter, T. (1999) Hybridisation in larch-alternating *Melampsora epitea* (*M. larici-epitea*). *Mycological Research* 103, 1440–1446.

Pinon, J., Newcombe, G., and Chastagner, G. (1994) Identification of races of *Melampsora larici-populina*, the Eurasian poplar leaf rust, on *Populus* species in California and Washington. *Plant Disease* 78, 1011.

Shain, L. (1988) Evidence for *formae speciales* in the poplar leaf rust fungus, *Melampsora medusae*. *Mycologia* 80, 729–732.

Shang, Y.Z. and Pei, M.H. (1984) Study on the leaf rust of Davids European Aspen caused by *Melampsora larici-tremulae* Kleb. [in Chinese]. *Journal of North-eastern Forestry Institute* 12 (1), 47–55.

Shang, Y.Z. and Pei, M.H. (1988) Different resistance to *Melamspora laricis* Hart. and *Melamspora larici-populina* Kleb. among three sections and hybrids of poplar [in Chinese]. *Journal of Northeast Forestry University* 16(4), 12–20.

Shang, Y.Z., Pei, M.H. and Yuan Z.W. (1986) A new rust fungus on poplars [in Chinese]. *Acta Mycologia Sinica. Supplement* 1, 180–184.

Spiers, A.G. (1990) *Melampsora* leaf rusts of poplar. *Forest Pathology in New Zealand* 20, 1–6.

Spiers, A.G. (1998) *Melampsora* and *Marssonina* pathogens of poplars and willows in New Zealand. *European Journal of Forest Pathology* 28, 233–240.

Spiers, A.G. and Hopcroft, D.H. (1994) Comparative studies of the poplar rusts *Melampsora medusae*, *M. larici-populina* and their interspecific hybrid *M. medusae-populina*. *Mycological Research* 98, 889–903.

Sydow, P. and Sydow, H. (1915) *Monographia Uredinearum III*. Bornträger, Leipzig, pp. 334–350.

Tian, C.M., Shang, Y.Z., Zhuang, J.W., Wang, Q. and Kakishima, M. (2004) Morphological and molecular phylogenetic analysis of *Melampsora* species on poplars in China. *Mycoscience* 45, 56–66.

Viljanson-Rollinson, S.L.H. and Cromey, M.G. (2002) Pathways of entry and spread of rust pathogens: Implications for New Zealand's biosecurity. *New Zealand Plant Protection* 55, 42–48.

Walker, J. and Hartigan, D. (1972) Poplar rust in Australia. *Australasian Plant Pathology Society Newsletter* 1, 3.

Wang, C. and Fang, C.F. (1984) *Salicaceae. Flora of the People's Republic of China*, Vol. 20 (2) [in Chinese]. Science Press, Beijing, pp. 2–78.

Wilkinson, A.G. and Spiers, A.G. (1976) Introduction of the poplar rusts *Melampsora larici-populina* and *M. medusae* to New Zealand and their subsequent distribution. *New Zealand Journal of Science* 19, 195–198.

Wilson, M. and Henderson, D.M. (1966) *The British Rust Fungi*. Cambridge University Press, Cambridge, UK.

Xu, M.Q. (1998) Leaf rust of *Populus* section *Leuce* in China. In: Jalkanen, R., Crane, P.E., Walla, J.A. and Aalto, T. (eds) *Proceedings of the First IUFRO Rusts of Forest Trees W.P. Conference*, 2–7 August, Saariselkä. Finnish Forest Research Institute, Rovaniemi, pp. 53–56.

Ziller, W.G. (1955) Studies of western tree rusts: II. *Melampsora occidentalis* and *M. albertiensis*, two needle rusts of Douglas fir. *Canadian Journal of Botany* 33, 177–188.

Ziller, W.G. (1974) *The Tree Rusts of Western Canada*. Publication No. 1329. Canadian Forestry Service, Environmental Canada, Ottawa, Canada.

5 Variability and Population Biology of *Melampsora* Rusts on Poplars

Pascal Frey[1], Pierre Gérard[2], Nicolas Feau[3], Claude Husson[1] and Jean Pinon[1]

[1]*UR Pathologie Forestière, INRA, F-54280 Champenoux, France;* [2]*Laboratoire Ecologie, Systématique et Evolution, UMR ENGREF-UPXI-CNRS 8079, Université Paris-Sud, 91405 Orsay, France;* [3]*Centre de Recherche en Biologie Forestière, Université Laval, Sainte-Foy (QC), G1K 7P4, Canada*

Background

There are eight *Melampsora* species infecting poplars in Europe (Pinon, 1973; Cellerino, 1999). Three of them, *M. larici-populina*, *M. allii-populina*, and *M. medusae* are pathogenic on the poplars of the sections *Aigeiros* and *Tacamahaca*, i.e. *P. nigra*, *P. deltoides*, *P. trichocarpa*, and their interspecific hybrids. Most of the commercial poplar cultivation in Europe is made up of *P.* × *euramericana* (*P. deltoides* × *P. nigra*) and *P.* × *interamericana* (*P. trichocarpa* × *P. deltoides*) hybrids. Rust was the most damaging disease of poplars in the past decade (Chapter 12, this volume). Five other *Melampsora* species, *M. pinitorqua*, *M. larici-tremulae*, *M. rostrupii*, *M. magnusiana*, and *M. pulcherrima*, are pathogenic on species of the *Populus* (formerly *Leuce*) section, i.e. *P. alba*, *P. tremula*, and their hybrids. Although these five species are difficult to distinguish, since they have in common evenly echinulated urediniospores, Pinon (1973) proposed a diagnosis key based on the morphology of the urediniospores and the paraphyses. Some authors follow the proposal of Wilson and Henderson (1966) to adopt *M. populnea* as a collective species to include four species which are difficult to distinguish in their morphology, *M. pinitorqua*, *M. larici-tremulae*, *M. rostrupii*, and *M. magnusiana* (Chapter 1, this volume). Since aspens and white poplars are not widely planted for commercial cultivation and exist only as wild and ornamental trees in Europe, the incidence of rust on these species has a much lower economic impact.

In this review, we will focus mainly on the three *Melampsora* species attacking poplars of the *Aigeiros* and *Tacamahaca* sections. Among these three species, *M. larici-populina* is responsible for most of the economic losses, causing the breakdown of several important resistance genes which had been released (Chapter 12, this volume). At present, *M. allii-populina* is much less frequent and its economic impact is negligible (Frey and Pinon, 1997). The North American species, *M. medusae*, present in Europe since the early 20th century, is very rare and restricted to south-western Europe (south-western France, Spain and Portugal; Pinon, 1986, 1991). Although registered as a quarantine pathogen in the EU, this species causes no economic losses in Europe.

Interspecific Variability

The advances of molecular biology have brought new approaches to investigations of rust fungi, and molecular tools are increasingly being used in phylogenetic studies, DNA-based diagnostics and the study of genes involved in pathogenicity.

Comparison of internal transcribed spacer sequences

The nuclear ribosomal DNA, and especially the internal transcribed spacer (ITS) regions, has widely been used to assess genetic variability at the species level in many fungal genera, including rusts (Zambino and Szabo, 1993; Nakamura *et al.*, 1998; Vogler and Bruns, 1998; Hantula *et al.*, 2002; Maier *et al.*, 2003). In order to determine the phylogenetic relationships among the poplar *Melampsora*, we studied the ITS sequence of nine isolates of *M. larici-populina*, *M. allii-populina* and *M. medusae*. Additional sequence information available from the GenBank database (http://www.ncbi.nlm.nih.gov/), including a sequence from *M. occidentalis*, was added to this study (Table 5.1). The full-length ITS sequences (ITS1, 5.8S, ITS2) were aligned using MULTALIN software (http://www.toulouse.inra.fr/multalin.html) (Corpet, 1988) and a phylogenetic tree was constructed using parsimony method in PAUP 4.0, using a heuristic search with the default settings (Swofford, 2002). The tree

Table 5.1. ITS sequences of *Melampsora* spp. used in this study.

Species	Isolate code	Original host	Year of collection	Collection site	GenBank accession	Source[a]
M. larici-populina	95XA1	*P.* × *interamericana* 'Beaupré'	1995	N France	AY375267	(1)
	93ID6	*P.* × *euramericana* 'I 45-51'	1993	NE France	AY375268	(1)
	97A1	*P.* × *euramericana*	1997	Morocco	AY375269	(1)
	97J10	*P.* × *euramericana* 'I 488'	1997	South Africa	AY375270	(1)
	9014-A	*P. maximowiczi* × *P. balsamifera*	2002	Québec, Canada	AY429656	(2)
	GHOYAH00-2	*P.* × *euramericana* 'Ghoy'	2000	SW England	AY444774	(3)
	TSH-R16975	*P. nigra* 'Italica'	2001	Jilin, China	AB116836	(4)
M. allii-populina	96B6-1	*Allium vineale*	1996	NE France	AY375271	(1)
	96M24-1	*Muscari comosum*	1996	NE France	AY375272	(1)
	BEAUPMK95	*P.* × *interamericana* 'Beaupré'	1995	N England	AY444773	(3)
	TSH-R04141	*P. laurifolia*	1986	Xinjiang, China	AB116875	(4)
M. medusae	88MM1	*P.* × *euramericana* 'Gaver'	1988	SW France	AY375273	(1)
	88MM2	*P.* × *interamericana* 'Beaupré'	1988	SW France	AY375274	(1)
	97CN1	*P.* × *interamericana* 'Boelare'	1997	SW France	AY375275	(1)
	97-4	*P. deltoides*	1997	South Carolina, USA	AF087711	(5)
M. occidentalis	97-13	*P. trichocarpa*	1997	Washington State, USA	AF087710	(5)

[a]Source: (1) This study, (2) Innes *et al.* (2004), (3) Pei *et al.* (2004), (4) Tian *et al.* (2004), (5) Newcombe *et al.* (2000).

obtained (Fig. 5.1) shows three clearly distinct clades: *M. larici-populina*, *M. allii-populina*, and one clade containing *M. medusae* and *M. occidentalis*. The interspecific polymorphism is relatively high (average identity 87–90%), whereas the intraspecific polymorphism within isolates of the same species is limited to a few point mutations (average identity 99–100%).

Recently, *M. larici-populina* was detected in north-eastern Canada (Innes *et al.*, 2004). Exactly the same ITS sequence was found in one isolate from France, two isolates from Canada (Innes *et al.*, 2004) and 16 isolates from China (Tian *et al.*, 2004), whereas a few single nucleotide polymorphisms (SNPs) were found in other isolates from France, Morocco, South Africa and England (Fig. 5.1). One surprising feature is the position of *M. occidentalis* in this tree: the ITS sequence of *M. occidentalis* differs from that of *M. medusae* only by one

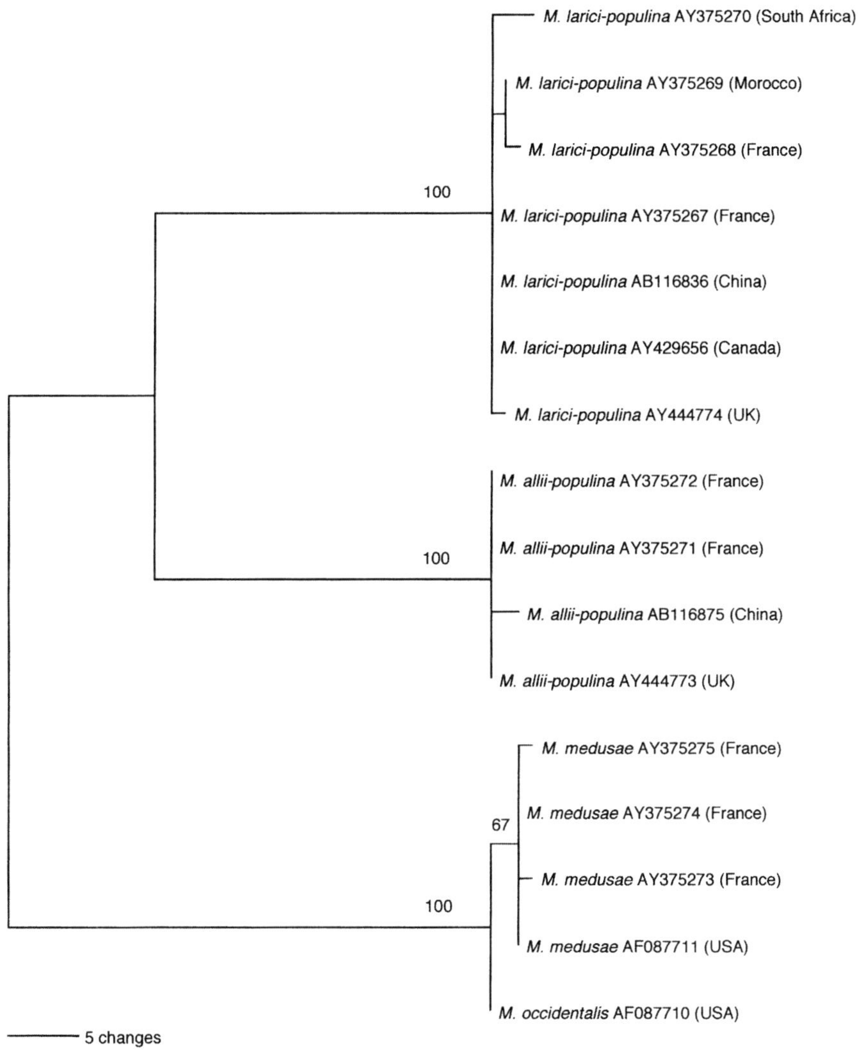

Fig. 5.1. Phylogenetic tree obtained with the aligned internal transcribed spacer (ITS) sequences of *Melampsora* spp. listed in Table 5.1. The numbers above the branches are the bootstrap values from 1000 replicates. GenBank accession numbers are indicated.

TT→CG substitution at position 78–79 (Newcombe *et al.*, 2000), although the two species differ clearly for several characteristics, such as the morphology of urediniospores and the host range. Clearly, the ITS region does not discriminate between these two species sufficiently.

Development of species-specific polymerase chain reaction primers and application to the detection of *M. medusae*

Although *M. larici-populina*, *M. allii-populina* and *M. medusae* are easily distinguishable under a light microscope at the uredinial stage, species-specific primers were developed in order to: (i) distinguish *M. larici-populina* and *M. medusae* at the aecial stage on larch; (ii) distinguish between the three species at the telial stage on fallen poplar leaves; and (iii) detect *M. medusae* when present at a low frequency mixed with *M. larici-populina* and/or *M. allii-populina*. From the alignment of the ITS sequences, three primer pairs were designed (unpublished data). The three primer pairs were tested on a worldwide collection of 110 isolates belonging to the three species. All the isolates resulted in a positive amplification with the corresponding primer pair. No cross-reaction was observed between the three species.

We also determined the detection threshold of the *M. medusae*-specific primers. By adding serial dilutions of *M. medusae* DNA into *M. larici-populina* DNA, the detection threshold was found to be 10^{-5} to 10^{-6}, which corresponds to less than 10 urediniospores of *M. medusae* within a sample containing 2 mg (i.e. 800,000) urediniospores. The same sensitivity was found by adding serial dilutions of *M. medusae* urediniospores into *M. larici-populina* urediniospores.

Since *M. medusae* is registered as a quarantine pathogen in the EU (Anon., 2000), the trade of poplar plants between and within EU countries is subjected to a phytosanitary passport which states the absence of *M. medusae* in the producer's nursery. Therefore, since 1993, all the commercial nurseries producing poplar saplings or cuttings in France have been inspected annually by the French Plant Protection Service (Laboratoire National de la Protection des Végétaux, LNPV) for the presence of *M. medusae*. The official detection method is based on observation of the morphological characters of the urediniospores. No *M. medusae* has been detected since 1998, even in south-western France. In collaboration with LNPV, we assessed the polymerase chain reaction (PCR)-based detection method, along with the official method. In 2002, 333 rust samples were collected in 154 commercial nurseries throughout France. All the samples were tested by PCR with the three primer pairs. The results of the molecular detection method were consistent with those of the morphological diagnosis: *M. larici-populina* was detected in 87% of the samples, *M. allii-populina* in 16% of the samples, and *M. medusae* was never detected. In 2003, although the number of samples collected was reduced greatly due to the exceptional drought that year, the results were similar to those in 2002, without any occurrence of *M. medusae*. In the future, the molecular detection method will be proposed as the official method for *M. medusae* detection, since the detection threshold is reduced by 10,000-fold, compared to the morphological method.

Interspecific Hybrids in Poplar Rusts

In the 1990s, two interspecific hybrids of *Melampsora* spp. were discovered in two different geographic areas. *M. medusae-populina*, a hybrid of *M. larici-populina* and *M. medusae*, was first discovered in New Zealand in 1991 (Spiers and Hopcroft, 1994) and was subsequently found in South Africa. *M.* × *columbiana*, a hybrid of *M. occidentalis* and *M. medusae*, was discovered in 1995 in the Pacific Northwest region of the USA (Newcombe *et al.*, 2000).

Melampsora medusae-populina

M. medusae-populina was first discovered in New Zealand in 1991 (Spiers and Hopcroft, 1994), where both *M. larici-populina* and *M. medusae* were introduced in 1973 (Van Kraayenoord *et al.*, 1974). This new rust taxon presents several morphological characters which are intermediate between the two supposed parental species, the most evident being the presence of both apical and equatorial smooth patches on the surface of the urediniospores. The rust was discovered after it attacked several 'rust-resistant' poplar cultivars (Spiers and Hopcroft, 1994). These authors hypothesized that this hybrid arose in Australia, where both supposed parental species are present, and that it may have been transported to New Zealand by wind currents. The mechanism of hybridization still remains an enigma. Spiers and Hopcroft (1994) seemed to favour the hypothesis of anastomosis at the uredinial, i.e. dicaryotic, stage on poplar, instead of a heterologous plasmogamy on larch, which is the common alternate host for both parental species.

Independently, we found *M. medusae-populina* in 1997 in samples originating from South Africa (unpublished data). Since both *M. larici-populina* and *M. medusae* have been described in South Africa since 1967 and 1985, respectively, *M. medusae-populina* in this country may have originated from: (i) transportation from New Zealand or Australia; or (ii) a new hybridization event. In order to confirm the hybrid status of *M. medusae-populina*, we studied the ITS region of four mono-uredinial isolates of this taxon. After amplification, the amplicons were cloned and several clones were sequenced. Within each *M. medusae-populina* isolate, two types of ITS sequences were found, one matching exactly that of *M. larici-populina* and the other matching that of *M. medusae* (Frey *et al.*, 1999). Further confirmation of the hybrid status of *M. medusae-populina* was obtained with random amplified polymorphic DNA (RAPD) fingerprinting. For most of the RAPD primers tested, the hybrid showed DNA bands from *M. larici-populina* and *M. medusae*.

Melampsora × *columbiana*

Melampsora × *columbiana*, a hybrid of *M. occidentalis* and *M. medusae*, was discovered in 1995 in the Pacific Northwest region of the USA by Newcombe *et al.* (2000). This new rust taxon was discovered after many *P.* × *interamericana* cultivars became infected by rust. This hybrid arose after *M. medusae* was introduced to the Pacific Northwest in 1991, where only *M. occidentalis* was present (Newcombe, 1998). Although less obvious than in the case of *M. medusae-populina*, the morphology of the urediniospores of *M.* × *columbiana* is intermediate between that of the parental species. The hybrid status also is evident from the host range: whereas *M. medusae* is mostly pathogenic to *P. deltoides* clones and *M. occidentalis* to *P. trichocarpa* clones, *M.* × *columbiana* combines both host ranges and is highly pathogenic on *P.* × *interamericana* hybrids. Thus the hybrid rust is particularly adapted to the hybrid poplars (Newcombe *et al.*, 2001).

Newcombe *et al.* (2000) also studied the ITS region of *M. medusae*, *M. occidentalis* and *M.* × *columbiana*. The ITS sequences of *M. medusae* and *M. occidentalis* were found to be 99.7% identical, as they only differ by two base pairs at one position (Fig. 5.1). Of six isolates of *M.* × *columbiana*, two were found containing ITS sequences from both parental species, whereas three presented only the ITS sequence of *M. medusae*, and one that of *M. occidentalis*. Therefore, the authors suggested that the hybrids may have produced advanced generations such as F_2 and/or backcrosses.

These two hybrids illustrate perfectly the risk of interspecific hybridization between fungal species after a geographical transposition of one (or both) species into a new region (Brasier, 1995; Newcombe and Frey, 2002). Phylogenetically close, but geographically distant, fungal species may hybridize when they come into contact in the same niche, and the resulting hybrids may become a 'superpathogen' in some cases (Brasier *et al.*, 1999).

Population biology of *Melampsora larici-populina*

Objectives

Since *M. larici-populina* is responsible for most of the economic losses in poplar cultivation in Europe, our efforts in population biology have focused mainly on this species. The populations of *M. larici-populina* have shown very rapid changes during the past two decades, through the rise of new virulence that overcame several resistance genes (Chapter 12, this volume). After the occurrence of a new virulence, a very rapid spread of the new virulent rust was observed, associated with the appearance of numerous new pathotypes through recombination (Pinon *et al.*, 1998). In the meantime, no obvious changes were noticed on the impact of rust on wild stands of *P. nigra*. Taking advantage of the presence in France of both wild stands of *P. nigra* and cultivated stands planted with exotic hybrids such as *P.* × *euramericana* and *P.* × *interamericana*, our studies were aimed mainly at comparison of the structure of *M. larici-populina* populations in both pathosystems (Pinon and Frey, 1997; unpublished data). Furthermore, some aspects of the biology and the epidemiology of this pathogen remained unknown. For instance, the relative role of sexual reproduction versus asexual survival remained unclear. Therefore, we examined more closely the effect of the presence of larch on the structure of *M. larici-populina* populations.

Markers used

We have described a universal differential set of poplar clones which discriminates eight virulences of *M. larici-populina* (Chapter 12). All the isolates studied were made from single uredinia and then inoculated on to leaf discs of the poplar differential set. Parameters studied were the frequency of each virulence, the frequency of each pathotype, the richness of the populations assessed with the Alpha, Shannon and Gleason indices (Pinon and Frey, 1997), and the complexity of the populations, defined as the mean number of virulences per isolate (Andrivon and de Vallavieille-Pope, 1995).

Along with these virulence markers, we also developed molecular markers. RAPD markers were used to compare the genetic structure of *M. larici-populina* populations. Microsatellites (or single-sequence repeat, SSR) markers are currently being developed. These co-dominant markers, suitable for automation, will be very useful tools for the study of migration, selection and genetic drift within and between populations of *M. larici-populina*.

Genetic structure of populations from wild versus cultivated poplar stands

In order to assess the effect of the nature of the host population, we compared the structure of *M. larici-populina* populations from wild and cultivated poplar stands. We selected wild riparian *P. nigra* stands along two rivers in the French Alps (Durance and Drac rivers), and cultivated stands in two main areas for poplar cultivation (Picardie, northern France, and Franche-Comté, north-eastern France). For each region, sites with and without larch (*Larix decidua*) in their vicinity (< 1 km) were selected in order to take the role of sexual reproduction on the structure of *M. larici-populina* populations into account. Pathotype analysis clearly distinguished the populations from wild and cultivated stands. Populations from cultivated stands comprised a significantly higher ($P = 0.0001$) number of pathotypes ($Np = 13.1 \pm 4.0$) than those from wild stands ($Np = 5.3 \pm 2.2$). Both the richness and the complexity of the populations were significantly higher ($P = 0.0001$) in cultivated than in wild stands (Fig. 5.2).

Populations from cultivated stands were composed of numerous pathotypes combining many virulences, such as pathotype 1–3–4–5–7, which has been the most frequent since 1994 (Pinon *et al.*, 1998; Miot *et al.*, 1999). In contrast, populations from the wild stands were composed of a few simple pathotypes, such as pathotypes 0, 2 and 4. These findings are consistent with those of previous studies (Pinon and Frey,

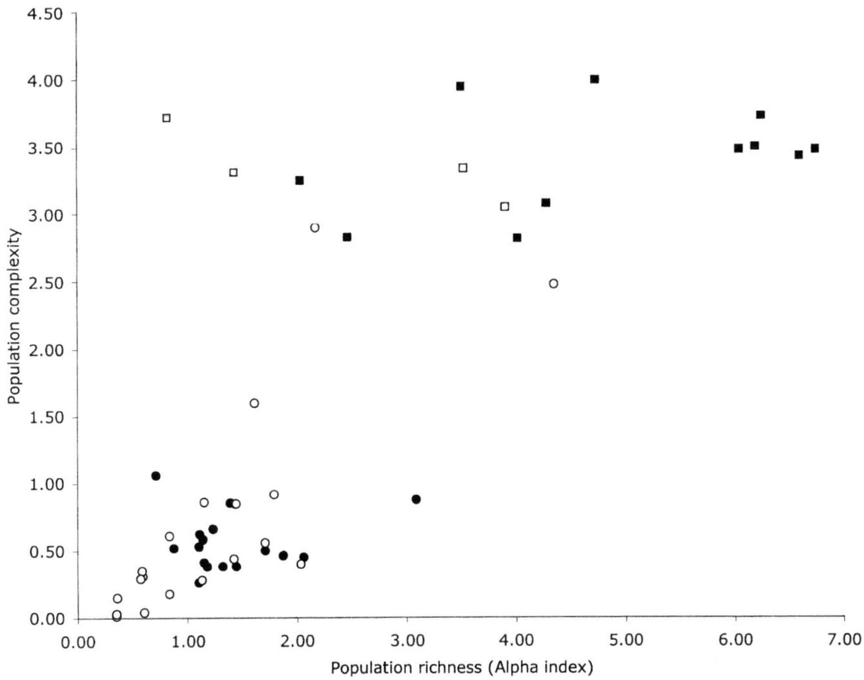

Fig. 5.2. Richness and complexity of 51 populations of *Melampsora larici-populina* collected from wild and cultivated poplar stands in 2001–2003. Circles, wild stands; squares, cultivated stands; black symbols, stands with larches in their vicinity (< 1 km); open symbols, stands without larches in their vicinity.

1997). The high complexity of *M. larici-populina* populations in cultivated stands raises the question of the cost of unnecessary virulence (Miot *et al.*, 1999; Chapter 12).

RAPD markers did not distinguish clearly populations of *M. larici-populina* from wild versus cultivated stands. The overall genetic distances between populations were low. Nei's unbiased genetic distance (Nei, 1978) between paired populations ranged from 0.01 to 0.07, suggesting a high gene flow at the country scale. Genetic distances were very low between pairs of populations from the same region. Similarly, *F*-statistics showed low values (F_{ST} ranging from 0.02 to 0.09), suggesting a low degree of differentiation between populations.

Effect of the presence of larch

The presence of larch in the vicinity (< 1 km) of the selected poplar stands significantly increased the number of pathotypes ($P = 0.001$) either for cultivated or for wild stands. It also increased the richness of the populations ($P = 0.001$), but did not increase the complexity ($P = 0.9$). Surprisingly, we did not observe a reduction of the linkage disequilibria in the presence of larch, neither for virulence markers nor for RAPD markers.

Furthermore, a more detailed study of the space–time dynamics of *M. larici-populina* populations along the Durance River gave us a better view of the natural pathosystem. Every year, rust epidemics seem to begin in the area where both poplar and larch occur, in the upstream of the river. During summer, the epidemics spread downstream following the continuous riparian forest of *P. nigra*. The dates of infection of *P. nigra* in the downstream part of the river clearly suggest that there is no asexual survival for urediniospores of *M. larici-populina*. Thus the only primary inoculum source under this climate is larch. These findings are consistent with previous observations in the cultivated stands (Frey *et al.*, 1998).

Comparison between regions with and without larch plants was also made in order to assess the role of sexual reproduction on the structure of populations of other larch-alternating *Melampsora* species. *M. medusae* populations from the areas in Canada where both poplar and larch occur, exhibited a higher genetic diversity compared to those from the areas in southern Canada and in the USA where only poplars occur (Bourassa *et al.*, 1998). Similar results were obtained by comparing sexual populations of *M. larici-epitea* on willow from Sweden and asexual populations from Northern Ireland (Samils *et al.*, 2001; Chapter 7).

Conclusions

Although *M. allii-populina* and *M. medusae* are present in some parts of Europe and can infect cultivated poplar stands (mainly *P.* × *euramericana* and *P.* × *interamericana* cultivars), most of the damage is caused by *M. larici-populina*. With its ability to overcome several pathotype-specific resistance genes, this species is presently the main concern for poplar breeders (Lefèvre *et al.*, 1998; Dowkiw *et al.*, 2003). There were great differences in the structure of *M. larici-populina* populations from cultivated and from wild poplar stands. The low complexity of *M. larici-populina* populations from wild stands may suggest a cost for unnecessary virulence, whereas the high complexity of populations from cultivated stands may suggest the opposite. This apparent contradiction may result from the difference in the rate of evolution of virulence markers versus molecular markers. Virulence markers have been shown to evolve very rapidly in the cultivated stands during the past 30 years. However, the genetic distances between populations from cultivated and wild stands were low, suggesting it is unlikely that there is a strong barrier between cultivated and wild stands. Anyway, the cultivated stands have existed only for about two centuries, whereas the fungus and its wild host, *P. nigra*, have coevolved for millions of years. Therefore, an in-depth study of the population genetics and population dynamics in the wild stands is useful, to have a better view of the cultivated pathosystem.

Acknowledgements

This research was partly funded by grants from GIP Ecofor and from INRA and DGAL (Ministry of Agriculture).

References

Andrivon, D. and de Vallavieille-Pope, C. (1995) Race diversity and complexity in selected populations of fungal biotrophic pathogens of cereals. *Phytopathology* 85, 897–905.

Anon. (2000) Council Directive 2000/29/EC of 8 May 2000 on protective measures against the introduction into the Community of organisms harmful to plants or plant products and against their spread within the Community. *Official Journal* L 169, 10/07/2000, 1–112.

Bourassa, M., Bernier, L., Milligan, B.G. and Hamelin, R.C. (1998) Sympatry between alternate hosts affects population structure of poplar rust. In: Jalkanen, R., Crane, P.E., Walla, J.A. and Aalto, T. (eds) *Proceedings of the First IUFRO Rusts of Forest Trees Working Party Conference*, 2–7 August 1998, Saariselkä. Finnish Forest Research Institute, Research papers 712, pp. 65–69.

Brasier, C.M. (1995) Episodic selection as a force in fungal microevolution, with special reference to clonal speciation and hybrid introgression. *Canadian Journal of Botany* 73, S1213–S1221.

Brasier, C.M., Cooke, D.E. and Duncan, J.M. (1999) Origin of a new *Phytophthora* pathogen through interspecific hybridization. *Proceedings of the National Academy of Sciences of the USA* 96, 5878–5883.

Cellerino, G.P. (1999) Review of poplar diseases. International Poplar Commission, FAO. Available at Website www.efor.ucl.ac.be/ipc/pub/celle01/celle01.htm (accessed 28 September 2004).

Corpet, F. (1988) Multiple sequence alignment with hierarchical clustering. *Nucleic Acids Research* 16, 10881–10890.

Dowkiw, A., Husson, C., Frey, P., Pinon, J. and Bastien, C. (2003) Partial resistance to *Melampsora larici-populina* leaf rust in hybrid poplars: genetic variability in inoculated excised leaf disk bioassay and relationship with complete resistance. *Phytopathology* 93, 421–427.

Frey, P. and Pinon, J. (1997) Variability in pathogenicity in *Melampsora allii-populina* expressed on poplar cutivars. *European Journal of Forest Pathology* 27, 397–407.

Frey, P., Husson, C., Pinon, J., Gumez, J. L., Munnier, P. and Servant H. (1998) Confirmation du rôle des mélèzes dans l'initiation des infections du peuplier par la rouille (*Melampsora larici-populina*). *Lettre du Département Santé des Forêts* 18, 2–3.

Frey, P., Gatineau, M., Martin, F. and Pinon, J. (1999) Molecular studies of the poplar rust *Melampsora medusae-populina,* an interspecific hybrid between *M. larici-populina* and *M. medusae*. International Poplar Symposium II, Orléans, France, 13–17 September 1999 [abstract].

Hantula, J., Kasanen, R., Kaitera, J. and Moricca, S. (2002) Analyses of genetic variation suggest that pine rusts *Cronartium flaccidum* and *Peridermium pini* belong to the same species. *Mycological Research* 106, 203–209.

Innes, L., Marchand, L., Frey, P., Bourassa, M. and Hamelin, R.C. (2004) First report of *Melampsora larici-populina* on *Populus* spp. in Eastern North America. *Plant Disease* 88, 85.

Lefèvre, F., Goué-Mourier, M.C., Faivre-Rampant, P. and Villar, M. (1998) A single gene cluster controls incompatibility and partial resistance to various *Melampsora larici-populina* races in hybrid poplars. *Phytopathology* 88, 156–163.

Maier, W., Begerow, D., Weiß, M. and Oberwinkler, F. (2003) Phylogeny of the rust fungi: an approach using nuclear large subunit ribosomal DNA sequences. *Canadian Journal of Botany* 81, 12–23.

Miot, S., Frey, P. and Pinon, J. (1999) Varietal mixture of poplar clones: effects on infection by *Melampsora larici-populina* and on plant growth. *European Journal of Forest Pathology* 29, 411–423.

Nakamura, H., Kaneko, S., Yamaoka, Y. and Kakishima, M. (1998) Differentiation of *Melampsora* rust species on willows in Japan using PCR-RFLP analysis of ITS regions of ribosomal DNA. *Mycoscience* 39, 105–113.

Nei, M. (1978) Estimation of average heterozygosity and genetic distance from a small number of individuals. *Genetics* 89, 583–590.

Newcombe, G. (1998) A review of exapted resistance to disease of *Populus*. *European Journal of Forest Pathology* 28, 209–216.

Newcombe, G. and Frey, P. (2002) Evolution of interspecific *Melampsora* hybrids exhibiting new host specificities. 7th International Mycological Congress, Oslo, Norway, 11–17 August [abstract].

Newcombe, G., Stirling, B., McDonald, S. and Bradshaw, H.D. Jr (2000) *Melampsora* × *columbiana,* a natural hybrid of *M. medusae* and *M. occidentalis*. *Mycological Research* 104, 261–274.

Newcombe, G., Stirling, B. and Bradshaw, H.D. Jr (2001) Abundant pathogenic variation in the new hybrid rust *Melampsora* × *columbiana* on hybrid poplar. *Phytopathology* 91, 981–985.

Pei, M.H., Bayon, C. and Ruiz, C. (2004) Phylogenetic relationships in some *Melampsora* rusts on *Salicaceae* assessed using rDNA sequence information (submitted).

Pinon, J. (1973) Les rouilles du Peuplier en France: systématique et répartition du stade urédien. *European Journal of Forest Pathology* 3, 221–228.

Pinon, J. (1986) Situation of *Melampsora medusae* in Europe. *EPPO Bulletin* 16, 547–551.

Pinon, J. (1991) Eléments de répartition des rouilles des peupliers cultivés en France. *Comptes-Rendus de l'Académie d'Agriculture de France* 77, 109–115.

Pinon, J. and Frey, P. (1997) Structure of *Melampsora larici-populina* populations on wild and cultivated poplar. *European Journal of Plant Pathology* 103, 159–173.

Pinon, J., Frey, P., Husson, C. and Schipfer, A. (1998) Poplar rust (*Melampsora larici-populina*): the development of E4 pathotypes in France since 1994. In: Jalkanen, R., Crane, P.E., Walla, J.A. and Aalto, T. (eds) *Proceedings of the First IUFRO. Rusts of Forest Trees Working Party Conference,* 2–7 August, Saariselkä. Finnish Forest Research Institute, Research papers 712, pp. 57–64.

Samils, B., Stepien, V., Lagercrantz, U., Lascoux, M. and Gullberg, U. (2001) Genetic diversity in relation to sexual and asexual reproduction in populations of *Melampsora larici-epitea*. *European Journal of Plant Pathology* 107, 871–881.

Spiers, A. and Hopcroft, D. (1994) Comparative studies of the poplar rusts *Melampsora medusae, M. larici-populina* and their interspecific hybrid *M. medusae-populina*. *Mycological Research* 98, 889–903.

Swofford, D.L. (2002) *PAUP**: phylogenetic analysis using parsimony (*and other methods).* Version 4. Sinauer Associates, Sunderland, Massachusetts.

Tian, C.M., Shang, Y.Z., Zhuang, J.Y., Wang, Q. and Kakishima, M. (2004) Morphological and molecular phylogenetic analysis of *Melampsora* species on poplars in China. *Mycoscience* 45, 56–66.

Van Kraayenoord, C.W.S., Laundon, G.F. and Spiers, A.G. (1974) Poplar rusts invade New Zealand. *Plant Disease Reporter* 58, 423–427.

Vogler, D.R. and Bruns, T.D. (1998) Phylogenetic relationships among the pine stem rust fungi (*Cronartium* and *Peridermium* spp.). *Mycologia* 90, 244–257.

Wilson, M. and Henderson, D.M. (1966) *The British Rust Fungi*. Cambridge University Press, Cambridge.

Zambino, P.J. and Szabo, L.J. (1993) Phylogenetic relationships of selected cereal and grass rusts based on rDNA analysis. *Mycologia* 85, 401–414.

6 Genetic Diversity of *Melampsora* Willow Rusts in Germany

Mirko Liesebach[1] and Irmtraut Zaspel[2]
[1]*Federal Office and Research Centre for Forests, Department of Forest Genetics, Hauptstrasse 7, A-1140 Vienna, Austria;* [2]*Federal Research Centre for Forestry and Forest Products, Institute for Forest Genetics and Forest Tree Breeding, Eberswalder Chaussee 3A, D-15377 Waldsieversdorf, Germany*

Background

Over 300 species are described in the genus *Salix* and hybridization between *Salix* species occurs commonly (Lautenschlager-Fleury, 1994; Zander, 2000; Newsholme, 2002). This provides a rich genetic resource with which to combat the threat of pest and disease attacks through breeding (Ahman and Larsson, 1994). Willows are hosts to a wide range of phytophagous insect species and fungal pathogens. Leaf rusts caused by *Melampsora* spp. are important fungal pathogens on willows in central Europe.

Willow rusts can cause severe damage to the leaves and shoots of their host plants and lead to leaf senescence and premature defoliation, thus debilitating the plant (Helfer, 1992; Pei *et al.*, 1999a). With the introduction of willow short-rotation coppice (SRC) plantations in northern and north-western Europe, rusts have become major pathogens in energy forestry, resulting in reduced growth rates and decreased yields by as much as 40% (Parker *et al.*, 1993; Tucker and Sage, 1999). Severe rust can predispose plants to infections by secondary pathogens, which often kill the plants.

More than 20 species of the genus *Melampsora* have been recognized on willows worldwide, and most of them have a complex life cycle and alternate on very different herbaceous and woody plants (Pei *et al.*, 1997). To different species and populations of rust pathogens, species and varieties of willows have different degrees of resistance and tolerance (Hunter *et al.*, 1996).

Three species of rusts occur on shrub willows for biomass. One of them is the highly heterogeneous collective species *M. epitea* (Kunze et Schm.) Thüm complex. Pei *et al.* (1999a) have defined eight putative pathotypes within this *M. epitea* complex. Five of these pathotypes belong to the *forma specialis* (f. sp.) *larici-epitea typica* (LET), two to f. sp. *larici-retusae* (LR) and one to f. sp. *larici-daphnoides* (LD). The groupings of willow rust by Gäumann (1959) and Butin (1960) were similar. Butin (1960) divided *M. epitea* into five subspecies, including *M. larici-epitea*. The *M. larici-epitea* in central Europe and Switzerland was further divided into six *formae speciales* (Gäumann, 1959), including LET, LR and LD. Much of the pioneering work on the species identity, host alternation and host range in willow rusts in central Europe was carried out in the late 19th to early 20th century (reviewed by Gäumann, 1959). Since then, little work has been done on willow rusts in central Europe. Klenke (1998) discussed the taxonomy of willow rusts but his work was restricted to Saxonia only.

There has been no recent investigation on the range of pathotypes of the different *formae speciales* in Germany. These different types cannot be distinguished by their morphology. They can only be identified through standardized pathotyping using a range of willow clones as differentials. Willow clones used for differentials may belong to several sections and species. An understanding of genetic variability in pathogens is thought to be highly important for the management of diseases of fast-growing tree species like willows. At the beginning of the 21st century, we conducted a new study to identify and characterize different rust species and forms posing threats to willows for renewable energy and conservation in central Europe, especially in Germany.

Melampsora Species on Willow in Germany and their Life Cycles

The majority of *Melampsora* species have a macrocyclic life cycle, during which they develop five spore stages and alternate on very different host species (alternate hosts). The most important alternate host of *Melampsora* species infecting willows and poplars in SRC plantations is larch. The life cycle of *M. larici-epitea* is illustrated in Fig. 6.1. During summer, rust can be observed mostly on the lower sides of willow leaves as yellow to orange pustules (uredinia). The urediniospores are responsible for many cycles of asexual reproduction and hence the spread of rust disease in a growing season. Probably as a reaction to low temperatures and senescence of leaves, uredinia are transformed into telia which serve as

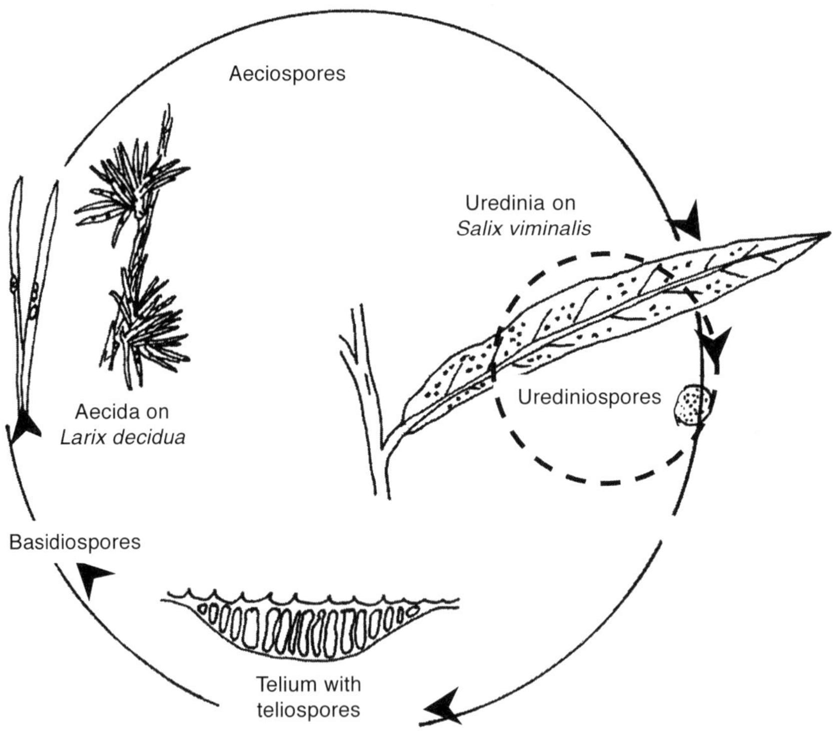

Fig. 6.1. Life cycle of *Melampsora epitea* on *Salix viminalis* and *Larix decidua*.

the overwintering form of the rusts after leaf fall. In spring, teliospores germinate to produce basidiospores which infect alternate hosts. Consequently, fertilization between spermatia and receptive hyphae results in formation of aecia. Inoculations with aeciospores from the larch hosts *Larix decidua*, *L. kaempferii* and *L.* × *eurolepis* on to *S. viminalis*, *S. daphnoides*, *S. caprea* and hybrid willows showed that all three species/hybrids of the genus *Larix* are capable of serving as alternative hosts of willow rusts.

According to the literature, important rust species and sub-species in Germany and their main willow hosts and alternate hosts, can be compiled as in Table 6.1. In contrast to Butin (1960), Klenke (1998) described ten *Melampsora* species on *Salix* and classified *M. repentis* as a distinct species, renamed as *M. arctica* Rostr. em. Braun. This rust occurs very rarely on *S. aurita*. He also placed *M. ribesii-epitea* Kleb. in the *M. epitea* complex.

Further work is needed to clarify the discrepancies between the systems of Butin (1960) and Klenke (1998) and the taxonomy of the willow rusts.

Variation in Size and Appearance of Uredinia

Recently an investigation (Liesebach and Zaspel, unpublished data) was conducted to evaluate the occurrence of *Melampsora* collected from 28 locations in Germany (Fig. 6.2). There were three different types of locations: willow collections (C, $n = 8$), SRC plantations (P, $n = 9$) and willow sites in the landscape (N, $n = 11$). For the purpose of comparison, samples from Sweden (3 sites), Northern Ireland (1), Belgium (1), France (1) and Georgia (1), and from *Populus* (6) were also investigated. In all, leaves bearing well-developed uredinia were collected from the 35 locations (Fig. 6.2). A large proportion of the assessed willows were *S. viminalis* and its hybrids. Altogether, more than 500 willow samples were studied and the specific phenotypic traits of all rusts, as well as the host clones, were recorded. During the period of study, the incidence and severity of rust changed from site to site in each year and between the years. The rust severity was assessed using a 5-step scale, separately for the

Table 6.1. Rust species and *formae speciales* described by Butin (1960) (B) and Klenke (1998) (K) in Germany.

Rust	Willow species	Alternate host species	Source
Melampsora allii-fragilis Kleb.	*S. fragilis*	*Allium*	B, K
Melampsora ari-sacina Raabe	*S. fragilis*	*Arum*	K
Melampsora galanthi-fragilis Kleb.	*S. fragilis*	*Galanthus*	K
Melampsora amygdalinae Kleb.	*S. amygdalina*, *S. triandra*, *S. pentandra*	*Autecious*	B, K
Melampsora caprearum (DC.) Thün.	*S. aurita*, *S. caprea*, *S. cinerea*	*Larix* (not obligate on alternate host species)	B, K
Melampsora epitea (Kunze et Schm.) Thüm.			
(a) *M. abieti-caprearum* Tub.	*S. caprea*	*Abies*	B, K
(b) *M. euonymi-caprearum* Kleb.	*S. cinerea*	*Euonymus*	B, K
(c) *M. larici-epitea* Kleb.	*S. viminalis*	*Larix*	B, K
(d) *M. repentis* Plowr.	*S. repens*	*Orchidaceae*	B
(e) *M. ribesii-purpureae* Kleb.	*S. purpurea*	*Ribes*	B, K
(f) *M. ribesii-epitea* Kleb.	*S. caprea*, *S. cinerea*	*Ribes*	K
Melampsora larici-pentandrae Kleb.	*S. pentandra*	*Larix*	B, K
Melampsora ribesii-viminalis Kleb.	*S. viminalis*	*Ribes*	B, K
Melampsora salicis-albae Kleb.	*S. alba*	*Allium* (not obligate on alternate host species)	B, K
Melampsora arctica Rostr. em. Braun	*S. repens*, *S. aurita*	*Orchidaceae*	K

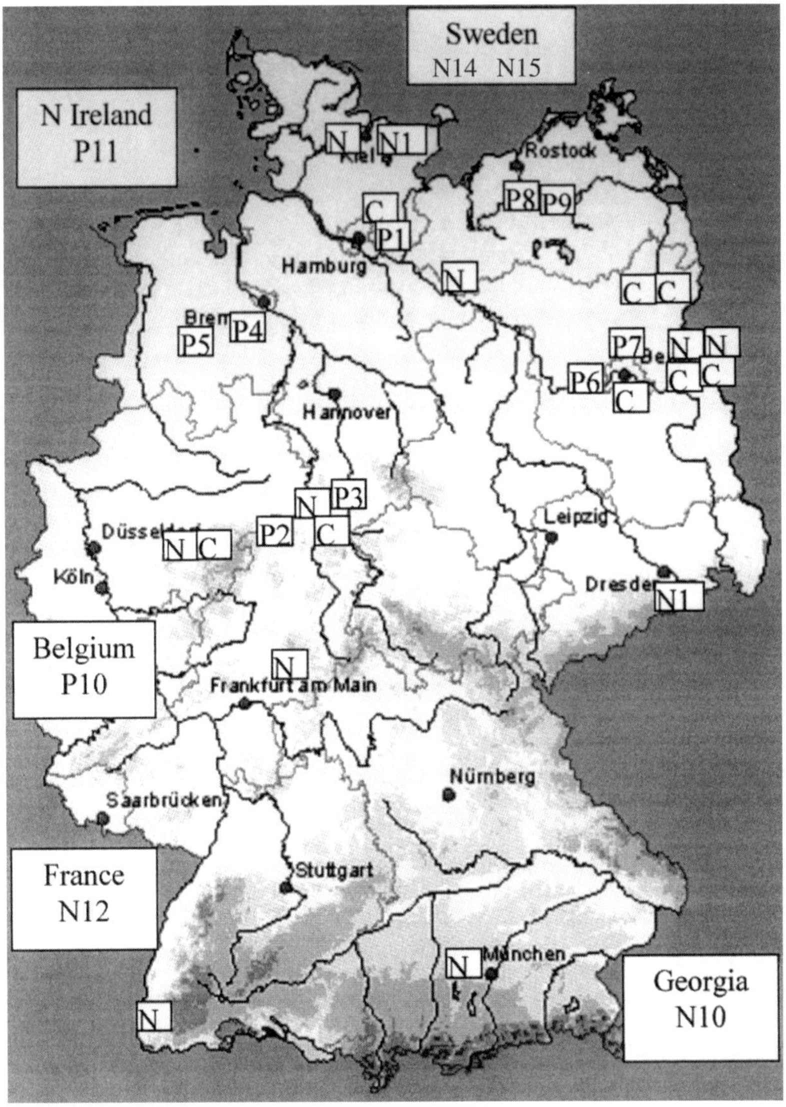

Fig. 6.2. Locations of sites where *Melampsora* rusts were assessed and collected. C, willow collections; P, short-rotation coppice (SRC) plantations; N, willow sites in the landscape.

expanded leaves and for the younger, partially expanded leaves (modified after McCracken and Dawson, 1992; Ahman, 1998; Hunter and Pei, unpublished) (Table 6.2).

On 55% of the samples, rust pustules occurred only on the expanded leaves, while on the other 45%, the younger, partially expanded leaves were infected. In many cases, the infection of the younger leaves occurred following the appearance of disease on the expanded leaves (Liesebach *et al.*, 2002). With some willow species, such as *S. alba* hybrids, *S. cinerea* hybrids, *S. daphnoides* hybrids, *S. nigricans* hybrids, *S. purpurea* hybrids, *S. schwerinii*, and *S.* × *undulata*, the disease was only observed on the expanded leaves. It was noted that, with the exception of *S. schwerinii*, only hybrids showed this type of infection. In some samples of the willow species *S. acutifolia*, *S. adenophylla*, *S. humboltiana*,

S. kazbekensis, *S. magnifica*, *S. megeriana* and *S.* × *sericans*, infections were observed only on younger leaves.

There were differences on which side of the leaf surface the pustules developed, i.e. whether the pustules were only on the lower or the upper sides, or on both sides, of the leaves. In some cases stem infections were also found. Table 6.3 lists rust samples grouped by willow genotype and the position of the leaves where the uredinia occurred. About 67% of the infections were observed only on the lower side of the leaves. In 31% of samples, the pustules were on both sides of the leaves. Very seldom (2%) did the pustules appear only on the upper side of the leaves.

According to the results from the assessment, the host clones could be grouped into five groups (Table 6.3). In one of these groups uredinial pustules were found only on the lower side of the leaves, and in a few cases on both sides. The most important biomass species *S. viminalis* and its hybrids, *S.* × *dasyclados* and *S.* × *aquatica*, were included in this group. *Salix caprea*, *S. daphnoides* and their hybrids were placed into another group which was characterized by having pustules only on the lower side in about 50% of the samples and on both sides of the leaves in the other 50% of the samples. The third main group included the species *S. alba*, *S. aurita*, *S. fragilis*, *S. nigra*, *S. repens*, *S. triandra* and *S.* × *sericans*, on which pustules were predominantly on both sides of the leaves. In the two remaining groups, the sample size was too small to give a general description.

Table 6.2. Rust disease assessment key for willow clones (Liesebach *et al.*, 2002).

Numerical scale	Category for the expanded leaves	Category for the younger, partially expanded leaves	Description
0	None		No rust can be detected
1	Very slight	–	Only one pustule or a single leaf is infected
2	Slight	Slight	Leaves bear a few conspicuous uredinia or more, but barely recognizable rust pustules
3	Moderate	Moderate	Leaves frequently bear rust pustules, up to an average of 1–5% leaf area covered by uredinia
4	Severe	Severe	Most leaves bear numerous pustules

Table 6.3. List of willow species/hybrids and the frequency of leaf sides on which rust pustules appeared.

	Frequency of leaf sides on which rust pustules appeared, and willow hosts				
Lower side:	100%	About 2/3	50%	About 1/3	–
Both sides:	–	About 1/3	50%	About 2/3	100%
Willow species/ hybrid	*S. schwerinii* *S.* × *undulata* *S. cineria* hybrids *S. purpurea* hybrids *S. triandra* hybrids *S. cordata*	*S. viminalis* *S. viminalis* hybrids *S.* × *aquatica* *S.* × *dasyclados* *S. cinerea* *S. pentandra* *S. fragilis* hybrids *S. nigricans* hybrids *S. repens* hybrids *S.* × *hirtei*	*S. alba* hybrids *S. aurita* hybrids *S. caprea* *S. caprea* hybrids *S. daphnoides* *S. daphnoides* hybrids *S. purpurea*	*S. triandra* *S. alba* *S. aurita* *S. repens* *S. fragilis* *S.* × *sericans* *S. nigra*	*S. humboltiana* *S. kazbekensis* *S. magnifica* *S. megeriana* *S. adenophylla* *S. acutifolia*

In addition, 60 willow clones (10.9%) showed stem infections. The stem infections were frequently found on the clones of *S.* × *sericans* and *S. nigra*. There were similar degrees of infection on leaves and stems in the clones of *S. purpurea* and *S. caprea* and their hybrids, and *S.* × *smithiana*, from which only small-size samples were collected. Stem infection was found only rarely on the clones of *S.* × *dasyclados*, *S.* × *aquatica*, *S. viminalis* and its hybrids, *S. triandra*, *S. daphnoides* and *S. aurita* and their hybrids. Furthermore, the size and colour of the pustules were estimated using a 5-step scale (Table 6.4). It was noted that small pustules were associated with the light-yellow colour, while larger pustules were associated with colours towards orange.

The results from the assessments of the pustule colours were not as clear-cut as those from the assessment of the pustule sizes. The colour of the uredinia varied between different willow species and species hybrids. Most clones of *S. viminalis* had pustules with light yellow or yellow colour (Fig. 6.3A). On *S. daphnoides* clones, pustules were predominantly orange, and light-yellow pustules were missing (Fig. 6.3B). The colour step '(yellow-) brown' was rarely found. This colour category might be associated with old pustules.

For clones of *S. viminalis* (grown mostly in SRC plantations) and *S. daphnoides*

Table 6.4. Scoring system for the size and the colour of the rust pustules.

Size of the rust pustules	Colour of rust pustules
Small	Light-yellow
Small–medium	Yellow
Medium	Yellow-orange
Medium–large	Orange
Large	(Yellow-) brown

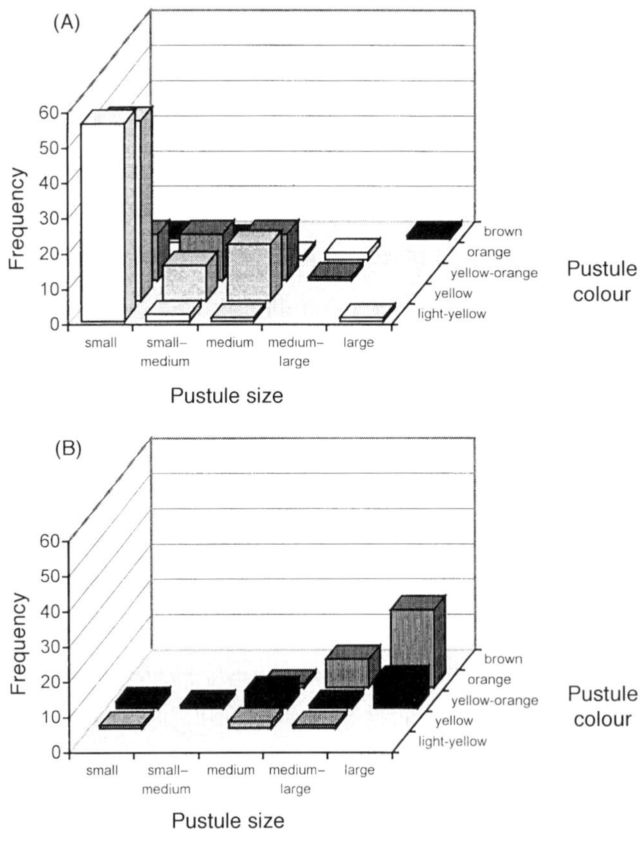

Fig. 6.3. The frequencies of pustule colours and sizes in *Melampsora* on (A) *Salix viminalis* and its hybrids (n = 191), and (B) *Salix daphnoides* and its hybrids (n = 52).

(clones with high salicylic acid content), the frequency of the colour and the size of the pustules are shown in Fig. 6.3. On both species, pustules were assessed from small to large. However, small pustules were predominant on *S. viminalis* (Fig. 6.3A) and large ones on *S. daphnoides* (Fig. 6.3B).

With selected rust samples, urediniospores were measured using a light microscope. Most samples had spores that were more or less round, with a diameter of 15–20 μm. Some samples from *S. fragilis* had a different spore size (35–40 μm long and 12.5–15 μm wide). All the *S. fragilis* grew in meadows. According to its morphology and collection sites, the rust was identified as *M. allii-fragilis*.

Development of Rust Infection

The development of rust infection was observed on a set of 35 clones and clones grown in a SRC trial over three consecutive years (Liesebach *et al.*, 2002; unpublished data). Rust severity was assessed using a 5-step scale (0–4 numerical scales) (Table 6.2). The assessment was carried out separately for the expanded leaves and for the younger, partially expanded leaves.

Differences were observed in the development of the rust disease between the years. The first appearance of disease varied by about 2 weeks, probably due to the weather conditions. According to the onset and the severity of rust disease, the clones assessed could be assigned to six groups. Between the years there were slight differences in the clones assigned to different groups.

A few resistant clones, such as *S. schwerinii* '77077', *S. fragilis* × *S. alba* '77802' and a few *S. viminalis* clones, did not have any rust pustules, or had only occasional pustules. Most clones had moderate rust infections on the expanded leaves (Fig. 6.4, top). The infection started between the 160th and 235th Julian day. A few clones had moderate rust infections on the partially expanded leaves and young shoots. The infections on partially expanded leaves on young shoots of clones appeared 10 or more days earlier than those on the clones having infections only on expanded leaves.

The hybrid clone *S. caprea* × *S. viminalis* '77666' was found to be highly susceptible. Disease started very early on the younger, partially expanded leaves (Fig. 6.4, bottom) and increased continuously to a high level. In the third year, two further clones, *S. caprea* × *S. viminalis* 'Coles' and *S. nigra* 'SN2', were clustered in this group. The remaining two groups were represented by one clone each. On these clones, rust development was different: the infection started earlier but was only temporary, and was concentrated on the expanded leaves.

In the newly established field trial at Hann. Münden and the clone collection Waldsieversdorf, *S. daphnoides* clones had hardly any rust in the year of establishment. Rust infection started during the following years and these clones were severely

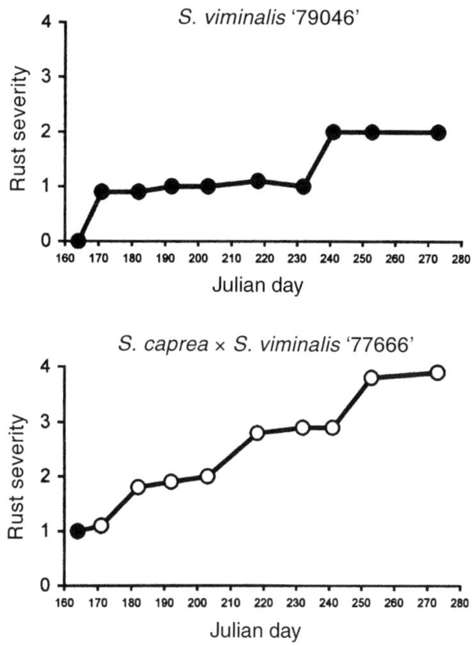

Fig. 6.4. Development of the *Melampsora* rust infection assessed on two clones: *S. viminalis* '79046' (top) and *S. caprea* × *S. viminalis* '77666' (bottom) in 2002. Closed circles indicate the infection on expanded leaves; open circles indicate the infection on partially expanded leaves. (From Liesebach *et al.*, 2002.)

affected in July/August and defoliated at the end of August. At the beginning of September, the plants flushed again and the tips of young shoots were immediately infected. Soon after, the plants showed dieback symptoms.

Pathogenicity Tests

Virulence patterns of the rusts from central Europe in relation to the known pathotypes were examined using the leaf disc method described by Pei *et al.* (1993, 1999a). After inoculation, the leaf discs were incubated at 16°C with 16 h day^{-1} illumination at an intensity of about 80 $\mu E\ m^{-2}\ s^{-1}$. The number of rust pustules on each leaf disc was observed from the 6th day after inoculation and counted until 2 weeks after inoculation. From these observations, 'infection types' were assigned as: 0 = no pustules; 1 = up to 5 pustules per leaf disc; 2 = 6–10 pustules per leaf disc; 3 = more than 10 pustules per leaf disc. In total, 76 rust samples were tested on 28 willow clones (13 *S. viminalis*, 7 *S.* × *aquatica* and *S.* × *dasyclados*, 2 *S. daphnoides*, 1 *S. alba*, 1 *S. nigra*, 1 *S.* × *stipularis*, 1 *S. caprea* hybrid, 1 *S.* × *hirtei*, 1 *S. fragilis* hybrid). Altogether, at least 43 virulence patterns were found.

The virulence patterns were determined mainly by the origin of the rust samples. Samples derived from *S. viminalis* caused infections mostly on the tested *S. viminalis* clones. Hence these 22 patterns showed only a slight variation. The samples derived from *S.* × *stipularis* also infected mostly *S. viminalis* clones. Those from *S.* × *aquatica* infected the tested clones of *S.* × *aquatica* and *S.* × *dasyclados*. Samples collected from *S. fragilis* caused infections only on *S. fragilis* or did not cause any infections. The infections observed on *S. daphnoides* were mostly caused by rusts collected from *S. daphnoides*. However, four more patterns were recorded from rusts collected from *S. daphnoides*. Also, only a slight infection was observed on *S. caprea* in the test using the rust collected from *S. caprea* in the field. Six infection types were unique to the rust samples from *S. caprea*. The rust samples derived from *S. purpurea*, *S.* × *sericans*, *S. nigra*, *S. aurita*, *S. acutifolia*, *S. cordata*, *S. pentandra*, *S. repens*, *S. triandra* and *S.* × *hirtei* only caused slight infections on some of the tested clones.

Most of the rust samples produced visible symptoms 8–10 days after inoculation on the leaf discs. But some developed faster: the first uredinia occurred 6 days after inoculation and produced a large number of pustules within several days. Those 'fast-infecting' samples were from *S. viminalis* clones as well as several hybrid clones.

To evaluate these results, a correspondence analysis was conducted using SAS (Rasch *et al.*, 1998). The result is shown in Fig. 6.5. The first two dimensions explain 61.3% of total inertia (Dimension 1: 33.5% and Dimension 2: 27.8%). In this analysis only the samples which had more or less round spores with a diameter between 15 and 20 µm were included. Samples of *M. allii-fragilis* from *S. fragilis*, which could be recognized easily by their large spore size were not included. In the plot, several groups can be distinguished. In one group all the tested rust samples from *S.* × *aquatica* and *S.* × *dasyclados* and the willow clones belonging to *S.* × *aquatica* and *S.* × *dasyclados* were clustered. They belonged to the *forma specialis larici-retusae*. A second group combined most samples collected from *S. daphnoides*. Rust samples of the *forma specialis laricii-daphnoides* were more or less clustered on *S. daphnoides*. However, some samples from *S. daphnoides* could not be placed into this *forma specialis*, indicating that there may be further genetic differentiation. A third group, which included all the rust samples from *S. viminalis* clones, was assigned to f. sp. *larici-epitea typica*. This *forma specialis* group also contained rust samples from different species of willows.

The samples from *S. caprea* could not be assigned to one of the known *formae speciales* of *M. epitea*. Possibly they might belong to another rust species (*M. caprearum*) or to another form or species (*sensu stricto*) of *M. epitea* that alternates on aecial hosts other than larch (such as *M.*

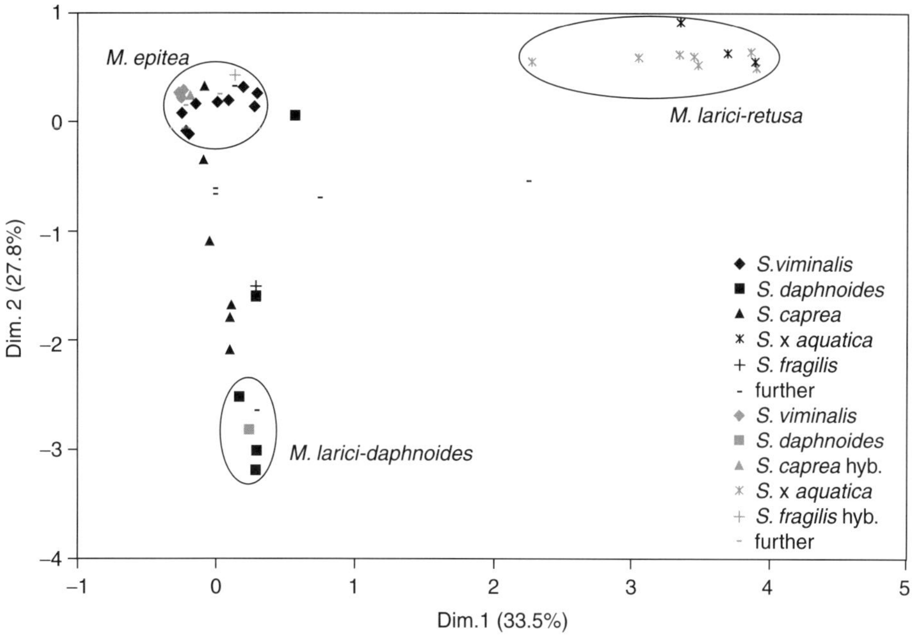

Fig. 6.5. Two-dimensional correspondence analysis for 28 willow clones (grey symbols) and for 75 rust samples coming from corresponding willow hosts (black symbols).

abieti-caprearum, *M. euonymi-caprearum* or *M. ribesii-epitea*). All other rust samples derived from willow species, which were only represented by a single or few samples, were not differentiated in Fig. 6.5 (they are indicated as –).

Differentiation Using ITS-RFLP and Sequencing of ITS unit and LSU

The ribosomal DNA (rDNA) of fungi contains one transcriptional unit with a cluster of genes coding for 18S, 5.8S and 28S rRNAs and two internal transcribed spacers, ITS1 and ITS2 (White *et al.*, 1990). The complex, non-coding and variable ITS regions and the coding and conserved 5.8S rRNA gene are useful in measuring close genealogical relationships, because they exhibit greater interspecies differences than the 18S and 28S genes. The coding regions within rDNA repeat units are known to be highly conserved in the non-transcribed spacer regions, both between and within species (Long and Dawid, 1980).

The ITS regions are a convenient target region for molecular identification of fungi, because in fungi the ITS regions can be readily amplified with universal primers that are complementary to sequences within the rRNA genes. Furthermore, the multi-copy nature of the rDNA repeat makes it easy to amplify from small, or diluted DNA samples. Several studies have demonstrated that the ITS regions are often highly variable among morphologically distinct fungal species, but intraspecific variation is low. Taxon-selective primers for the ITS regions in the nuclear ribosomal repeat unit are available for rusts (Gards and Bruns, 1993; Vogler, 1995).

The use of polymerase chain reaction (PCR)-based techniques, such as ITS-restriction fragment length polymorphism (ITS-RFLP) and sequencing of DNA fragments, made it possible to study genetic diversity at the molecular level and the phylogenetic relationships of isolates of *Melampsora* spp. In our study, DNA was extracted from a collection of 205 isolates (Liesebach and Zaspel, unpublished data), of which willow rust isolates accounted for 199. For comparison, six *Melampsora* isolates from poplar were also included.

Protocols for PCR amplification of ITS regions and RFLP analysis are described by Liesebach and Zaspel (2004). The primer pair ITS1 (5′ TCC GTA GGT GAA CCT GCG G 3′) and ITS4 (5′ TCC TCC GCT TAT TGA TAT GC 3′), permitted the amplification of a single DNA fragment of *Melampsora* containing ITS1, ITS2, 5.8S and partial sequences of 18S and 25S rDNA genes, of a total size between 625 and 628 bp. RFLPs of the 5.8S-ITS region were generated from the amplified products by restrictions with *Taq*I, *Eco*RI, *Mbo*I or *Ssp*I. Using the four restriction enzymes in combination, a total of 12 RFLP patterns could be distinguished clearly. Table 6.5 shows the frequency of the multi-locus patterns.

A set of 40 DNA samples, representing most of the patterns detected by RFLP and the resulting subgroups, were sequenced. Table 6.6 gives the length of the uncut amplicon (a PCR product that is a copy of the original DNA or RNA target region) using the primer pair ITS1 and ITS4, as well as the numbers of the fragments generated by restriction digestion with *Taq*I, *Eco*RI, *Mbo*I and *Ssp*I. The restriction sites are shown in Fig. 6.6. The 5.8S-ITS region of willow rusts was in a range from 625 bp to 628 bp. The sequences of all the tested willow rusts could be aligned in 630 nucleotide positions. The uncut amplicon of *M. larici-populina* had a length of 644 bp.

The analysis of the sequences of ITS1/ITS4 units confirmed most of the patterns obtained from the RFLP analyses. Based on the sequence information, some main groups can be distinguished. The nucleotide differences between the patterns of the *Melampsora* samples from *Salix* (38) and those from *Populus* (2) were from 6.7 to 8.5% (Table 6.7). The variation between the three main groups from *Salix* was lower (≤ 4.1%). There were two patterns (I, II) within the first main group. The nucleotide differences between the two patterns were 0.3 and 2.2%.

Table 6.5. RFLP patterns of ITS region, frequency and host species of isolates, and information on the collection.

RFLP pattern	Frequency	*Melampsora* rusts from species/hybrids (number of isolates)	Isolated from collection sites (Fig. 6.2)
I	14	*S. alba* (2); *S. fragilis* (8); *S. purpurea* (1); *S. triandra* (3)	C6, C8, N1, N2, N3
II	3	*S. purpurea* (3)	C4
III	40	*S. adenophylla* (1); *S. alba* (2); *S. aurita* (3 + 1 hybrid); *S. caprea* (14 + 9 hybrids); *S.* × *sericans* (9); *S.* × *stipularis* (1)	C1, C3, C4, C5, C6, N1, N3, N4, N5, N6, N8, N11, N12, N13, N15, P1, P7
IV	97	*S. acutifolia* (3); *S. daphnoides* (9); *S. fragilis* (2 + 1 hybrid); *S. magnifica* (1); *S. pentandra* (1); *S. purpurea* (2); *S. triandra* (5); *S. repens* (2 hybrid); *S. viminalis* (50 + 1 hybrid); *S.* × *aquatica* (7); *S.* × *dasyclados* (5); *S.* × *hirtei* (2); *S.* × *stipularis* (5); *S.* × *calodendron* (1)	C1, C2, C3, C4, C5, C6, C8, N1, N2, N3, N5, P1, P2, P3, P4, P5, P6, P7, P11
V	1	*S. triandra* (1)	P11
VI	1	*S. aurita* (1)	N5
VII	1	*S. daphnoides* (1)	C4
VIII	20	*S. daphnoides* (20)	C1, C3, C4, C5, P3
IX	3	*S. daphnoides* (3)	C4, C5
X	6	*S. nigra* (6)	C1
XI	12	*S. alba* (1 + 1 hybrid); *S. aurita* (1 + 1 hybrids); *S. caprea* (3 + 2 hybrids); *S. cinerea* (2 + 1 hybrid)	C1, C3, C4, N1, N5
XII	1	*S. caprea* (1)	N10
Po	6	*Populus* spp. (6)	C1, C8, N1, N8, P2, P10

RFLP, restriction fragment length polymorphism; ITS, internal transcribed spacer.

Table 6.6. Lengths of the uncut amplicons of ITS region and number of the fragments generated by restriction digestion with *Taq*I, *Eco*RI, *Mbo*I and *Ssp*I.

Pattern	Frequency	Amplicon (bp)	Fragments			
			*Taq*I	*Eco*RI	*Mbo*I*	*Ssp*I
I	6	625	4	2	4	1
II	2	626	4	2	3	1
III	4	626	3	2	3	2
IV	17	626	3	1	3	2
V	1	626	5	1	3	1
VI	1	627	3	1	3	2
VII	1	628	3	1	3	1
VIII	4	626	3	1	3	1
IX	1	627	3	1	3	1
X	1	626	2	1	3	2
Po	2	644	3	2	3	1

ITS, internal transcribed spacer.
*Including one fragment of 3 bp in size, which was not visible on the gel.

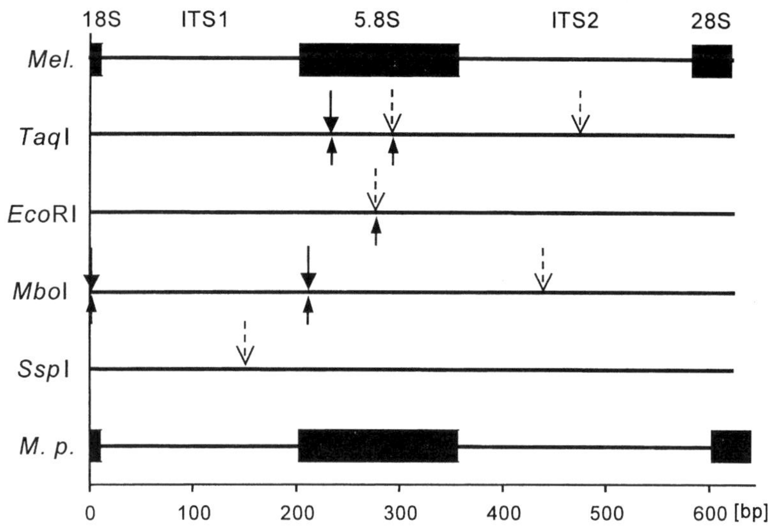

Fig. 6.6. Map of restriction sites along the rDNA-ITS region for the *Melampsora* collection (*Mel.* = *Melampsora* sp. from *Salix*; *M. p.* = *Melampsora larici-populina*). Solid arrows indicate invariable sites, dotted arrows indicate variable sites; arrows from below belong to *M. larici-populina*.

The variation within pattern I was 1% or less. The second main group comprised seven patterns (III, IV, VI, VII, VIII, IX, X) which differed in less than 1.8% of their nucleotide positions. Between these samples the differences were very small (< 2%). Sample 5.1 (pattern V) represented a distinct group with a 3% nucleotide difference to other *Melampsora* samples from willow.

The ITS1/ITS4 sequences of *Melampsora* isolates were used to infer phylogenetic relationships among these isolates. The sequence data were aligned using the CLUSTALX program (Thompson *et al.*, 1997). For calculating the tree, the neighbour-joining (N-J) method was used (Saitou and Nei, 1987). Bootstrap sampling was carried out using 1000 replications to assess the

Table 6.7. Percentage divergence between 18 *Melampsora* isolates. (Each sequence is only listed once. The sample numbers are a combination of the RFLP patterns and the different sequence types.)

Sample	1.1	1.2	1.3	1.4	2.1	3.1	3.2	3.3	10.1	8.1	8.2	7.1	9.1	8.3	4.1	6.1	5.1	Po
1.2	0.3			**I**														
1.3	0.5	0.8																
1.4	0.6	1	1															
2.1	1.9	2.2	2.2	1.6	**II**													
3.1	2.9	3.2	2.9	2.9	2.9				**III, X**									
3.2	3	3.4	3	3	3	0.2												
3.3	2.7	3	2.9	3	3	0.2	0.3											
10.1	3.4	3.7	3.4	2.7	3.1	0.5	0.6	0.6										
8.1	3.4	3.7	3.4	3	3	1.1	1.3	1.3	1.1				**IV, VI, VII, VIII, IX**					
8.2	3.5	3.8	3.5	3.2	3.2	1.3	1.4	1.4	1.3	0.2								
7.1	3.7	4.1	3.7	3.1	3.4	1.6	1.8	1.8	1.3	0.5	0.7							
9.1	3.5	3.9	3.5	2.9	3.2	1.3	1.4	1.4	1	0.2	0.3	0.3						
8.3	3.5	3.8	3.5	3.2	3.2	1.3	1.4	1.4	1.3	0.2	0.3	0.7	0.3					
4.1	3.7	4	3.7	3	3.4	1.1	1.3	1.3	0.8	0.6	0.8	0.8	0.5	0.5				
6.1	3.7	4	3.7	3	3.4	1.1	1.3	1.3	0.8	0.6	0.8	0.8	0.5	0.5	> 0			
5.1	3	3.4	3	2.7	3.4	2.9	3	3	2.9	2.4	2.6	2.9	2.6	2.6	2.7	2.7	**V**	
Po	7.2	7.5	7.5	7.2	7.4	6.7	6.9	6.9	7.2	6.9	6.7	7.3	7.2	6.7	6.9	6.9	8.5	**Po**

confidence limits of the branches (Felsenstein, 1985). The program TREEVIEW (Page, 1996) was used to construct the phylogenetic tree, which was rooted using the sequences ITS1/ITS4 units of *Melampsora larici-populina* (GenBank accession: AY375267) as the outgroup.

The phylogenetic tree constructed based on the results is shown in Fig. 6.7. The two *Melampsora* samples from *Populus* (Po) formed a distinct group. The sequences of these samples (*M. larici-populina*) were confirmed by that from the database (GenBank accession: AY375267). The analysis showed two clear lineages in willow rusts. The willow rust samples differed between these lineages in about 4% of their nucleotides. There was a close phylogenetic relationship among the *Melampsora* isolated from the rusts on *S. fragilis*, *S. triandra* and *S. purpurea* (RFLP patterns I, II, V). This main group can be divided into three subgroups. Two samples from *S. purpurea* were placed in one of these subgroups (pattern II). The samples in pattern I were mostly derived from *S. fragilis*.

The second main group comprised the samples from *S. viminalis*, *S. daphnoides*, *S.* × *dasyclados*, *S.* × *aquatica*, *S. aurita*, *S. nigra*, *S.* × *sericans* and *S.* × *stipularis*. Four subgroups can be distinguished. One subgroup contained only the samples from *S. daphnoides* (patterns VII, VIII, IX). The sequences of 18 samples, mostly from *S. viminalis*, *S.* × *aquatica*, *S.* × *dasyclados*, *S.* × *stipularis*, *S. fragilis*, *S. daphnoides* and *S. triandra*, were closely related to a sample from *S. aurita* (pattern VI). The next branch was formed by the samples from *S.* × *sericans* and *S. aurita* (pattern III), and one sample from *S. nigra* (pattern X).

The large subunit (LSU) ribosomal DNA sequence data were also used to infer phylogenetic relationships of *Uredinales* (Maier *et al.*, 2003). We used the primer pair NL1 (5′ GCA TAT CAA TAA GCG GAG GAA AGG 3′) and NL4 (5′ GGT CCG TGT TTC AAG ACG G 3′) to amplify the 28S rDNA (nuclear large subunit rDNA) (Guadet *et al.*, 1989). A set of 32 samples, including three rust isolates from *Populus* sp., were sequenced. The identified 28S region of *Melampsora* isolates was used to infer phylogenetic relationships among these isolates. The sequence data were aligned using the CLUSTALX program (Thompson *et al.*, 1997). The primer pair NL1 and NL4 amplified a single DNA fragment of the 28S rDNA genes of *Melampsora*,

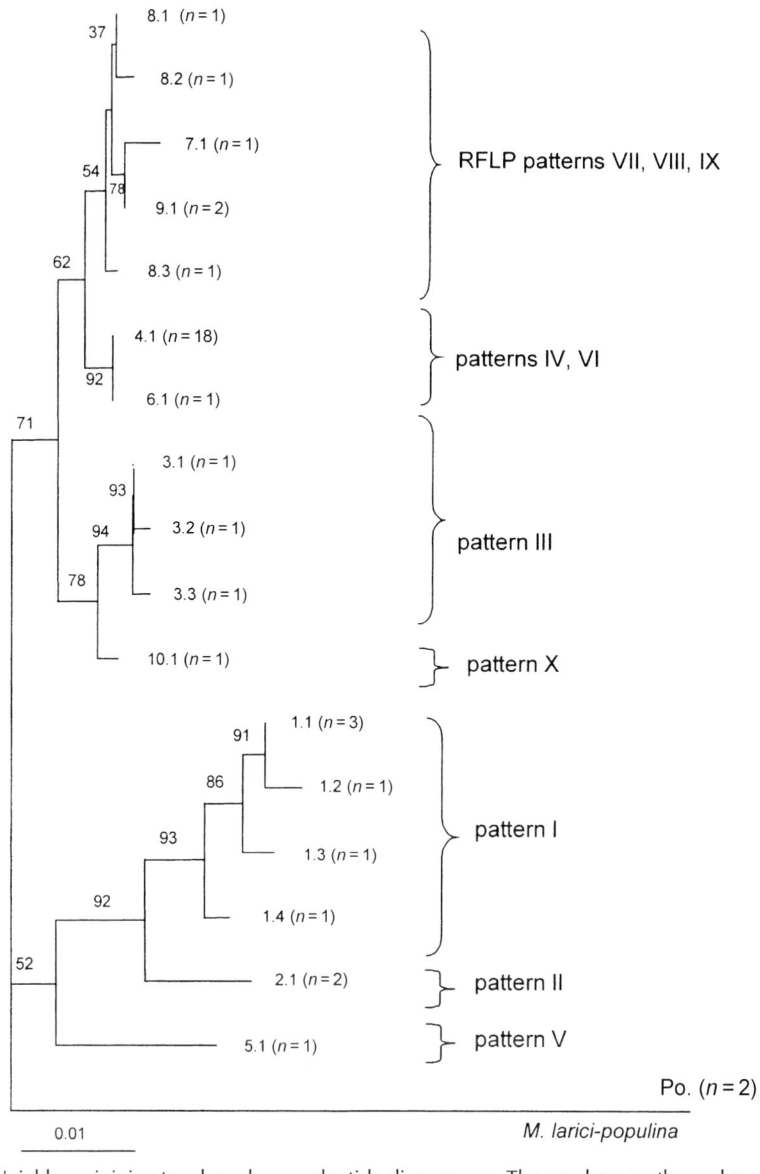

Fig. 6.7. Neighbour-joining tree based on nucleotide divergences. The numbers on the nodes are the frequencies (in per cent) with which a cluster appears in a bootstrap test of 1000 runs. The tree is rooted with *M. larici-populina* (Po) as outgroup. For an explanation of sample numbers see Table 6.7.

of a total size between 597 and 600 bp. The maximum nucleotide difference between the 29 samples from *Salix* and the three from *Populus* was only 3%. A neighbour-joining (N-J) tree was calculated according to the methods mentioned above (Fig. 6.8).

The tree was clearly composed of five branches. One branch was formed by the *M. larici-populina* samples. Another branch contained rust samples from *S. purpurea* and assigned to *M. ribesii-purpurea*. In the bottom left of the tree the samples from *S. fragilis* and *S. triandra* were grouped together and they were assigned to *M. amygdalinae*. *Melampsora allii-fragilis* was placed in a separate branch. The branches in

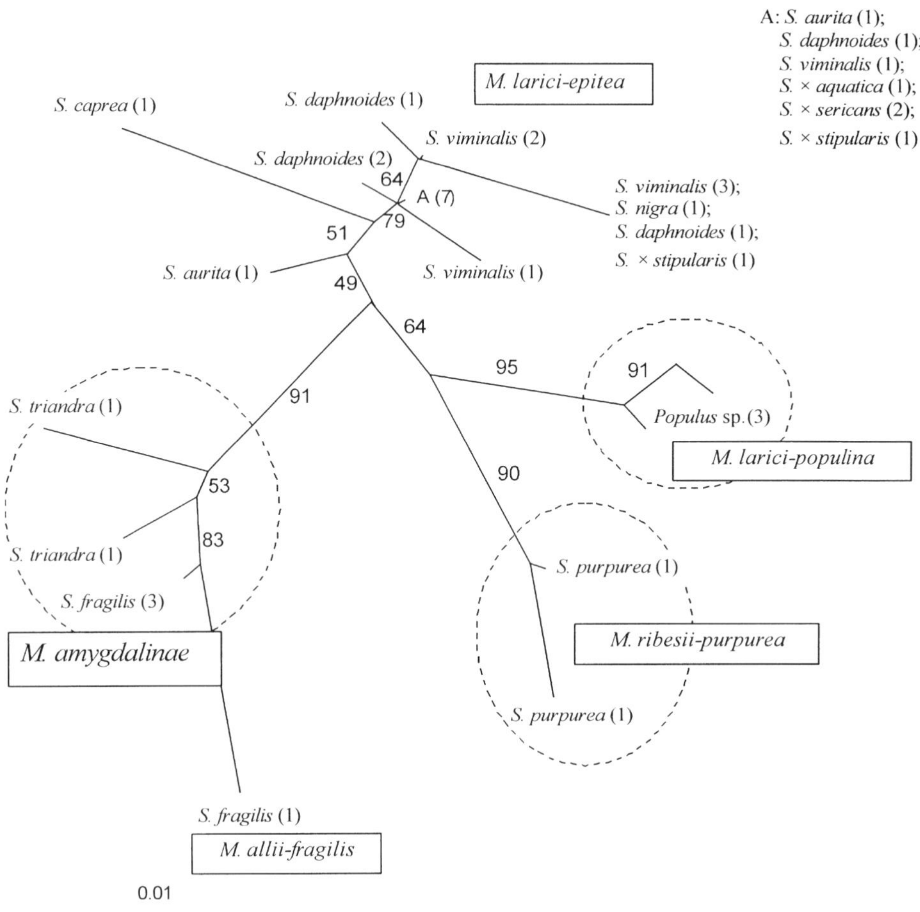

Fig. 6.8. Neighbour-joining tree based on nucleotide divergences of the large subunit 28S region. Bootstrap values for 1000 re-samplings are indicated on each branch. Willow hosts and putative *Melampsora* species are shown.

the upper part of the tree, which appeared to be predominated by *M. larici-epitea*, did not reveal the different *formae speciales*, i.e. *larici-epitea typica*, *larici-retusae*, *larici-daphnoides* and others. Overall, the results showed that there is a close phylogenetic relationship between *Melampsora* spp. isolates from rusts on *Salix* spp. and *Populus* spp. (*M. larici-populina*).

Concluding Remarks

The present study has demonstrated that, in central Europe, especially in Germany, *Melampsora* willow rusts are highly variable in their species and pathotype composition. The results are broadly in line with the taxonomic systems by Gäumann (1959) and Butin (1960). Klenke (1998) indicated a higher degree of variation in the willow rust complex, but his work was confined to a restricted region of Germany. Previous work referred to rusts occurring in natural willow stands, and those in SRC plantations were not implicated.

M. epitea has been treated as a species complex because of different spermagonial and aecial hosts and/or uredinial and telial hosts in Europe (Hylander *et al.*, 1953; Gäumann, 1959; Wilson and Henderson,

1966) and North America (Ziller, 1974). Under existing conditions in Germany, the genetic structure of the *M. larici-epitea* population, one of the subspecies of *M. epitea*, is of special interest. This rust plays an important role in rust disease epidemics on biomass willow clones in countries with SRC plantations, such as Sweden and the UK. It can be concluded that the three major *formae speciales*, out of the six listed by Gäumann (1959), occurred on biomass clones in trial plantations and clone collections in Germany.

The willow rust population comprises a number of pathotypes which vary between sites and geographical regions. Using leaf disc pathogenicity tests, Pei *et al.* (1999b) identified 12 pathotypes within *M. larici-epitea* in the UK. Using the same method, Ramstedt (1999) found 13 types and predicted about 20 types in Sweden. Our investigation revealed a distinctly higher number of rust types among the samples we tested, but the samples included, in addition to the *M. epitea* complex from cultivated clones, several other *Melampsora* species from natural stands. The stem-infecting rusts were recorded frequently but cannot yet be assigned to a distinct pathotype.

The application of molecular tools using PCR-based methods of ITS-RFLP, random amplified polymorphic DNA (RAPD) and amplification fragment length polymorphism (AFLP) markers revealed a high degree of genetic diversity among willow *Melampsora* (Nakamura *et al.*, 1998; Liesebach *et al.*, 2001; Samils *et al.*, 2001; Pei *et al.*, 2002). Because our molecular method was not exactly the same as that of the others, our results cannot be compared directly to the results by the cited authors. However, the extent of variation in willow rusts shows the potential of the rusts to act as aggressive pathogens under suitable conditions.

The clonal nature of willows as short-rotation coppice crops makes them especially vulnerable to the build-up of pests and diseases when grown in monocultures. Also, the intensity of pest damage and disease outbreaks can differ markedly between *Salix* species, clones and geographical locations (Hunter *et al.*, 1996). This can be seen by the different reactions of willow clones – from resistant to highly susceptible to rust infection. *S. viminalis* and some *S. daphnoides* clones were infected but without serious damage. Other clones, especially those belonging to *S.* × *dasyclados* or *S.* × *aquatica*, suffered severe attacks and showed dieback symptoms.

The pathotype composition of an SRC plantation depends on the species and the number of willow clones planted, and is further influenced by their age. The development of aggressive pathotypes endangering the sustainability of the whole plantation is more likely when only one, or a few, susceptible host clones have been grown. Therefore, the concept of multi-clonal plantations is creating a higher diversity of rust populations, resulting in a lower infection level and expected higher stability of yields (McCracken *et al.*, 2000). Information on the resistance of willow species/clones and genetics of the rust populations is important in the management of willow leaf rust in SRC coppice plantations, and in defining strategies in resistance breeding.

Acknowledgements

This research was funded by the EC and is part of the joint project 'Integrated, non-fungicidal control of *Melampsora* rusts in renewable energy willow plantations' (QLK5-1999-01585). We wish to thank Ms H. Mattauch and E. Ewald for technical assistance.

References

Ahman, I. (1998) Rust scorings in a plantation of *Salix viminalis* clones during ten consecutive years. *European Journal of Forest Pathology* 28, 251–258.

Ahman, I. and Larsson, S. (1994) Genetic improvement of willows (*Salix*) as a source of bioenergy. *Norwegian Journal Agricultural Sciences* 18(Suppl.), 47–56.

Butin, H. (1960) Die Krankheiten der Weide und deren Erreger. *Mitteilungen aus der Biologischen Bundesanstalt für Land- und Forstwirtschaft* 98.

Felsenstein, J. (1985) Confidence limits on phylogenies: an approach using the bootstrap. *Evolution* 39, 783–791.

Gards, M. and Bruns, T.D. (1993) ITS primers with enhanced specificity for basidiomycetes – application to the identification of mycorrhizae and rusts. *Molecular Ecology* 2, 113–118.

Gäumann, E. (1959) *Die Rostpilze Mitteleuropas. Beiträge zur Kryptogamenflora Schweiz 12.* Buchdruckerei Büchler and Co, Bern, Germany, pp. 144–177.

Guadet, J., Julien, J., Lafay, J.F. and Brygoo, Y. (1989) Phylogeny of some *Fusarium* species, as determined by large-subunit rRNA sequence comparison. *Molecular Biology and Evolution* 6, 227–242.

Helfer, S. (1992) The rust diseases of willows in Britain. *Proceedings of the Royal Society of Edinburgh* 98B, 119–134.

Hunter, T., Royle, D.J. and Arnold, G.M. (1996) Variation in the occurrence of rust (*Melampsora* spp.) and other diseases and pests, in short-rotation coppice plantations of *Salix* in the British Isles. *Annals of Applied Biology* 129, 1–12.

Hylander, N., Jorstad, I. and Nannfeldt, J.A. (1953) Enumeratio uredinearum scandinavicarum. *Opera Botanica* 1, 1–102.

Klenke, F. (1998) Sammel- und Bestimmungshilfen für phytoparasitische Kleinpilze in Sachsen. *Berichte der Arbeitsgemeinschaft sächsischer Botaniker* 16.

Lautenschlager-Fleury, D. (1994) *Die Weiden von Mittel- und Nordeuropa*. Birkhäuser Verlag, Basel.

Liesebach, M. and Zaspel, I. (2004) Genetic diversity of the mycoparasite *Sphaerellopsis filum* on *Melampsora* willow rusts. *Forest Pathology* 34, 292–305.

Liesebach, M., Zaspel, I. and Stauber, T (2001) Monitoring of *Melampsora* rusts in *Salix*. *Journal of Forest Sciences* 47 (Special Issue No. 2), 119–122.

Liesebach, M., Zaspel, I. and Stauber, T. (2002) Growth of willows and the influence of *Melampsora* infection on the establishment of biomass plantations. In: Ercan, M., Diner, A., Birler, A.S., Goulding, C. and Zoralioglu, T. (eds) *Management of Fast Growing Plantations*. International IUFRO Meeting, 11–13 September, Izmit/Turkey, pp. 150–160.

Long, E.O. and Dawid, I.B. (1980) Repeated genes in eukaryotes. *Annual Review of Biochemistry* 49, 727–764.

Maier, W., Begerow, D., Weiß, M. and Oberwinkler, F. (2003) Phylogeny of the rust fungi: an approach using nuclear large subunit ribosomal DNA sequences. *Canadian Journal of Botany* 81, 12–23.

McCracken, A.R. and Dawson, M. (1992) Clonal response in *Salix* to *Melampsora* rusts in short rotation coppice plantations. *European Journal of Forest Pathology* 22, 19–28.

McCracken, A.R., Dawson, W.M., Watson, S. and Allen, C.Y. (2000) Pathotype composition in *Melampsora epitea* populations occurring on willow (*Salix*) grown in mixed and monoculture plantations. *European Journal of Plant Pathology* 106, 879–886.

Nakamura, H., Kaneko, S., Yamaoka, Y. and Kakishima, M. (1998) Differentiation of *Melampsora* rust species on willows in Japan by using PCR-RFLP analysis of ITS regions of ribosomal DNA. *Mycoscience* 39, 105–113.

Newsholme, C. (2002) *Willows of the Genus* Salix. Batsford Ltd, London.

Page, R.D.M. (1996) TREEVIEW: An application to display phylogenetic trees on personal computers. *Computer Applications in the Biosciences* 12, 357–358.

Parker, S.R., Royle, D.J. and Hunter, T. (1993) Impact of *Melampsora* rust on yield of biomass willows. *Sixth International Congress of Plant Pathology, Montreal, Canada, 28 July–6 August*. National Research Council Canada, Ottawa, p. 117 [abstract].

Pei, M.H., Hunter, T. and Royle, D.J. (1993) Variation of *Melampsora* rusts occurring on biomass willows in the UK. *Proceedings IEA Task VIII Joint Meeting, Northern Ireland, September 1992.*

Pei, M.H., Parker, S.R., Hunter, T. and Royle, D.J. (1997) Variation in populations of *Melampsora* willow rust and the implications for design of short rotation coppice plantations. In: Bullard, M.J., Ellis, R.G., Heath, M.C., Knight, J.D.D. and Parker, S.R. (eds) *Aspects of Applied Biology* Vol. 49, *Biomass and Energy Crops*. The Association of Applied Biologists, Wellesbourne, UK, pp. 91– 96.

Pei, M.H., Royle, D.J. and Hunter, T. (1999a) Hybridization in larch-alternating *Melampsora epitea* (*M. larici-epitea*). *Mycological Research* 103, 1440–1446.

Pei, M.H., Hunter, T. and Ruiz, C. (1999b) Occurrence of *Melampsora* rusts in biomass willow plantations for renewable energy in the United Kingdom. *Biomass and Bioenergy* 17, 153–163.

Pei, M.H., Bayon, C., Ruiz, C., Yuan, Z.W. and Hunter, T. (2002) Genetic variation in *Melampsora larici-epitea* on biomass willows assessed using AFLP. *European Journal of Forest Pathology* 108, 229–236.

Ramstedt, M. (1999) Rust disease on willows – virulence variation and resistance breeding strategies. *Forest Ecology and Managment* 121, 101–111.

Rasch, D., Herrendörfer, G., Bock, J., Victor, N. and Guiard, V. (eds) (1998) *Verfahrensbibliothek. Versuchsplanung und -auswertung*, Vol. II. R. Oldenbourg Verlag, Munich, pp. 208–216.

Saitou, N. and Nei, M. (1987) The neighbor-joining method: a new method for reconstructing phylogenetic trees. *Molecular Biology and Evolution* 4, 406–425.

Samils, B., Lagercrantz, U., Lascoux, M. and Gullberg, U. (2001) Genetic structure of *Melampsora epitea* populations in Swedish *Salix viminalis* plantations. *European Journal of Plant Pathology* 107, 399–409.

Thompson, J.D., Gibson, T.J., Plewniak, F., Jeanmougin, F. and Higgins, D.G. (1997) The ClustalX windows interface: flexible strategies for multiple sequence alignment aided by quality analysis tools. *Nucleic Acids Research* 24, 4876–4882.

Tucker, K. and Sage, R. (1999) *Integrated Pest Management in Short Rotation Coppice for Energy – a Grower's Guide*. Game Conservancy Ltd, Fordingbridge, UK.

Vogler, D.R. (1995) Use of molecular techniques in rust systematics. In: Kaneko, S., Katsuya, K., Kakishima., M. and Ono, Y. (eds) *Proceedings of 4th IUFRO Rusts of Pines Working Party Conference, Tsukuba*, pp. 9–15.

White, T.J., Bruns, T., Lee, S. and Taylor, J. (1990) Amplification and direct sequencing of fungal ribosomal RNA genes for phylogenetics. In: Innis, M.A., Gelfand, D.H., Sninsky, J.J. and White, T.J. (eds) *PCR Products: a Guide to Methods and Applications*. Academic Press, San Diego, California, pp. 315–322.

Wilson, M. and Henderson, D.M. (1966) *British Rust Fungi*. Cambridge University Press, Cambridge, UK.

Zander, M. (2000) Untersuchungen zur Identifizierung ausgewählter Vertreter der Gattung *Salix* L. im NO-deutschen Tiefland, unter besonderer Berücksichtigung des *Salix repens*-Komplexes. *Mitteilungen und floristische Karten Sachsen-Anhalt* (Halle) 5, 3–137.

Ziller, W.G. (1974) *The Tree Rusts of Western Canada*. Publication No. 1329. Canadian Forestry Service, Environmental Canada, Ottawa, Canada.

7 Genetic Structure of *Melampsora larici-epitea* Populations in North-western Europe

Berit Samils
Department of Plant Biology and Forest Genetics, Swedish University of Agricultural Sciences, Uppsala, Sweden

Introduction

The population genetic structure of a pathogen reflects its evolutionary history and its potential to evolve. By making inferences from the population structure we can get an understanding of the evolutionary mechanisms that lead to adaptations of the pathogen, and this knowledge can contribute to a more durable resistance through improved breeding programmes and strategies for the deployment of resistance (Leung *et al.*, 1993; Wolfe and McDermott, 1994; McDonald and Linde, 2002). Many mechanisms interact in the genetic changes and evolution of populations. The important ones are mutation, reproduction mode, gene flow, selection and random genetic drift.

The recent development of molecular markers for *Melampsora larici-epitea* has provided new opportunities to investigate genetic variation of the fungus. RAPD (random amplified polymorphic DNA) markers were used to distinguish between stem- and leaf-infecting forms of *Melampsora* on willow (Pei *et al.*, 1997). More recently, AFLP (amplified fragment length polymorphism) was applied to willow rust (Pei and Ruiz, 2000; Samils *et al.*, 2001a) and has since been used in a number of population studies. AFLP is a polymerase chain reaction (PCR)-based DNA fingerprint technique (Vos *et al.*, 1995) that has produced highly polymorphic and robust markers, which makes it a powerful tool in genotyping and examination of genetic variation (Majer *et al.*, 1996).

This chapter reviews the studies of genetic variation in populations of *M. larici-epitea* in biomass willow plantations in north-western Europe (mostly in Sweden). From these studies some inferences on the evolutionary potential of the pathogen will be made.

Biology of *M. larici-epitea*

M. larici-epitea has a complex life history. The fungus alternates between two hosts (willow and larch) and has five distinct kinds of spores (Fig. 7.1). The urediniospores start to form on the lower sides of willow leaves in the beginning of the summer. These spores are windspread and serve as the inoculum in repeated asexual infection cycles throughout the growing season. In autumn, telia develop on infected leaves and the fungus overwinters

 Rust Diseases of Willow and Poplar (eds M.H. Pei and A.R. McCracken)

in this form on fallen leaves. In the spring, basidiospores are produced and they infect larch (*Larix*), the alternate host, where sexual reproduction takes place (i.e. mating between spermatia and receptive hyphae). The sexual phase results in the formation of recombined aeciospores, which infect new willow leaves early in the summer. Thereafter, the asexual uredinial phase follows and the cycle is completed. The sexual phase on larch is presumed to be essential for winter survival of *M. larici-epitea*, but it has been discussed whether overwintering in the uredinial form might also occur. So far, however, evidence is lacking.

Spore dispersal

In rust fungi, wind-transported urediniospores (and possibly also aeciospores) are generally believed to be capable of long-distance migration (Roelfs, 1986; Aylor, 1990; Nagarajan and Singh, 1990). This capability has also been demonstrated for some *Melampsora* species on poplars and willows (Latch, 1980). The dispersal range of the drought- and ultraviolet (UV)-sensitive basidiospores is much more limited, probably to around a few hundred metres (Roelfs, 1985), although under favourable conditions basidiospores have been reported to move as far as 8 km (van Arsdel, 1967). This implies that, in the case of *M. larici-epitea*, the distance between willow plantations and larch stands may be important for the rate of winter survival of the fungus. Only those basidiospores that in spring manage to get to a larch plant and find a mating partner will contribute to the following season's initial inoculum in the form of recombined aeciospores.

Host specificity

Like other rust fungi, *M. larici-epitea* is highly specialized to its host, and genetically differentiated forms have been defined based on the host range (Gäumann, 1959; Pei *et al.*, 1996). The three most common form species (*formae speciales*) of *M. larici-epitea* on biomass willow are f. sp. *larici-epitea typica* (LET; infecting, for example, *S. viminalis*), f. sp. *larici-retusae* (LR; infecting, for example, *S. dasyclados*/*S. burjatica*) and f. sp. *larici-daphnoides* (LD; infecting, for example, *S. daphnoides*). These *formae speciales* probably constitute genetically separated populations, since low fertility in crossing experiments suggests that they do not normally interbreed (Pei *et al.*, 1999b). This has also been confirmed in extensive field surveys where no genotypes in rust samples expressed virulence to both of the two host species *S. viminalis* (infected by LET) and *S. dasyclados* (infected by LR) (Pei *et al.*, 1999a; Ramstedt, 1999). Further evidence of the *formae speciales* being genetically separate groups has been provided by DNA marker analyses (Pei *et al.*, 2002; Samils *et al.*, 2002).

The population studies of *M. larici-epitea* reviewed in this chapter relate mainly to f. sp. *larici-epitea typica* (LET), causing leaf rust on *S. viminalis*, one of the most important biomass willow species. All studies were conducted in biomass willow plantations or experimental fields.

Genetic Diversity Within Populations of *M. larici-epitea*

Genetic structure relates to the amount and distribution of genetic diversity within and among populations. It is useful to distinguish between the two components of genetic diversity, namely gene diversity and genotypic diversity. Gene diversity refers to the number of alleles at a locus in the population, while genotypic diversity refers to the number of genetically distinct individuals, or genotypes, in a population. It should be kept in mind that the absolute number of individuals of the rust fungus in the form of uredinia in the epidemic phase can be incredibly large. Even a moderately infected willow stand can harbour millions of individual pustules per hectare. But since *M. larici-epitea* reproduces asexually in repeated infection cycles during the

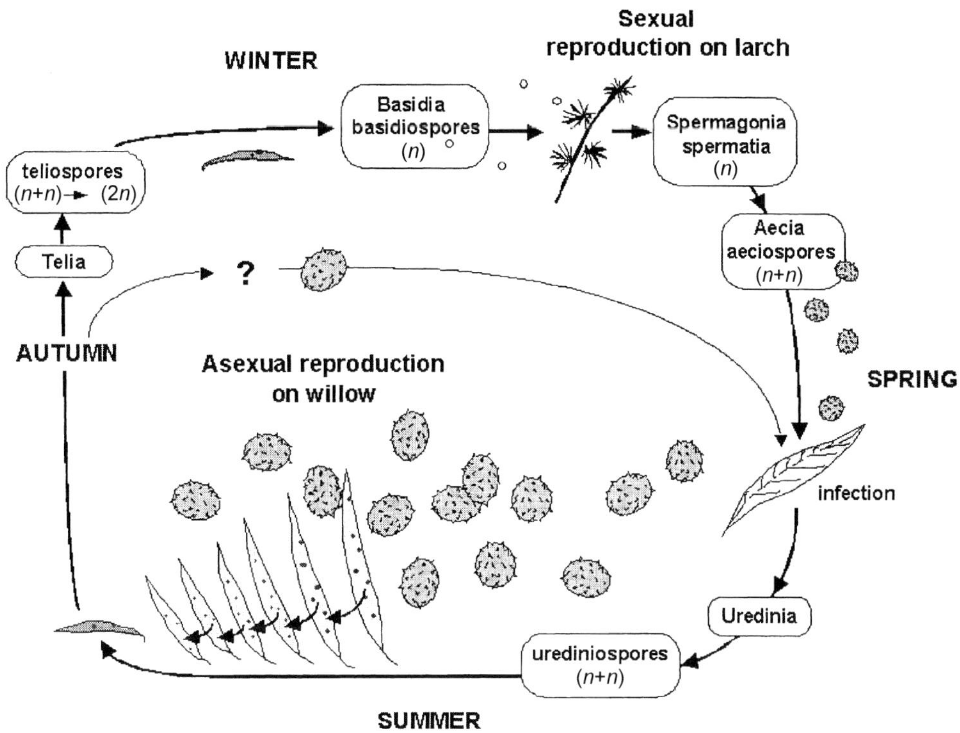

Fig. 7.1. The life cycle of *M. larici-epitea* (from Samils *et al.*, 2001). The nuclear condition of the fungus in each of the five spore stages is indicated.

summer, many of the uredinia can belong to the same genotype (i.e. they belong to the same clone).

Studies of Swedish *M. larici-epitea* populations show high levels of genetic diversity, both with regard to gene and genotypic diversity. Samils *et al.* (2001a) investigated three *M. larici-epitea* populations in biomass willow field trials in middle and southern Sweden. The three fields were located in Uppsala, Bränna and Svalöv, 300–600 km apart (Fig. 7.2). A high genotypic diversity was demonstrated in all populations by AFLP fingerprinting. Most urediniospore isolates (191 out of 197 investigated) were unique multilocus phenotypes, and only a few clonemates were detected. This means that the effective population size is very large. A high gene diversity was also indicated in these populations, with 96% of the AFLP loci being polymorphic.

Variation Among Willow Clones

There is little or no evidence of race-specific resistance towards *M. larici-epitea* within species of *Salix*, probably because breeding of biomass willow is still at an early stage in this relatively new crop. However, some degree of specificity has been indicated. For example, a study by McCracken *et al.* (2000) showed differences in pathotype composition between two *S. viminalis* varieties in a mixture experiment. To evaluate whether the genetic composition of *M. larici-epitea* populations differs between clones of *S. viminalis*, a comparison of rust populations on three willow clones (*S. viminalis* '78112', '78183' and 'Rapp') was made in three Swedish locations (Samils *et al.*, 2001a) (Fig. 7.2). In general, *M. larici-epitea* populations on the three *S. viminalis* clones were genetically similar. An exceptional finding was the occurrence of five contrasting isolates from *S. viminalis* 'Rapp' in Svalöv in southern Sweden. Although these isolates could not be separated by morphology, they had a markedly different AFLP pattern compared to the prevalent type of the rust. Since these isolates might belong to another species, they were excluded from the genetic analyses in the paper. This odd type (designated the 'B-type') has also been found on other *S. viminalis* clones in southern Sweden and its genetic relationship to other *Melampsora* on *Salix*, as revealed by AFLP markers, has been described (Samils *et al.*, 2002).

In contrast to the results from Sweden, a clear genetic difference was found between *M. larici-epitea* populations on two *S. viminalis* clones ('78118' and '78183') in a mixture experiment in Castlearchdale in Northern Ireland (Samils *et al.*, 2003) (Fig. 7.2). An analysis of molecular variance based on AFLP showed a relatively large differentiation (19.8%, $\phi_{ST} = 0.198$; Table 7.1) between the two willow varieties, and differences in genotypic composition were indicated by the non-random distribution of clonal rust isolates between the varieties. It seemed that the pathogen population was composed both of genotypes that are more adapted to one or other host clone and those that are well adapted to both clones. The finding of differences between willow clones in Castlearchdale, but not in Sweden, could simply be because different sets of

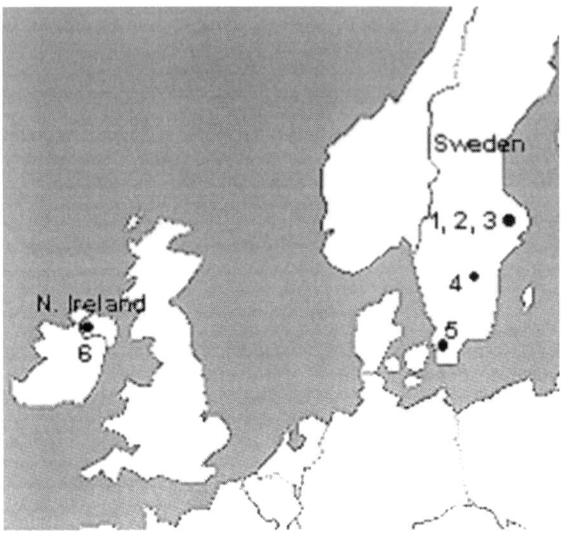

Fig. 7.2. Map showing the location of *Melampsora larici-epitea* populations in studies by Samils *et al.*: 1, Sätuna; 2, Uppsala; 3, Djurby; 4, Bränna; 5, Svalöv; and 6, Castlearchdale.

Table 7.1. Summary of results from analysis of molecular variance (AMOVA) for *Melampsora larici-epitea* isolates in studies by Samils *et al.* (The publication where the complete result is presented is given.)

Source of variation	Populations considered	ϕ_{ST}[a]
Between Sweden and N. Ireland	Djurby and Sätuna (Sw), and Castlearchdale (NI)[b]	0.089***
Among provinces in Sweden	Uppsala, Bränna and Svalöv[c]	0.025***
Between populations within a province	Djurby and Sätuna[b]	–0.003
Among willow clones in Sweden	*S. viminalis* '78112', '78183' and 'Rapp'[c]	0.001
Between willow clones in Castlearchdale (N. Ireland)	*S. viminalis* '78183' and '78118'[d]	0.198***

***$P < 0.001$ (determined by 1000 randomizations of the data set).
[a]ϕ_{ST} (analogous to F_{ST}) is an estimate of population subdivision (Excoffier *et al.*, 1992).
[b]Samils *et al.* (2001b), [c]Samils *et al.* (2001a), [d]Samils *et al.* (2003)

willow clones were investigated in the two countries (e.g. '78118' was not included in Sweden). Another explanation could be that detection of virulence differences with molecular markers is less likely in a population with frequent sexual reproduction, as seems to be the case in Sweden (Samils *et al.*, 2001a), than in a population where asexual reproduction is predominant, as might be the case in Castlearchdale in Northern Ireland (Samils *et al.*, 2001b).

Influence of Larch – the Alternate Host

Since larch is the host for sexual reproduction of *M. larici-epitea*, it is also the place where new, genetically recombined genotypes are created in each spring–early summer. Recombination is an important feature for the evolution of fungal genotypes. With new combinations of virulence genes, recombinants are better adapted to the resistances of the host. To investigate whether the presence of larch will increase the genetic diversity of willow rust populations, Samils *et al.* (2001b) studied the genetic structure of *M. larici-epitea* populations in two plantations of *S. viminalis* in the province of Uppland in central Sweden. On one of the plantations, a larch stand was less than 100 m from the willow field, whereas no larch plants were present within 1.5 km of the other plantation. A comparison was also made with a population in an experimental trial in Northern Ireland, where larch grew less than 500 m from the willow field. In order to analyse the spatial distribution of genetic variation within the fields, rust samples were collected in a hierarchical pattern, and the rust isolates were analysed with AFLP markers. Comparing the two Swedish populations, it turned out that the proximity of the alternate host had no apparent effect on the genetic structure, as there was no genetic differentiation between the two sites and the level of genotypic diversity was similarly high for both populations. Most of the molecular variation was present on a very fine spatial scale, with no detectable within-field clustering of clonal isolates. A single leaf was normally infected by more than one *M. larici-epitea* genotype. In fact, when two uredinia from each of 56 leaves were investigated, it turned out that the two uredinia had different genotypes in all cases. The occurrence of sexual reproduction in the Swedish populations was supported by tests of non-random association among loci, which revealed no significant departure from random mating in either of the two populations. According to weekly records of disease severity in the two fields, disease development in the beginning of the summer started a few weeks earlier in the plantation close to larch. The infection level was also significantly higher than in the non-larch field during a large part of the growth period.

For the purpose of the study, the leaf rust present within a willow field was regarded as a separate population. In reality, populations might extend over larger areas, and the two Swedish populations (located 45 km apart) might perhaps better be considered as belonging to the same population, as they were shown to be genetically similar. A plausible scenario would be that urediniospores will spread from field to field during the repeated cycles of asexual production during the epidemics. The sexual reproduction in spring might occur only at those locations where willow and larch grow close enough to allow basidiospore transmission (i.e. probably less than a few hundred metres apart). From these larch plants the recombined aeciospores can spread to nearby willow fields, and the subsequently produced urediniospores can be dispersed throughout the population range.

The *M. larici-epitea* population in Castlearchdale in Northern Ireland showed a different structure compared to the two Swedish populations. As in the Swedish populations, genetic variation was randomly distributed within the field, and no spatial clustering of clonal isolates was indicated. However, the level of clonality was much higher, with the three most frequent clones representing 70% of the total sample (16, 11 and 7 isolates, respectively, out of 47). The genotypic diversity calculated by a normalized Shannon's index was 0.54 in this population, while it was 0.95 and 1.0, respectively, in the two Swedish populations. Further, the occurrence of multilocus associations (linkage disequilibrium) indicated non-random mating, which is a characteristic of a predominantly asexual population.

The overall results suggest that while sexual reproduction seems to be frequent and regular in Swedish *M. larici-epitea* populations, it may be less frequent or less regular in Castlearchdale. This could be due to a seasonal bottleneck during winter survival and sexual reproduction, resulting in a limited amount of source inoculum for the next season and thus leading to limited genotypic diversity. Alternatively, the milder climate in Northern Ireland might allow for winter survival of urediniospores and sexual reproduction may thus not be obligatory, which would also result in lower genotypic diversity.

Genetic Variation Among Populations of *M. larici-epitea*

The amount of genetic variation among populations can give us indirect measures of the amount of gene flow, since little variation among populations typically implies high rates of gene flow and vice versa. Migration of wind-transported urediniospores and aeciospores will be the primary way of gene flow in *M. larici-epitea*. The rate of gene flow is of importance from a breeding point of view since it will give an indication on how fast virulence genes will spread among regions. It should be noted that other factors, such as selection and genetic drift, especially in small populations, also affect genetic variation among populations.

Generally there seems to be relatively little variation among geographically separated willow rust populations in north-western Europe. An analysis of molecular variance (AMOVA) based on AFLP showed only a few per cent differentiation among three Swedish *M. larici-epitea* populations, located 300–600 km apart (Samils *et al.*, 2001a); and, as already mentioned, no genetic differentiation was found between two populations located 45 km apart in the province of Uppland in Sweden (Samils *et al.*, 2001b) (Table 7.1). When these two populations were compared to a population in Northern Ireland, analysis of molecular variance showed a differentiation of 8.9% ($\phi_{ST} = 0.089$; Table 7.1) between the two countries.

In contrast to the low levels of geographic differentiation in the above studies, a case with a high degree of subdivision among *M. larici-epitea* populations was observed in Great Britain. Pei *et al.* (2000) investigated *M. larici-epitea* populations on *Salix* × *mollisima* (*S. triandra* × *S. viminalis*) during the first year of rust outbreak (in 1992) on this previously highly resistant clone. The investigation included three plantations in the British Isles: one in Long Ashton and one in Markington (both in England, roughly 300 km apart) and the third in Loughall in Northern Ireland. In an analysis of molecular variance based on AFLP, as much as 86% of the variation was due to differences among populations. The authors suggested that the high degree of differentiation could be due to inbreeding in small, newly established populations. They further suggested that the previously unknown pathotype was not spread from a common source but was spread from different sources.

Even though isolated diverged populations may temporarily occur in *M. larici-epitea*, the overall results, with small to moderate differentiation, suggest relatively high levels of gene flow between populations.

Conclusions

The genetic structure of *M. larici-epitea* populations, with large amounts of both gene and genotypic variation present within most populations, and little geographic differentiation, suggests high rates of genetic recombination through sexual reproduction and high rates of gene flow. Such qualities are likely to allow the pathogen populations to readily adapt to a changing environment, such as new resistances in the host population. In other words, *M. larici-epitea* has a high evolutionary potential and can be rated as a 'high-risk' pathogen (McDonald and Linde, 2002). This means that when resistant *Salix* clones are introduced and cultivated over large areas, they run the risk of losing resistance as *M. larici-epitea* can rapidly acquire and spread adaptive genotypes with matching virulences. Such losses of resistance have already been experienced in the *Salix*/*Melampsora* pathosystem, where resistances of the willow clones *S. burjatica* 'Korso' and *S.* × *mollisima* 'Q83' were broken after 8–10 years of cultivation (McCracken and Dawson, 1998; Pei *et al.*, 2000). Similar breakdowns have also occurred in the closely related *Populus*/*Melampsora* pathosystem (Pinon and Frey, 1997). Therefore, finding ways to slow host adaptation in *M. larici-epitea* populations is a major concern. One promising approach is the cultivation of host mixtures, which are treated in detail in Part 4 of this book. Another option is breeding for partial resistance (race-non-specific resistance) (Lascoux *et al.*, 1996; Ramstedt, 1999), which is believed to be more durable than complete resistance (race-specific resistance). Partial resistance is thought to be the combined effect of a number of resistance components and may be characterized by a long latent period, low spore production and low infection frequency (Parlevliet, 1993). A large amount of heritable variation for resistance components has been detected among *S. viminalis* clones (Lascoux *et al.*, 1996), which raises the prospects for breeding for partial resistance in biomass willows.

References

Aylor, D.E. (1990) The role of intermittent wind in the dispersal of fungal pathogens. *Annual Review of Phytopathology* 28, 73–92.

Excoffier, L., Smouse, P.E. and Quattro, J.M. (1992) Analysis of molecular variance inferred from metric distances among DNA haplotypes: application to human mitocondrial DNA restriction sites. *Genetics* 131, 479–491.

Gäumann, E. (1959) *Die Rostpilze Mitteleuropas. Beiträge zur Kryptogamenflora Schweiz 12.* Buchdruckerei Büchler and Co., Bern, Germany.

Lascoux, M., Ramstedt, M., Åström, B. and Gullberg, U. (1996) Components of resistance of leaf rust (*Melampsora laricii-epitea* Kleb./*Melampsora ribesii-viminalis* Kleb.) in *Salix viminalis* L. *Theoretical and Applied Genetics* 93, 1310–1318.

Latch, B.J. (1980) Weeping willow rust in New Zealand. *New Zealand Journal of Agricultural Research* 23, 535–538.

Leung, H., Nelson, R.J. and Leach, J.E. (1993) Population structure of plant pathogenic fungi and bacteria. *Advances in Plant Pathology* 10, 157–205.

Majer, D., Mithen, R., Lewis, B.G., Vos, P. and Oliver, P. (1996) The use of AFLP fingerprinting for the detection of genetic variation in fungi. *Mycological Research* 100, 1107–1111.

McCracken, A.R. and Dawson, W.M. (1998) Short rotation coppice willow in Northern Ireland since 1973: development of the use of mixtures in the control of foliar rust (*Melampsora* spp.). *European Journal of Forest Pathology* 28, 241–250.

McCracken, A.R., Dawson, W.M., Watson, S. and Allen, C.Y. (2000) Pathotype composition in *Melampsora epitea* populations occurring on willow (*Salix*) grown in mixed and monoculture plantations. *European Journal of Plant Pathology* 106, 879–886.

McDonald, B.A. and Linde, C. (2002) Pathogen population genetics, evolutionary potential, and durable resistance. *Annual Review of Phytopathology* 40, 349–379.

Nagarajan, S. and Singh, D.V. (1990) Long-distance dispersion of rust pathogens. *Annual Review of Phytopathology* 28, 139–153.

Parlevliet, J.E. (1993) What is durable resistance, a general outline. In: Jacob, T. and Parlevliet, J.E. (eds) *Durability of Disease Resistance.* Kluwer Academic Publishers, Dordrecht, The Netherlands, pp. 23–39.

Pei, M.H. and Ruiz, C. (2000) AFLP evidence of the distinctive patterns of life-cycle in two forms of *Melampsora* rust on *Salix viminalis*. *Mycological Research* 104, 937–942.

Pei, M.H., Royle, D.J. and Hunter, T. (1996) Pathogenic specialization in *Melampsora epitea* var. *epitea* on *Salix*. *Plant Pathology* 45, 679–690.

Pei, M.H., Parker, S.R., Hunter, T. and Royle, D.J. (1997) Variation in populations of *Melampsora* willow rust and the implications for design of short rotation coppice plantation. *Aspects of Applied Biology* 49, 91–96.

Pei, M.H., Hunter, T. and Ruiz, C. (1999a) Occurrence of *Melampsora* rusts in biomass willow plantations for renewable energy in the United Kingdom. *Biomass and Bioenergy* 17, 153–163.

Pei, M.H., Royle, D.J. and Hunter, T. (1999b) Hybridization in larch-alternating *Melampsora epitea* (*M. larici-epitea*). *Mycological Research* 103, 1440–1446.

Pei, M.H., Yuan, Z.W., Hunter, T. and Ruiz, C. (2000) Heterogeneous nature of a 'new' pathotype of *Melampsora* rust on *Salix* revealed by AFLP. *European Journal of Plant Pathology* 106, 771–779.

Pei, M.H., Bayon, C., Ruiz, C., Yuan, Z.W. and Hunter, T. (2002) Genetic variation in *Melampsora larici-epitea* on biomass willows using AFLP. *European Journal of Plant Pathology* 108, 229–236.

Pinon, J. and Frey, P. (1997) Structure of *Melampsora larici-populina* populations on wild and cultivated poplar. *European Journal of Plant Pathology* 103, 159–173.

Ramstedt, M. (1999) Rust disease on willows – virulence variation and resistance breeding strategies. *Forest Ecology and Management* 121, 101–111.

Roelfs, A.P. (1985) Wheat and rye stem rust. In: Roelfs, A.P. and Bushnell, W.R. (eds) *The Cereal Rusts*, Vol. II, *Diseases, Distribution, Epidemiology, and Control*. Academic Press, Orlando, Florida.

Roelfs, A.P. (1986) Development and impact of regional cereal rust epidemics. In: Leonard, K.J. and Fry, W.E. (eds) *Plant Disease Epidemiology: Population Dynamics and Management.* Macmillan, New York, pp. 129–150.

Samils, B. (2001) Population genetic structure of *Melampsora larici-epitea,* a willow leaf rust fungus. Doctoral thesis. Acta Universitatis Agriculturae Sueciae, Agraria 292. SLU Servive/Repro, Uppsala, Sweden. ISSN 1405-6249.

Samils, B., Lagercrantz, U., Lascoux, M. and Gullberg, U. (2001a) Genetic structure of *Melampsora epitea* populations in Swedish *Salix viminalis* plantations. *European Journal of Plant Pathology* 107, 399–409.

Samils, B., Stepien, V., Lagercrantz, U., Lascoux, M. and Gullberg, U. (2001b) Genetic diversity in relation to sexual and asexual reproduction in populations of *Melampsora larici-epitea. European Journal of Plant Pathology* 107, 871–881.

Samils, B., Lagercrantz, U. and Gullberg, U. (2002) Genetic relationships among genetically distinct forms of *Melampsora larici-epitea* and related species based on AFLP data. *Forest Pathology* 32, 379–386.

Samils, B., McCracken, A.R., Dawson, W.M. and Gullberg, U. (2003) Host specific genetic composition of *Melampsora larici-epitea* populations on two *Salix viminalis* varieties in a mixture trial. *European Journal of Plant Pathology* 109, 183–190.

Van Arsdel, E.P. (1967) The nocturnal diffusion and transport of spores. *Phytopathology* 57, 1221–1229.

Vos, P., Hogers, M.B., Reijans, M., van de Lee, T., Hornes, M., Frijters, A., Pot, J., Peleman, J., Kuiper, M. and Zabeau, M. (1995) AFLP: a new technique for DNA fingerprinting. *Nucleic Acids Research* 23, 4407–4414.

Wolfe, M.S. and McDermott, J.M. (1994) Population genetics of plant pathogen interactions: The example of the *Erysiphe graminis–Hordeum vulgare* pathosystem. *Annual Review of Phytopathology* 32, 89–113.

8 Current Taxonomic Status of *Melampsora* Species on Poplars in China

Cheng-Ming Tian[1] and Makoto Kakishima[2]

[1]*College of Resource and Environment, Beijing Forestry University, 35 Tsinghua Eastern Road, Beijing 100083, China;* [2]*Institute of Agriculture and Forestry, University of Tsukuba, Tsukuba, Ibaraki 305-8572, Japan*

Melampsora Rusts on Populus

The genus *Melampsora* was established by Castagne in 1843 based upon *M. euphorbiae* (Schub.) Cast. About 90 species of *Melampsora*, showing either an autoecious or a heteroecious life-cycle pattern, have been described worldwide (Kirk *et al.*, 2001). Most *Melampsora* species occur on poplars and willows, and not all are morphologically distinct (Cummins and Hiratsuka, 2003).

A worldwide literature survey by Van Kraayenoord *et al.* (1974) showed that 32 *Melampsora* species names had been described on *Populus*. Of these, 11 species could be recognized and others treated as synonyms. Shang *et al.* (1986b) reviewed 34 species of poplar rusts reported worldwide and recognized 12 species based on the host range and the characteristics of uredinia and telia. In a numerical taxonomical study, Dai (1989) examined *Melampsora* species on poplars using 24 characters derived from urediniospores and teliospores, and aecial and telial hosts. As a result, 14 species could be recognized. Bagyanarayana (1998) studied the morphology of *Melampsora* species on *Populus* species and recognized nine species and five *formae speciales* under *M. populnea*. Cellerino (1999) listed 14 species of *Melampsora* on poplars.

The classification of *Melampsora* species on poplars is based mainly on the morphology of urediniospores and teliospores, and their host range. However, some criteria often overlap and the taxonomy of *Melampsora* on various poplar species is not entirely clear at the present time. With many poplar rusts, identification of species can be very difficult for several reasons. First, almost all the morphological characteristics of teliospores are very similar, and teliospores of some species do not appear during the growing season. Secondly, most species infecting poplars are heteroecious, having *Abies*, *Allium*, *Arum*, *Chelidonium*, *Corydalis*, *Fumaria*, *Larix*, *Mercurialis*, *Papaver*, *Picea*, *Pinus* and *Tsuga* as alternate hosts, but information on the alternate hosts is often unavailable. Some species are also capable of overwintering as mycelia in dormant buds of *Populus* and continue their existence without infecting alternate hosts. Thirdly, the same species of *Populus* can be infected by different *Melampsora* species. Furthermore, non-host poplars can sometimes be infected in artificial inoculations.

 Rust Diseases of Willow and Poplar (eds M.H. Pei and A.R. McCracken)

Record of Poplar Rusts in China

The earliest report on *Melampsora* species on poplars in China was made by Miyake (1914), based on the specimens collected in north-eastern China. In the following years, the species of poplar rusts were complied in publications by other authors, such as Miura (1928), Huo and Wang (1934) and Liu and Wang (1935). Tai (1979) reviewed earlier studies of *Melampsora* spp. on poplars in China and listed five species, *M. larici-populina* Kleb., *M. laricis* Hart., *M. magnusiana* Wagn., *M. rostrupii* Wagn. and *M. pruinosae* Tranz. Yuan (1984) examined the morphology of *Melampsora* species on poplars and recorded three more species in China, i.e. *M. abietis-canadensis* (Farl.) Ludw., *M. allii-populina* Kleb. and *M. occidentalis* Jacks. Shang *et al.* (1986a) described a new species, *M. multa* Shang, Pei and Yuan on *P.* × *euramericana* Moench. collected in Liaoning, China. *Melampsora magnusiana* and *M. rostrupii* were treated by Shang *et al.* (1986b, 1990) as synonyms of *M. aecidioides* Plowr.

Recently, *M. abietis-populi* Imai was reported on *Populus wilsonii* Schneid. in Shaanxi and Hubei provinces (Guo, 1989; Zhang *et al.*, 1997; Cao and Li, 1999). Previously, *M. abietis-populi* had only been recorded in Japan (Hiratsuka *et al.*, 1992; Bagyanarayana, 1998). Zhuang and Wei (1994) reported *M. populnea* (Pers. ex Pers.) Karst., for the first time in China, on *P. pseudoglauca* Wang and Fu., and *P. szechuanica* Schneid. var. *tibetica* Schneid. In their report, *M. laricis* was treated as a synonym of *M. populnea*. Subsequently, some specimens from different regions were also recognized as *M. populnea* (Cao and Li, 1999). In total, 11 *Melampsora* species names have been recorded on poplars in China (Table 8.1). In this chapter, we present a review of the current taxonomic status of *Melampsora* species on *Populus* in China.

Melampsora abietis-populi Imai, in Ito et Murayama, *Trans. Sapporo Nat. Hist. Soc.* 17, 164; 1943

= *Caeoma abietis-mayrianae* S. Imai, 1942.
Host plant: II, III on *Populus cathayana* Rehd. (Shaanxi: Cao and Li, 1999) and *P. wilsonii* Schneid. (Hubei: Guo, 1989; Shaanxi: Cao and Li, 1999; Zhang *et al.*, 1997).
Distribution: China, Japan, Russia.

This species resembles *M. magnusiana* and *M. laricis* morphologically (Fig. 8.1A), but can be distinguished from the two species by the following features: (a) its urediniospores have thin walls (Fig. 8.1B); (b) the distance between spinules is less compared with that in *M. magnusiana* and *M. laricis* (Fig. 8.1C); and (c) it alternates on *Abies*.

Zhuang and Wei (1994) identified some specimens (HMAS 67387, 67388) occurring on *P. pseudoglauca* Wang et Fu collections from Tibet as *M. populnea* (Per.) Karst. We also saw these specimens and found that, morphologically, they differed from *M. populnea* in having urediniospores with thin walls (Fig. 8.1B). In our phylogenetic analyses, these specimens were clearly separated from other groups such as *M. laricis* Hartig. or *M. populnea* (Tian *et al.*, 2004). Thus, we consider that these specimens belong to *M. abietis-populi*. We also recorded this species for the first time on *P. yunnanensis* Dode collected in Yunnan Province (Tian *et al.*, 2004).

This species was established by Imai (1942) based on a rust fungus on *P. nigra* var. *italica* (Moench) Koehne in Hokkaido, Japan, and the alternation of *M. abietis-populi* between *Populus* and *Abies* was proven by Imai (1942) and Hiratsuka *et al.* (1992). In China, no specimens of spermogonial and aecial stages have been collected and no inoculation experiments conducted.

We consider that the thin wall is a stable feature in distinguishing this species from other *Melampsora* species, based on our morphological and molecular phylogenetic studies (Tian *et al.*, 2004).

Table 8.1. *Melampsora* spp. reported in China (Tai, 1979; Yuan, 1984; Shang *et al.*, 1986b; Guo, 1989; Zhuang and Wei 1994; Cao and Li, 1999).

	Urediniospores					
Species	Size (μm)	Walls (μm)	Smooth parts (μm)	Teliospore size (μm)	Host section	Alternate hosts (genus)
M. larici-populina	23–46 × 16–25	2–3 (5.1–6.8)	Apical parts	37–60 × 8–12	*Tacamahaca*, *Aigeiros*	*Larix*
M. allii-populina	26–38 × 17–20	2–3	Apical parts	38–56 × 9–13	*Tacamahaca*, *Aigeiros*	
M. magnusiana (*M. aecidioides*)	19–26 × 16–21	2.8–3.5	No	37–51 × 9–13	*Leuce*	
M. populnea	16–28 × 14–17	2.5–3.5	No	30–60 × 8–13	*Leuce*	
M. rostrupii	15–28 × 14–18	2–3	No	27–44 × 7–12	*Leuce*	
M. laricis (*M. larici-tremulae*)	17–27 × 13–20	2–2.8	No	27–48 × 8–14	*Leuce*	*Larix*
M. pruinosae	18–27 × 17–20	2.7–3.5	No	38–59 × 9–13	*Turanga*	
M. abietis-canadensis	17–25 × 13–20	1.7–2.7	Equatorial parts	27–48 × 8–12	*Leuce*, *Tacamahaca*	
M. occidentalis	27–41 × 20–27	2–3 (5.1–6.8)	No	30–48 × 9–12	*Tacamahaca*	
M. abietis-populi	17–32 × 11–22	2–3	No	15–32 × 5–10	*Leucoides*	
M. multa	26–44 × 16–25		Apical parts	30–45 × 6–13	*Aigeiros*	

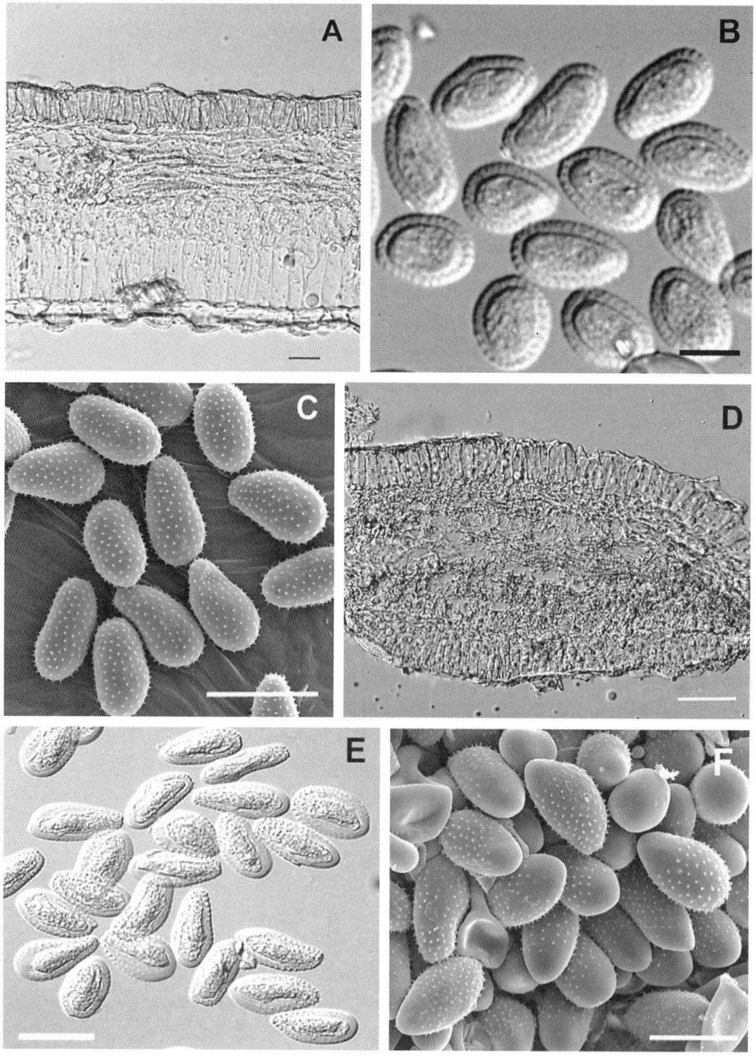

Fig. 8.1. Urediniospores and teliospores of *Melampsora* spp. on *Populus* observed by light microscopy and scanning electron microscopy. (A)–(C) *M. abietis-populi* on *P. wilsonii.* (A) Hypophyllous telia; (B) globose or ellipsoid urediniospores with thin walls; (C) urediniospores with echinulate surface. (D)–(E) *M. allii-populina* on poplars. (D) Amphigenous telia on *P. laurifolia;* (E) oblong urediniospores with uniformly thickened wall on *P. laurifolia;* (F) urediniospores with a smooth apex on *P. talassica*. Bars: (A) 30 μm; (B) 15 μm; (C) 20 μm; (D) 55 μm; (E) 35 μm; (F) 20 μm.

Melampsora abietis-canadensis (Farl.) Ludw. *Phytop.* 5, 279; 1915

= *Caeoma abietis-canadensis* Farl., *Proc. Am. Acad.* 20, 323; 1885.
= *Melampsora populi-tsugae* Havis, *Trans. Wis. Acad.* 19, 676; 1919.

Host plant: II, III on *P. davidiana* Dode (Shanxi and Jilin: Yuan, 1984), *P. cathayana* Rehd. (Shanxi: Yuan, 1984).
Distribution: China, Canada, USA.

Yuan (1984) described *M. abietis-canadensis* for the first time in China based on the specimens on *P. davidiana* Dode and

P. cathayana Rehd. He considered that 'a few urediniospores with a smooth area at the equator' is an important feature in distinguishing this species from others. Although we were not able to see the same specimens that Yuan examined, we noticed a smooth area at the equator of urediniospores in some other specimens, such as the rust on *P. davidiana* (Fig. 8.2D). These specimens were grouped together with other specimens of *M. laricis* in our phylogenetic analyses (Tian *et al.*, 2004). In Japan, Ito (1938) and Hiratsuka and Karube (1969) also described some specimens of *M. laricis* that had urediniospores with slightly thickened side walls, and some urediniospores that had a smooth area at the equator. Thus, we consider that the previously reported *M. abietis-canadensis* (Farl.) Ludwig in China is a consequence of misidentification. *Melampsora abietis-canadensis* is regarded as a synonym of *M. medusae* (Cellerino, 1999).

Melampsora allii-populina Kleb. *Zeit. Pfl. – Krankh.* 12, 25; 1902

= *Caeoma alliorum* Link, sp. Pl. 2, 7; 1825.
= *Caeoma allii-ursini* Winter, *Pilze Deutschl.* I, 255; 1881.
= *Uredo allii-populina* Arth., *Rèsult Scient. Congr. Intern. Bot. Vienne* 338; 1905, 1906.
= *Melampsora populnea* f. sp. *allii-populina* (Kleb.) Bagyanarayana, *Proc. 1st IUFRO Fore. Trees Working Party Conf.* 47; 1998.

Host plants: II, III on *P. talassica* Komarov (Xinjiang: Yuan, 1984).

Distribution: China, southern Europe, Russia.

Morphological characteristics of urediniospores and teliospores of *M. allii-populina* are somewhat similar to those of *M. larici-populina*, but it is distinguishable by having urediniospores with uniformly thickened (2–4 μm) walls (Fig. 8.1E), paraphyses with uniformly thin walls, and amphigenous telia (Fig. 8.1D). Although *M. allii-populina* completes its full life cycle by infecting herbaceous plants belonging to *Allium*, *Arum* and *Muscari*, it can perpetuate infection on poplar as the uredinial stage. Bagyanarayana (1998) placed *M. allii-populina* together with other four species, as *formae speciales* of *M. populnea* (Pers. ex Pers.) Karst., based on different aecial hosts. However, *M. allii-populina* can be distinguished from *M. populnea* by the urediniospores with a smooth apex (Fig. 8.1F), and its amphigenous telia. Furthermore, it differs from other *Melampsora* species in phylogeny (Tian *et al.*, 2004).

Previously, *M. allii-populina* was known in China by the only specimen collected from *P. talassica* Komarov from Xinjiang Province (Yuan, 1984). In this chapter we record this species for the first time from Inner Mongolia on two new poplar hosts, *P. opera* of section *Tacamahaca*, and *P. davidiana* Dode of section *Leuce*, and also from Xinjiang on *P. laurifolia* Ledebour.

Melampsora laricis Hartig, *Allg. Forest-Jagdzeit.* 61, 326; 1885

= *Uredo laricis* Arth. *Result. Scient. Congr. Intern. Bot. Vienne* 338; 1905, 1906.
= *M. larici-tremulae* Kleb., *Forestl. Nat. Zeits.* 468; 1897.
= *M. tremulae* f. sp. *laricis* Hartig, *Lehrb. Baumkrankh. Ed.* 2, 14; 1889.
= *M. populnea* f. sp. *laricis* (Hart.) Boer. and Verh., *Neth. J. Pl. Path.* 78 (1), 82–201; 1972.

Host plants: 0, I on *Larix olgensis* Henry, *L. gmelinii* (Rupr.) Rupr., *L. kaempferi* (Sieb. et Zucc.) Gord., *L. principis-ruppechtii* Mayr. and *L. sibirica* Ledeb. (Heilongjiang: Shang and Pei, 1984). II, III on *Populus alba* L. × *P. davidiana* Dode (Heilongjiang: Yuan, 1984), *P. adenopoda* Maxim. (Guangxi: Tai, 1979; Hubei: Guo, 1989; Shaanxi: Cao and Li, 1999), *P. davidiana* Dode (Heilongjiang and Shaanxi: Tai, 1979; Shanxi: Yuan, 1984; Inner Mongolia: Shang *et al.*, 1990; Shaanxi: Cao and Li, 1999; Inner Mongolia: Zhuang and Wei, 2002), *P. pseudoglauca* Wang et Fu (Tibet: Zhuang, 1986), *P. tomentosa* Carr. (Shaanxi: Tai, 1979), *P. tremula* L. (Tibet: Zhuang, 1986), *P.* sp. (Henan: Tai, 1979), *P. wilsonii* Schneid (Hubei: Guo, 1989).

Distribution: China, Japan, Europe, Russia, Africa, USA.

Shang and Pei (1984) were the first to establish the full life history of *M. laricis* (= *M. larici-tremulae* Kleb.) in China, when they produced uredinia and telia on *Populus* spp. in artificial inoculations with aecidiospores from *Larix* spp. It was proven that basidiospores of *M. laricis* can infect *L. olgensis*, *L. gmelinii*, *L. kaempferi*, *L. principis-rupprechtii* and *L. sibirica*, and that its aeciospores can infect the species of section *Leuce*. It also infected the hybrids having section *Leuce* as female parents, but not the species of sections *Aigeios* and *Tacamahaca*, and the hybrids between these two sections.

Wilson and Henderson (1966) adopted *M. populnea* as a complex species to include 'various races' which are similar in morphology, but different in aecial hosts. Boerema and Verhoeven (1972) also treated *M. laricis* as a *forma specialis* (f. sp. *laricis*) within *M. populnea*, based on the aecial host. In China, *M. laricis* has recently been regarded as synonymous with *M. populnea* by Zhuang and Wei (1994) and others (Cao and Li, 1999), but these workers also have accepted *M. magnusiana* as a distinct taxon. Although *M. laricis* is similar to *M. magnusiana* in the shape and size of urediniospores, it differs from *M. magnusiana* by having urediniospores with slightly thickened lateral walls (Fig. 8.2B) and sometimes with a smooth area at the equator (Fig. 8.2D). Furthermore, the uredinia of *M. laricis* are hypophyllous (Fig. 8.2A), scattered and pale-yellow, while *M. magnusiana* often produces uredinia on leafstalks and young buds. The uredinia of *M. magnusiana* are epiphyllous, often aggregated in groups and yellow (Shang *et al.*, 1990; Y.Z. Shang, 2002, personal communication). Recently, our phylogenetic analyses placed the *M. laricis* specimens on *P. adenopoda* Maxim., *P. davidiana* Dode, and *P. tomentosa* Carr. collected from China into one group, and showed that this rust has a distant relationship with other *Melampsora* species occurring on section *Leuce* (Tian *et al.*, 2004). Therefore, we consider that *M. laricis* is a distinct taxon.

Melampsora larici-populina Kleb., *Zeit. Pfl. –Krankh.* 12, 43; 1902

= *Caeoma laricis* Hart. *Wichtige Krankh. Wäldbaume* 93; 1847.
= *Uredo larici-populina* Arth., *Rèsult. Scient. Congr. Intern. Bot. Vienne* 338; 1905, 1906.
= *Melampsora populina* (Jacq.) Lev., *Ann. Sci. Nat.*, 375; 1847.
= *Melampsora populina* Auct. pr. P

Host plants: 0, I on *Larix gmelinii* (Rupr.) Kuz. (Heilongjiang: Tai, 1979; Pei and Shang, 1984; Zhuang and Wei, 2002), *L. kaempferi* (Sieb. et Zucc.) Gord. (north-east China: Tai, 1979; Heilongjiang: Pei and Shang, 1984), *L. olgensis* Henry (Heilongjiang: Pei and Shang, 1984; Shaanxi: Cao and Li, 1999), *L. principis-ruppechtii* Mayr. (Heilongjiang: Pei and Shang, 1984; Shaanxi: Cao and Li, 1999), *L. sibirica* Ledeb. (Heilongjiang: Pei and Shang, 1984), *L.* sp. (Qinghai and Heilongjiang: Tai, 1979). II, III on *Populus adenopoda* Maxim. (Shaanxi: Wei and Zhuang, 1997), *P.* × *berolinensis* Dippel (Inner Mongolia, Liaoning, Jilin and Heilongjiang: Tai, 1979; Shang *et al.*, 1990), *P.* × *beijingensis* Hsu (Beijing, Inner Mongolia and north-east China: Tai, 1979; Yuan, 1984; Shang *et al.*, 1990), *P. cathayana* Rehd. (Hebei, Liaoning: Tai, 1979; Beijing: Yuan, 1984; Inner Mongolia: Shang *et al.*, 1990; Shaanxi: Wei and Zhuang, 1997; Cao and Li, 1999), *P.* × *euramericana* (Dode) Guinier (Inner Mongolia, Liaoning and north-east China: Tai, 1979; Beijing, and Jilin: Yuan, 1984; Inner Mongolia: Shang *et al.*, 1990), *P.* × *euramericana* var. *regenerata* (Henry) Rehd. (Inner Mongolia and north-east China: Tai, 1979), *P. davidiana* Dode (Hebei: Tai, 1979; Tibet: Zhuang, 1986), *P. deltoides* Marsh (Beijing: Yuan, 1984; Inner Mongolia: Shang *et al.*, 1990), *P. euphratica* Oliv. (Liaoning: Tai, 1979; Inner Mongolia: Yuan, 1984), *P. harbinensis* Wang et Skv. (Inner Mongolia and north-east China: Tai, 1979; Beijing: Yuan, 1984), *P.* × *gansuensis* Wang et Yang (Inner Mongolia: Shang *et al.*, 1990), *P. koreana* Rehd. (Liaoning and Heilongjiang: Tai, 1979; Inner Mongolia: Shang *et al.*, 1990), *P. laurifolia* Ledebour

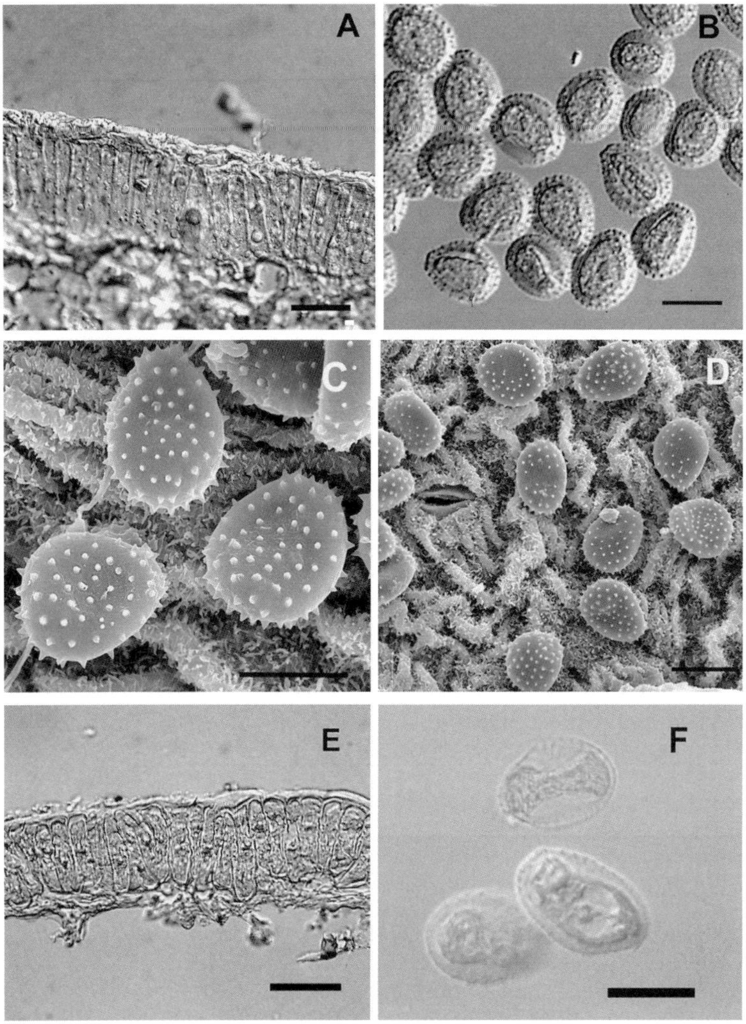

Fig. 8.2. Urediniospores and teliospores of *Melampsora* spp. on *Populus* observed by light microscopy and scanning electron microscopy. (A)–(D) *Melampsora laricis* on *Populus davidiana*. (A) Teliospores; (B) ovoid or ellipsoid urediniospores with walls uniformly thickened or slightly thickened laterally; (C) urediniospores with echinulate surface; (D) occasionally, urediniospores have a smooth area at the equator. (E)–(F) *M. multa* on *P.* × *euramericana*. (E) Telia; (F) ellipsoid or oblong urediniospores with walls uniformly thickened or slightly thickened laterally. Bars: (A) 35 μm; (B) 20 μm; (C) 12 μm; (D) 20 μm; (E) 30 μm; (F) 30 μm.

(Inner Mongolia and north-east China: Tai, 1979; Yuan, 1984; Shang *et al.*, 1990; Zhuang, 1999), *P. lasiocarpa* Oliv. (Shaanxi: Wei and Zhuang, 1997), *P. manshurica* Nakai (Liaoning: Tai, 1979), *P. maximowiczii* Henry (north-east China: Tai, 1979; Jilin: Yuan, 1984), *P. nigra* L. (Inner Mongolia, Liaoning and north-east China: Tai, 1979; Shang *et al.*, 1990; Beijing and Jilin: Yuan, 1984) *P. nigra* var. *italica* (Moench) Koehne (Hebei, Inner Mongolia, Heilongjiang and north-east China: Tai, 1979; Shang *et al.*, 1990; Liaoning: Yuan, 1984; Shaanxi: Wei and Zhuang, 1997; Cao and Li, 1999), *P. nigra* var. *thevestina* (Dode) Bean (Inner Mongolia: Yuan, 1984; Shang *et al.*, 1990; Shaanxi: Wei and Zhuang, 1997), *P. nigra* L. × *P. hopeiensis* Hu et Chow (Shaanxi: Wei and Zhuang, 1997), *P. opera* Hsu (Beijing: Yuan, 1984;

Inner Mongolia: Shang *et al.*, 1990), *P. pioneer* (Heilongjiang: Yuan, 1984), *P. purdomii* Rehd. (Shaanxi: Wei and Zhuang, 1997; Cao and Li, 1999), *P. pseudoglauca* Wang et Fu (Tibet: Zhuang, 1986; Zhuang and Wei, 1994), *P. pseudo-simonii* Kitag. (Beijing, Inner Mongolia, Liaoning and north-east China: Tai, 1979; Yuan, 1984; Shang *et al.*, 1990), *P. pseudo-simonii* × *P. deltoides* (Beijing: Yuan, 1984), *P.* × *euramericana* 'Robusta' (Inner Mongolia and north-east China: Tai, 1979), *P. simonii* (Beijing: Liu and Wang, 1939; Hebei, Shanxi, Inner Mongolia, Liaoning, Jilin, Heilongjiang and Henan: Tai, 1979; Yuan, 1984; Inner Mongolia: Shang *et al.* 1990; Shaanxi: Wei and Zhuang, 1997; Cao and Li, 1999), *P. simonii* var. *shomliflia* (Beijing: Yuan, 1984), *P. simonii* × *P. davidiana* (Inner Mongolia: Shang *et al.*, 1990), *P.* sp. (Tibet: Zhuang, 1986), *P. szechuanica*, *P. trichocarpa* (Beijing and Inner Mongolia: Yuan, 1984), *P. ussuriensis* Kom. (Inner Mongolia, Heilongjiang and north-east China: Tai, 1979; Yuan, 1984; Shang *et al.*, 1990), *P.* × *xiaohei* Hwang et Liang (Inner Mongolia: Shang *et al.*, 1990), *P.* × *xiaozuanica* Hsu et Liang (Inner Mongolia: Shang *et al.*, 1990), *P. yunnanensis* Dode (Yunnan: Tai, 1979), and most cultivars of *Argeiros* and *Tacamahaca* and some hybrids between *Argeiros* and *Tacamahaca* (Tai, 1979; Yuan, 1984; Shang *et al.*, 1990).

Distribution: China, Europe, Japan, India, Korea, Mongolia, New Zealand, Australia, South America, USA and Russia.

M. larici-populina is similar to *M. allii-populina* in morphology (Figs 8.1F, 8.3C), but readily distinguishable from other poplar rusts, such as *M. allii-populina*, by having epiphyllous telia (Fig. 8.3A), elongate urediniospores with distinct laterally thickened walls (Fig. 8.3B), and paraphyses with apical walls thickened up to 25 μm. In China, this species is the most prevalent and widely distributed of all the poplar pathogens. It also has the greatest range of telial hosts and is one of the most damaging forest tree diseases in man-made forests in northern China. In inoculation experiments (Pei and Shang, 1984), its aeciospores infected all the tested species of section *Tacamahaca* and most of the species of section *Aigeiros*, and the hybrids within or between these two sections. *M. larici-populina* did not infect the species of section *Leuce* or section *Turanga* under natural conditions. Urediniospores of *M. larici-populina* infected *P. euphratica* in inoculation experiments when urediniospores collected from *Populus* × *gansuensis* Wang et Yang were used as an inoculum (Yuan, 1984). However, the urediniospores produced had uniformly thickened walls, and only a few urediniospores had slightly laterally thickened walls, differing from the spores used for inoculation.

Melampsora magnusiana Wagner, *Zeit. Pfl. –Krankh.* 7, 340; 1897

= *M. aecidioides* Plowr., *Brit. Ured. Ustil.* 241; 1889.
= *M. aecidioides* (DC) Schroet., Cohn, *Krypt. Fl. Schles.* 3 (1), 362; 1989.
= *M. rostrupii* Wagner, *Oest. Bot. Zeit.* 46, 274; 1896 (*nomen nudum*).
= *M. pulcherrima* Maire, *Myc. Bor. Afr. Fass.* 5, 108; 1914.
= *M. populnea* (Pers.) Karst., *Brdr. Kanned. Finl. Nat. Folk.* 31, 53; 1879.
= *M. tremulae* Tulasne, *Ann. Sci. Nat.* 4 (2), 95; 1854.
= *M. populnea* f. sp. *magnusiana* Bagyanarayana, *Proc. IUFRO Rust Forest Tree Working Party Conf.* 48; 1998.
= *M. populnea* f. sp. *rostrupii* Bagyanarayana, *Proc. IUFRO Rust Forest Tree Working Party Conf.* 50, 1998.

Host plants: 0, I on *Corydalis* sp. (Hebei and north-east China: Tai, 1979). II, III on *Populus alba* L. (Liaoning and Jiangsu: Tai, 1979; Shaanxi: Ge *et al.*, 1964; Beijing: Yuan, 1984; Inner Mongolia: Shang *et al.*, 1990), *P. alba* L. var. *pyramidalis* Bge. (Xinjiang: Yuan, 1984; Inner Mongolia: Yuan, 1984; Shang *et al.*, 1990), *P. davidiana* Dode (Hebei, Xinjiang and north-east China, Tai, 1979; Guizhou: Yuan, 1984; Shaanxi: Zhang *et al.*, 1997), *P. hopeiensis* Hu et Chow (Inner Mongolia: Shang *et al.*, 1990), *P. shanxiensis* Wang

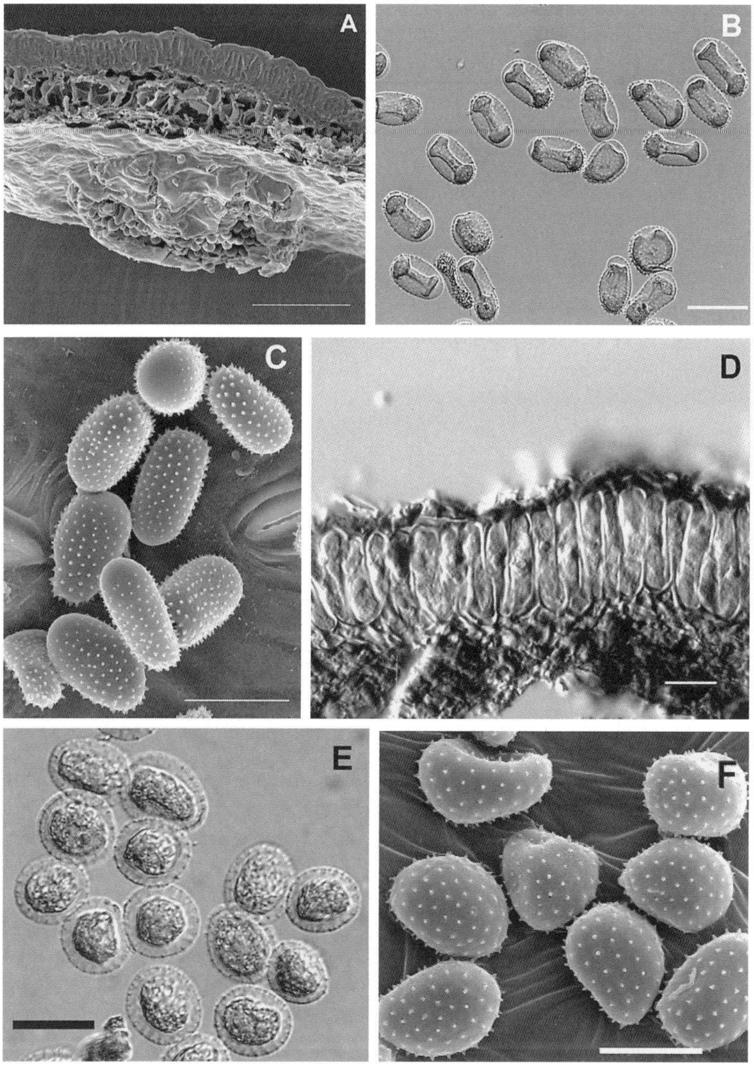

Fig. 8.3. Urediniospores and teliospores of *Melampsora* on poplars observed by light microscopy and scanning electron microscopy. (A)–(C) *Melampsora larici-populina* on *Populus* spp. (A) Hypophyllous uredinia and epiphyllous telia on *Populus opera*; (B) ellipsoid or oblong urediniospores with equatorially thickened walls on *P. laurifolia*; (C) echinulate urediniospores with a smooth apex on *P. simonii* × *P. nigra* var. *italica*. (D)–(F) *Melampsora magnusiana* Wagn. on *Populus* spp. (D) Teliospores on *P. alba* var. *pyramidalis*; (E) globose or ovate urediniospores with uniformly thickened walls on *P. tomentosa*; (F) urediniospores with echinulate surface on *P. tomentosa*. Bars: (A) 120 μm; (B) 35 μm; (C) 21 μm; (D) 35 μm; (E) 14 μm; (F) 20 μm.

et Tung (Inner Mongolia: Shang *et al.*, 1990), *P. tomentosa* Carr. (Shandong, Shaanxi, north-western China and Shanxi: Tai, 1979; Shaanxi: Wei and Zhuang, 1997; Cao and Li, 1999; Inner Mongolia: Shang *et al.*, 1990), and *Populus* spp. of section *Leuce*.

Distribution: China, eastern Europe, Russia, USA.

According to Wilson and Henderson (1966), *M. magnusiana* is morphologically very similar to *M. rostrupii* and both have the uredinial and telial stages on the same species of *Populus*. The only difference is in their aecial hosts. *M. rostrupii* was a *nomen nudum* of *M. aecidioides* and is placed under *M. populnea*. Boerema and Verhoeven (1972) reduced *M. rostrupii* to a *forma specialis* of the species *M. populnea*. On a similar basis, Bagyanarayana (1998) considered five species as *formae speciales* of *M. populnea* and treated two as *formae speciales* within *M. populnea*, i.e. f. sp. *magnusiana* and f. sp. *rostrupii*. He described the length of urediniospores of *M. populnea* as over 30 µm. However, urediniospores of *M. populnea* or *M. magnusiana* occurring on the section *Leuce* are less than 30 µm long (Wilson and Henderson, 1966).

Three rust species have been reported on *Populus tomentosa* in China: *M. magnusiana* (Wang, 1949; Zhou *et al.*, 1979), *M. rostrupii* (Ge *et al.*, 1964) and *Uredo tholopsora* (Cummins, 1951). The former two species are common in China and differ from each other only in their aecial hosts. *M. magnusiana* alternates on *Corydalis* and *Chelidonium* whereas *M. rostrupii* alternates on *Mercurialis*. In Japan, Hiratsuka (1927, 1932) established that *M. magnusiana* alternates between *Populus sieboldii* and *Chelidonium majus*. According to the records, spermogonia and aecia of *M. magnusiana* have been found on *Corydalis* in Hebei province, China (Tai, 1979). This record is doubtful as Miura (1928) reported a rust species, *Caeoma fumariae* on *Corydalis* spp. in Caohekou of Liaoning Province, but he did not establish the relationship between *M. magnusiana* and *Caeoma fumariae*. *M. yezoensis* Miyebe et Matsumoto also forms aecia on *Corydalis*. At present, host alternation in *M. magnusiana* and *M. rostrupii* in China is not clear. Field observations and results from inoculation experiments have shown that *M. magnusiana* overwinters on poplar as uredinial mycelium. Thus, it can survive and propagate itself in many areas where its alternate host is absent. In fact, *Corydalis* and *Chelidonium* are rarely infected and the teliospores of *M. magnusiana* may not be important in the life cycle.

Although *M. magnusiana* is very similar to *M. laricis* in urediniospore shape and size, uredinia of this species are epiphyllous (Fig. 8.3D), often found on leafstalks and young buds, aggregated in groups and yellow. Urediniospores of *M. magnusiana* have uniformly thickened walls (Fig. 8.3E) and the distance between spinules is 2–3 µm (Fig. 8.3F).

Zhuang (1999) reported *M. aecidioides* Plowr. (on *Populus* collected from Xinjiang province) as well as *M. magnusiana* in China. But many authors considered that these two species names refer to the same species and *M. magnusiana* was a *nomen nudum* of *M. aecidioides* (Shang *et al.*, 1990). Recently, morphological and molecular studies suggested that there are no obvious differences among specimens collected in China which had been identified as *M. rostrupii* and *M. magnusiana*, and *M. aecidioides* (Tian *et al.*, 2004).

Melampsora multa Shang, Pei et Yuan, *Acta Mycol. Sin.* suppl. I, 180–184; 1986

Host plant: II, III on *P.* × *euramericana* cv. 'Robusta', *P.* × *euramericana* cv. 'Serotina', *P.* × *euramericana* cv. 'I-214' (Liaoning: Shang *et al.*, 1986a).

Distribution: China (Liaoning).

This species was established by Shang *et al.* (1986a) based on a rust fungus on *P.* × *euramericana* which occurred in Liaoning province. The same authors reported that this species did not infect *Larix principis-ruppechtii* Mayr., *Picea wilsonii* Mast., *Picea meyeri* Rehd., *Pinus tabulaefomis* Carr. and *Sabina chinensis* (L.) Antonine in artificial inoculation experiments. Morphologically, this species resembles *M. larici-populina*, but differs from other poplar rusts by having single or 2–3-layered telia and being amphigenous (Shang *et al.*, 1986a). Our examination of the isotype of *M. multa* suggested that the morphology of this species is almost the same as that of *M. larici-populina*, although *M. multa* occasionally has two- or three-layered telia. It is possible that some specimens of *M. larici-populina* Kleb. occasionally produce two-layered telia. Further studies are needed to clarify the taxonomic status of these closely related rust fungi.

Melampsora occidentalis Jacks., *Phytopath.* 7, 354; 1917

= *Caeoma occidentalis* Arth., *Bull. Torrey Bot. Club* 34, 591; 1907.

Host plant: On *P. cathayana* Rehd. (Heilongjiang: Yuan, 1984), *P. koreana* Rehd. (Heilongjiang: Yuan, 1984), *P. yunnanensis* Dode (Yunnan: Yuan, 1984) and *P.* × *euramericana* cv. 'Marilandica' (Jilin: Yuan, 1984).

Distribution: Canada and USA.

This species was reported for the first time in China by Yuan (1984) based on three specimens on *Populus*. The fungus is characterized by having spinules at the apex of urediniospores. Usually, the smooth area at the apex of urediniospores is observed using a light microscope. However, scanning electron microscopy revealed that in some specimens of *M. larici-populina* Kleb., such areas are either smooth or have some spinules varying in size (Gao, 1982; Shang *et al.*, 1990). We also found that, in some specimens of *M. larici-populina*, many urediniospores have spinules at the apex and only a few urediniospores have a smooth area at the apex. In our phylogenetic analyses, these specimens were grouped together with other specimens of *M. larici-populina* (Tian *et al.*, 2004). Thus, we consider that the distribution of *M. occidentalis* in China is doubtful.

Melampsora populnea (Pers.) Karst., *Bidr. Kammed. Finl. Nat. Folk* 31, 53; 1879

Host plant: *Populus adenopoda* Maxim (Shaanxi: Cao and Li, 1999), *P. davidiana* Dode (Shaanxi: Cao and Li, 1999), *P. pseudoglauca* Wang et Fu (Tibet: Zhuang and Wei, 1994), *P. szechuanica* Schneid. var. tibetica Schneid. (Tibet: Zhuang and Wei, 1994), *Populus* sp. (Tibet: Zhuang and Wei, 1994).

Distribution: China, Japan, Europe.

Recently, rust species occurring on *Populus adenopoda* and *P. davidiana* were treated as *M. populnea* (Pers. ex Pers.) Karst. in China. Many authors adopted *M. populnea* as a complex species and some poplar rusts, such as *M. laricis*, *M. magnusiana*, *M. allii-populina* were included under this species (Wilson and Henderson, 1966; Boerema and Verhoeven, 1972; Bagyanarayana, 1998).

Zhuang (1986) identified some specimens occurring on *Populus pseudoglauca* collections from Tibet as *M. laricis*, and the same author also treated some other specimens on *P. pseudoglauca* collected from Tibet as *M. populnea* (Zhuang and Wei, 1994). We were not able to clarify whether these two species on *P. pseudoglauca* were different or the same in their papers. But these specimens on *P. pseudoglauca* were identified as *M. abietis-populi* based our studies (Tian *et al.*, 2004).

Zhang *et al.* (1997) reported *M. magnusiana* on *P. davidiana*, but this species cannot be distinguished from *M. populnea* in reality, and specimens on *P. davidiana* may be *M. laricis*.

Melampsora pruinosae Tranz., Tranzxchel et Serebriannikov, *Mycotheca Rossica* No. 265; 1912

Host plant: II, III on *P. euphratica* Oliv. (Inner Mongolia: Shang *et al.*, 1990; Xinjiang and Inner Mongolia: Yuan, 1984), *P. nigra* var. *italica* (Moench) Koehne (Shaanxi: Wei and Zhuang, 1997), *Populus* sp. (Xinjiang: Tai, 1979), *P. tomentosa* Carr. (Shaanxi: Wei and Zhuang, 1997), *P.* × *xiaozuanica* Hsu et Liang (Inner Mongolia: Shang *et al.*, 1990).

Distribution: China (Xinjiang, Ningxia, Inner Mongolia), Russia.

This species is similar to *M. magnusiana* in morphology, but easily distinguishable by its urediniospores, which have very small spinules, a short distance between spinules (< 2 μm; Fig. 8.4C) and thick walls (Fig. 8.4B). Geographically, this rust is distributed in China's north-western desert areas, i.e. the provinces of Xinjiang, Ningxia and Inner Mongolia. In nature, it has only been recorded on *Populus euphratica* and is non-pathogenic to poplars of the section *Leuce* and their hybrids. This rust can cause slight infection on *P. tomentosa* Carr. in artificial inoculations in nurseries, but not under natural conditions (Yuan, 1984). No alternate hosts have been found with *M. pruinosae*.

Wei and Zhuang (1997) identified several specimens on *P. tomentosa* Carr., and *P. nigra* var. *italica* (Moench) Koehne from Shaanxi province as *M. pruinosae*. However, our work suggested that these specimens should be identified as *M. magnusiana* as their urediniospores have larger spinules, the distance between spinules is longer (> 2 μm) and both uredinia and telia are amphigenous (Fig. 8.4A, C).

Melampsora rostrupii Wagn., *Oest. Bot. Zeit.* 46, 274; 1896 (*nomen nudum*)

Host plant: *Populus alba* L. (Liaoning and Jiangsu: Tai, 1979), *P. candicans* Ait (Liaoning: Tai, 1979), *P. rotundifolia* Griff. (Yunnan: Tai, 1979), *P. tomentosa* Carr. (Shaanxi: Ge *et al.*, 1964; Henan: Tai, 1979), *P. tremula* L. (Hebei: Tai, 1979), *P. tremula* L. var. *villosa* Weam (Hebei: Tai, 1979), *Populus* sp. (Yunnan: Tai, 1979).

Distribution: China, Europe, Russia.

Melampsora rostrupii has been recorded together with *M. magnusiana* on the same *Populus* species in China, such as *P. tomentosa*. This species is morphologically similar to *M. magnusiana*; the only difference being in the aecidial host. However, nobody can distinguish these two species by their morphology. Zhou *et al.* (1979) investigated the distribution of alternate hosts, i.e.

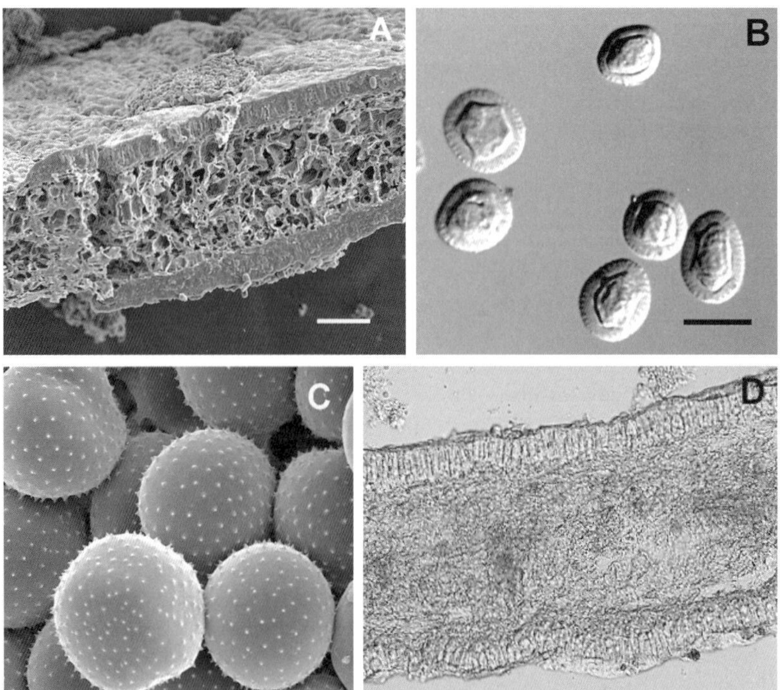

Fig. 8.4. Urediniospores and teliospores of *Melampsora pruinosae* on *Populus euphratica* observed by light microscopy and scanning electron microscopy. (A) Amphigenous uredinia and telia. (B) Globose or ovate urediniospores with uniformly thickened wall. (C) Urediniospore surface with small spinules. (D) Amphigenous telia. Bars: (A) 100 μm; (B) 20 μm; (C) 8.4 μm; (D) 65 μm.

Mercurialis; he considered that the rust on *P. tomentosa* in China is *M. magnusiana* (see the entry on *Melampsora magnusiana*).

Acknowledgements

We thank Professor Y.Z. Shang (College of Forestry, Inner Mongolia Agricultural University, China) for constructive review comments and providing many herbarium specimens. We are also grateful to Dr J.Y. Zhuang (Institute of Microbiology, Chinese Academy of Sciences), Dr Z.M. Cao and Y.M. Liang (Northwest Sci-Tech University of Agriculture and Forestry, China), Q. Wang (College of Chinese Medicinal Materials, Jilin Agricultural University, China), Professor Z.Y. Yue (Xinjiang Academic of Forestry, China) and Professor Y.D. Mo (University of Qinghai, China) for providing herbarium specimens. This work was supported by the Japan Society for the Promotion of Science (JSPS).

References

Bagyanarayana, G. (1998) The species of *Melampsora* on *Populus* (*Salicaceae*). In: Jalkanen, R., Crane, P.E., Walla, J.A. and Aalto, T. (eds) *Proceedings of the First IUFRO Rusts of Forest Trees Working Party Conference*, 2–7 August, Saariselkä. Finish Forest Research Institute, Rovaniemi, Finland, pp. 37–51.

Boerema, C.H. and Verhoeven, A.A. (1972) Check-list for scientific names of common parasitic fungi. Series la: Fungi on trees and shrubs. *Netherlands Journal of Plant Pathology* 78, 24–25.

Cao, Z.M. and Li, Z.Q. (1999) *Rust Fungi of Qinling Mountains* [in Chinese]. China Forestry Publishing House, Beijing, pp. 29–41.

Cellerino, G.P. (1999) Review of poplar diseases. (3. Disease by fungi. 3.4.1. Rusts caused by *Melampsora* spp.) Grugliasco. Available at website http://www.efor.ucl.ac.be/ipc/pub/celle01/celle01.htm (accessed 27 September 2004).

Cummins, G.B. (1951) Uredinales of continental China collected by S.Y. Cheo II. *Mycologia* 43, 78–98.

Cummins, G.B. and Hiratsuka, Y. (2003) *Illustrated Genera of Rust Fungi*, 3th edn. APS Press, Minnesota, p. 74.

Dai, Y.C. (1989) Study on poplar rusts (*Melampsora* spp.) by numerical taxonomic method [in Chinese]. *Scientia Silvae Sinicae* 25, 87–91.

Gao, Y. (1982) Scanning of the urediniospore of poplar leaf rust *Melampsora larici-populina* [in Chinese]. *Acta Microbiologica Sinica* 22, 26–30.

Ge, G.P., Jing, Y., Shen, M.M. and Shi, Q. (1964) Observations on the development and morphology of *Melampsora rostrupii* Wagner on *Populus tomentosa*. *Scientia Silvae Sinicae* 9, 221–232.

Guo, L. (1989) Uredinales of Shennongjia, China. In: Mycological and Lichenological Expedition to Shennongjia, Academia Sinica (eds) *Fungi and Lichens of Shennongjia* [in Chinese]. Beijing World Publishing Co, Beijing, p. 113.

Hiratsuka, N. (1927) Notes on Japanese species of *Melampsora* parasitic on species of *Larix* [in Japanese]. *Journal Society of Agricultural and Forestry, Sapporo* 19, 180–194.

Hiratsuka, N. (1932) Inoculation experiments with some heteroecious species of the *Melampsoraceae* in Japan. *Japanese Journal of Botany* 6, 1–33.

Hiratsuka, N. and Karube, Y. (1969) Studies on poplar leaf rusts in Japan [in Japanese]. *Bulletin Sugino Women's College* 6, 31–56.

Hiratsuka, N., Sato, S., Katsuya, K., Kakishima, M., Hiratsuka, Y., Kaneko, S., Ono, Y., Sato, T. and Harada, Y. (1992) *The Rust Flora of Japan*. Tsukuba Shuppankai, Tsukuba, Ibaraki, Japan, pp. 271–272.

Huo, J.F. and Wang, M.D. (1934) A preliminary list of fungi in north China. *Annals of the Research Council National University of Peking* 1, 1–22.

Imai, S. (1942) Damage caused by *Caeoma abietis-mayrianae* on todo-fir seedlings and the life-cycle of the pathogen. *Annals of the Phytopathological Society of Japan* 12, 68–69.

Ito, S. (1938) Uredinales-Melampsoraceae. In: *Mycological Flora of Japan*. Vol. 2. *Basidiomycetes. 2* [in Japanese]. Yokendo, Tokyo, pp. 113–117.

Kirk, P.M., Cannon, P.F., David, J.C. and Stalpers, J.A. (2001) *Ainsworth and Bisby's Dictionary of the Fungi*, 9th edn. CAB International, Wallingford, UK, pp. 311.

Liu, T.N. and Wang, Y.Z. (1935) Materials for study on rusts of China (III). *Contributions from the Institute of Botany, National Academy of Peking* 3, 353.

Miura, M. (1928) *Flora of Manchuria and East Mongolia. Part. III, Cryptogams, Fungi* [in Japanese]. South Manchuria Railway Co, pp. 230, 393.

Miyake, I. (1914) Über Chinensiche Pilze. *Botanical Magazine* 28, 37–56.

Pei, M.H. and Shang, Y.Z. (1984) Study on the leaf rust of cathay poplar caused by *Melampsora larici-tremulae* Kleb [in Chinese]. *Journal of Northeastern Forestry Institute* 12 (4), 40–49.

Shang, Y.Z. and Pei, M.H. (1984) Study on the leaf rust of Davids European Aspen caused by *Melampsora larici-tremulae* Kleb. *Journal of Northeastern Forestry Institute* 12 (3), 47–55.

Shang, Y.Z., Pei, M.H. and Yuan, Z.W. (1986a) A new rust fungus on poplars [in Chinese]. *Acta Mycologica Sinica* Supplement 1, 180–184.

Shang, Y.Z., Yuan, X.Y. and Hao, J.Z. (1986b) Taxonomy of *Melampsora* on *Populus* [in Chinese]. *Journal of Inner Mongolia Forestry College* 8, 126–133.

Shang, Y.Z., Hao, J.Z. and Yuan, X.Y. (1990) *Melampsora* on poplars in Nei Monggol [in Chinese]. *Acta Agriculturae Boreali-Sinica* 5, 86–92.

Tai, F.L. (1979) *Sylloge Fungorum Sinicorum* [in Chinese]. Science Press, Academia Sinica, Beijing, pp. 537–539.

Tian, C.M., Shang, Y.Z., Zhuang, J.Y., Wang, Q. and Kakishima, M. (2004) Morphological and molecular phylogenetic analysis of *Melampsora* species on poplars in China. *Mycoscience* 45, 56–66.

Van Kraayenoord, C.W.S., Laundon, G.F. and Spiers, A.G. (1974) Poplar rusts invade New Zealand. *Plant Disease Reporter* 58, 423–427.

Wang, Y.C. (1949) Uredinales of Shensi. *Contributions from the Institute of Botany, National Academy of Peking* 6, 224–225.

Wang, Y.C., Zang, M., Ma, Q.M. *et al.* (ed.) *Fungi of Xizang (Tibet)* [in Chinese]. Science Press, Academia Sinica, Beijing, pp. 39.

Wei, S.X. and Zhuang, J.Y. (1997) Species checklist of uredinales of the Qinling Mountains [in Chinese]. In: Mao, X.L. and Zhuang, J.Y. (ed.) *Fungi of the Qinling Mountains*. China Agricultural Science and Technology Publishing House, Beijing, pp. 37–38.

Wilson, M. and Henderson, D.M. (1966) *The British Rust Fungi*. Cambridge University Press, London, pp. 64–93.

Yuan, Y. (1984) On identification of some species of *Melampsora* infecting poplar trees in China and a test of host range of *M. pruinosae* [in Chinese]. *Journal of Beijing Forestry College* 6, 48–82.

Zhang, N., Zhuang, J.Y. and Wei, S.X. (1997) Fungal flora of the Daba Mountains: Uredinales. *Mycotaxon* 61, 59–61.

Zhou, Z.M., Shen, R.X., Huo, Z. and Lei, Z.P. (1979) Study on the tolerance of *Populus tomentosa* to *Melampsora magnusiana* [in Chinese]. *Journal of Beijing Forestry College* 1, 61–66.

Zhuang, J.Y. (1986) Uredinales from East Himalaya. *Acta Microbiologica Sinica* 5, 77.

Zhuang, J.Y. (1999) Rust fungi from the Altai. *Journal of Anhui Agricultural University* 26, 261–265.

Zhuang, J.Y. and Wei, S.X. (1994) An annotated checklist of rust fungi from the Mt. Qomolangma region (Tibetan Everest Himalaya). *Mycosystema* 7, 45.

Zhuang, J.Y. and Wei, S.X. (2002) A preliminary checklist of rust fungi in the Greater Khingan Mountains. *Journal of Jilin Agricultural University* 24, 6.

9 Current Status of Poplar Leaf Rust in India

R.C. Sharma[1], S. Sharma[1] and K.R. Sharma[2]
[1]*Department of Mycology and Plant Pathology,* [2]*Department of Forest Products, Dr Y.S. Parmar University of Horticulture and Forestry, Nauni, Solan-173 230 Himachal Pradesh, India*

Poplars and Poplar Rusts in India

The genus *Populus* represents a significant component of the world's potential renewable resources for the 21st century. In their natural range, poplars occur interspersed throughout the forests of temperate regions. Six species of *Populus* are indigenous to India, of which *P. ciliata* Wall. ex Royle is fast growing and offers a potential to meet the increasing demand of wood for the plywood industry, pulp industry, packaging and other industrial uses. This species is well distributed in temperate and sub-temperate regions of the Himalayas between 1200 and 3500 m above mean sea level. *Pinus roxburghii, Quercus* spp., *Cedrus deodara, Pinus wallichiana, Abies pindrow, Picea smithiana, Pinus gerardiana, Salix* spp. and *Hippophae* spp. are the other important forest tree species in this zone. In addition to indigenous poplar species, a number of exotic hybrids/clones developed from *P. deltoides, P. alba, P. nigra, P. tremula, P.* × *euramericana, P. yunnanensis* and *P. maximowiczii* are also grown in the Himalayas and northern Indian plains.

The most predominant *Melampsora* rust species causing poplar leaf rust in India is *M. ciliata* Barclay, which was first reported from the vicinity of Shimla in 1891 (Barclay, 1891). Another species, *M. populnea* (Pers.) Karst. (syn. *M. aecididoides* (DC) Schroet, *M. rostrupii* Wagner) on *P. alba* was first reported in 1943 from the north-west Himalayas (Cummins, 1943). *M. larici-populina* Kleb. was first recorded on exotic poplars in 1992 from Uttar Pradesh (Pandey, 1992). Recently, *M. populina* (Jacq.) Lev. in its uredinial stage was recorded on six poplar species from Himachal Pradesh and Jammu and Kashmir, although this still needs to be confirmed (Vattiprollu and Agarwal, 2002). In India, major work has been done on *M. ciliata*, which will be reviewed in this chapter.

Distribution and Hosts

M. ciliata is a microcyclic rust species recorded from the Indian and the Nepal Himalayas only (Khan, 1994; Vannini *et al.*, 1995; Sharma and Sharma, 2000). This rust species is widely distributed in the north-west Himalayas and causes moderate to severe damage (Bakshi and Singh, 1961; Singh *et al.*, 1983). In the north-west Himalayas, *M. ciliata* is prevalent from an altitude of 700 m to 3130 m above mean sea level. The rust has not been recorded below 700 m or above 3130 m. The climate at this altitude varies from sub-temperate with 600–1000 mm annual rainfall to dry, cold desert, where 3–5 m snow is experienced during winters. Defoliation usually occurs

 Rust Diseases of Willow and Poplar (eds M.H. Pei and A.R. McCracken)

when 50% or more of leaf area is covered by the rust pustules. In nurseries, leaf shedding due to rust infection is 1–2 months earlier than the normal leaf shedding period (Sharma and Sharma, 2000). Only two stages of the pathogen (uredinial and telial) are known. *M. ciliata* has been recorded on *P. ciliata*, *P. deltoides*, *P. alba*, *P. yunnanensis*, *P.* × *euramericana*, *P. nigra*, *P. tremula* and *P. trichocarpa*, and hybrids developed from *P. maximowiczii* × *P. berolinensis* and *P. nigra* × *P. trichocarpa* (Sharma and Sharma, 2000, 2002; Sharma *et al.*, 2001a). This rust occurs in native forests of *P. ciliata*, nurseries and plantations of indigenous and exotic poplars.

Another poplar rust recorded on *P. alba* in India is caused by *M. populnea* (Pers.) Karst (Bakshi and Singh, 1966). Pandey (1992) recorded the occurrence of *M. larici-populina* Kleb. on *P. deltoides* clones. The uredinial stage of *M. populina* (Jacq.) Lev. has been reported on *P. capsica*, *P. casale*, *P. ciliata*, *P. eugenei*, *P. oxfordii* and *P. robusta* from Jammu and Kashmir and Himachal Pradesh.

Epidemiology

At 1300 m above mean sea level, at Nauni, rust was first recorded on fully grown trees of *P. deltoides* 'Lux' in April. However, at the same site in the nursery, rust was first observed on *P. ciliata* 'Theog'. At 1500 m above mean sea level in Shilly nursery, rust was recorded in July on *P. ciliata*. The initial appearance of the rust disease varied according to the altitude and plants grown (Khan, 1994; Sharma and Khan, 2000).

The disease remains static during the summer months (May and June) and spreads rapidly during August, September and October, when the relative humidity is high and the temperature starts decreasing. Development of leaf rust (*M. ciliata*) in different species, hybrids and cultivars of poplar in the nursery revealed that the disease index of 9.4% on 10 August increased to 70.0% on 10 October. The lowest disease index (5.5%) was recorded on *P. yunnanensis*, whereas the maximum was recorded on *P.* × *euramericana* 'Rubra Poiret' (59.5%). The apparent infection rate per unit area per day was highest in *P. nigra* × *P. trichocarpa* (9.7×10^{-2}), whereas the minimum occurred in *P. maximowiczii* × *P. berolinensis* 'Oxford' (3.3×10^{-2}). The area under the disease progress curve was maximum in *P.* × *euramericana* 'Rubra Poiret' (26.7) and minimum in *P. yunnanensis* (1.6). Inoculum production was highest in *P. nigra* × *P. trichocarpa* and lowest in *P. maximowiczii* × *P. berolinensis* 'Oxford' (Sharma *et al.*, 2001a). *P. ciliata* clones, which remained disease free during July, became infected in the month of August. Hybrids of *P. ciliata* × *P. maximowiczii* also differed in their susceptibility to *M. ciliata* (Sharma *et al.*, 2002).

Between six temperature ranges and five levels of relative humidity, a temperature of 20°C and 100% relative humidity were best for the germination of urediniospores of *M. ciliata*. Low (5°C) and high (30°C) temperature and relative humidity of 90% did not favour the germination of urediniospores. Inoculum production (urediniospore/uredinium and urediniospores/unit leaf area) was affected adversely by maximum and minimum temperature but favourably by higher relative humidity and rainfall. Viability of urediniospores stored at 20°C or higher was lost within 10 days. Temperature and relative humidity adversely affected urediniospore viability (Sharma and Sharma, 2001).

The intensity of leaf rust is influenced by the stomatal density in poplar leaves. A study on the relationship between stomatal density/pore size and the incidence of leaf rust in 12 *P. ciliata* hybrids and clones with varying degrees of susceptibility revealed that stomatal density was positively and significantly correlated with disease incidence, whereas pore size was not (Sharma *et al.*, 2001b). Variation in the timing of initiation of leaf rust (*M. ciliata*) was correlated with variation in the leaf flushing time of poplar genotypes (Sharma and Sharma, 2004).

Variability in *Melampsora ciliata*

It has been observed that there is variation in the dimensions of urediniospores of *M. ciliata* collected from various poplar species grown under different environmental conditions. The smallest urediniospores (20.88 × 15.43 μm) produced on *P. ciliata* were recorded from a site of wet, temperate, high hills, while the largest (36.30 × 19.97 μm) were from sub-temperate, sub-humid, mid hills. Urediniospores produced on *P. deltoides* and *P. nigra* were smaller in a sub-temperate, sub-humid, mid hill zone than in a wet, temperate, high hill zone. Variations have also been recorded in the dimensions of urediniospores produced on *P. alba* in a dry, temperate zone (Sharma *et al.*, 2001c).

Influence of Host Genotype and Pathogen Isolate on Disease Development

In vitro inoculation tests using field collections of *M. ciliata* revealed that there are significant interactions between host genotype and pathogen isolate in determining the infection type and disease level. Shorter incubation period for flecking (IPF), latent period for eruption of first uredinium (LPU) and maximum uredinia per leaf disc (ULD) were associated with *P. deltoides* 'Lux' and *P. deltoides* '19', while longer IPF and LPU and minimum ULD were associated with *P. ciliata* × *P. maximowiczii* 'CMZ-26'. Flecking was not observed on *P.alba*, while on *P. yunnanensis*, only flecking appeared but no uredinia erupted. IPF was positively correlated (0.866) with the time periods required for eruption of first uredinium and eruption of 50% uredinia (LPU_{50}) while it was negatively correlated with ULD. Rust isolates I_1 and I_2 produced only flecking, while I_3 and I_4 could produce well-developed pustules. Time taken to show flecking, eruption of first and 50% uredinia by isolate I_4 was shorter when compared to isolate I_3, which produced more uredinia per leaf disc. Variation in the dimensions of urediniospores of various isolates was also observed. The smallest urediniospores were recorded in isolate I_4, while the largest were in I_2 (Sharma and Sharma, 2003).

Resistance

Sixty-nine clones and 84 hybrids of *P. ciliata* (UHF Selections), 141 clones, 7 hybrids and 48 families of *P. deltoides*, 6 clones of *P.* × *euramericana*, 11 cultivars and 6 species of *Populus* were screened under natural epiphytotic conditions at two locations. In *P. ciliata* all the clones were susceptible, while 14 hybrids remained disease-free. One hundred and fifteen clones, all hybrids and families of *P. deltoides* and all clones of *P.* × *euramericana* remained rust-free. Eight cultivars, *P. trichocarpa* and *P. nigra* × *P. trichocarpa* also remained disease-free (Sharma, 2000).

Biological Control

Control of plant diseases by the use of biocontrol agents, rather than with chemical sprays alone, is a supplementary measure to reduce impacts of disease. With a forest crop, a long rotation and low economic returns, biological control is a particularly attractive option. Fungal antagonists, *Alternaria alternata* and *Cladosporium oxysporum* were isolated from the phylloplane of poplar leaves infected with *M. ciliata*. Pre-inoculation with these antagonists resulted in minimum germination of urediniospores, minimum eruption of uredinia per unit leaf area and maximum parasitization of urediniospores and uredinia. *A. alternata* was found to be more parasitic than *C. oxysporum*. Reduction in viability of urediniospores was more obvious with *A. alternata* than *C. oxysporum*, and this potential may be exploited in the biocontrol of this disease (Sharma *et al.*, 2002).

Chemical Control

Successful control of leaf rust (*M. ciliata*) of *P. ciliata* in nursery seedlings was achieved by spraying carbendazim and mancozeb (Khan *et al.*, 1988). Post-symptom sprays of difenoconazole (0.02%), penconazole (0.06%) and carbendazim (0.05%) exhibited excellent eradicant activity and resulted in minimum production of urediniospores per uredinium and per unit leaf area when applied on nursery-grown poplar seedlings. When fungicides were tested as preventive and post-symptom sprays at fortnightly intervals, difenoconazole, penconazole and hexaconazole resulted in the least disease incidence and uredinia pustules per leaf, and minimum rate of spread of disease and area under disease progress curve (Sharma, 2000).

Conclusions

Leaf rust caused by an autoecious, microcylic *M. ciliata* is an important disease of *Populus* species, especially on the Himalayan poplars (*P. ciliata* in particular) in India. In the north-west Himalayas, this rust species is prevalent from an altitude of 700 m to 3130 m above mean sea level, where the climate varies from sub-tropical to temperate, cold desert. This rust species also infects a number of other exotic hybrids and clones of poplar. The disease initiation and incidence is influenced by the altitude and type of poplar species/hybrids/clones raised. A temperature of 20°C and 100% relative humidity is most favourable for urediniospore germination. The extent of disease is also influenced by the stomatal density. Differential interaction between host genotypes and pathogen isolates reveals the existence of races in *M. ciliata*. Difenoconazole, penconazole and hexaconazole fungicides are most effective in checking the disease in nursery-grown poplars.

References

Bakshi, B.K. and Singh, S. (1961) New and noteworthy records of some mildews and rusts on Indian trees. *Indian Forester* 87, 547.

Bakshi, B.K. and Singh, S. (1966) Rusts on Indian forest trees. *Indian Forest Record (New Series)* 2, 139–204.

Barclay, A. (1891) Additional urediniae from the neighbourhood of Simla. *Journal of Asiatic Society of Bengal* 60, 211–230.

Cummins, G.B. (1943) Uredinales from north west Himalayas. *Mycologia* 35, 446–458.

Khan, S.N., Rehill, P.S., Tiwari, R.K., Rawat, D.S. and Misra, B.M. (1988) Control of poplar rust, *Melampsora ciliata* in nurseries. *Indian Journal of Forestry* 11, 253–255.

Khan, Y. (1994) Studies on *Melampsora* leaf rust of *Populus* species. M.Sc. thesis, University of Horticulture and Forestry, Solan, India.

Pandey, P.C. (1992) Occurrence of destructive rust parasite on exotic poplars in India. *Indian Forester* 118, 168.

Sharma, R.C. and Khan, Y. (2000) Incidence and severity of *Melampsora* leaf rust of poplar. In: *Proceedings of Indian Phytopathological Society – Golden Jubilee International Conference on Integrated Plant Disease Management for Sustainable Agriculture*, IPS, New Delhi, India, pp. 970–971.

Sharma, R.C. and Sharma, S. (2000) Status and distribution of foliar diseases of poplar in Himachal Pradesh. *Indian Phytopathology* 53, 57–60.

Sharma, R.C., Khan, Y., Sharma, S. and Malhotra, R. (2001a) Development of *Melampsora ciliata* rust on nursery-grown poplars in north-western Himalayas. *Forest Pathology* 31, 313–319.

Sharma, R.C., Sharma, S. and Gupta, A.K. (2001b) Stomatal characters of *Populus ciliata* in relation to leaf rust and growth parameters. *Phytomorphology* 51, 199–205.

Sharma, S. (2000) Studies on epidemiology and management of poplar leaf rust. Ph.D. thesis, University of Horticulture and Forestry, Solan, India.

Sharma, S. and Sharma, R.C. (2001) Epidemiology of *Melampsora ciliata* leaf rust of poplars in India. *Zeitschrift für Pflanzenkrankheiten und Pflanzenschutz* 108, 337–344.

Sharma, S. and Sharma, R.C. (2002) Prevalence of poplar leaf rust in Himachal Pradesh. *Indian Phytopathology* 55, 81–83.

Sharma, S. and Sharma, R.C. (2003) Influence of host genotype and pathogen isolate in the development of poplar leaf rust. *Zeitschrift für Pflanzenkrankheiten und Pflanzenschutz* 110, 359–365.

Sharma, S. and Sharma, R.C. (2004) Relationship between leaf flushing time and initiation of poplar leaf rust. *Journal of Tropical Forest Science* 16, 369–372.

Sharma, S., Sharma, R.C. and Sharma, J.N. (2001c) Morphological variability in *Melampsora ciliata* – The incitant of poplar leaf rust. *Indian Forester* 127, 242–248.

Sharma, S., Sharma, R.C. and Malhotra, R. (2002) Effect of saprophytic fungi *Alternaria alternata* and *Cladosporium oxysporum* on germination, parasitism and viability of *Melampsora ciliata* urediniospores. *Zeitschrift für Pflanzenkrankheiten und Pflanzenschutz* 109, 291–300.

Singh, S., Khan, S.N. and Misra, B.M. (1983) Status of *Melampsora* rusts of poplars in India. *Indian Forester* 109, 743–747.

Vannini, A., Cecco, D., Monaci, N. and Anselmi, N. (1995) Main tree pathogens of western Himalayan forests in Nepal: Description and risk of introduction. *Bulletin OEPP* 25, 455–461.

Vattiprollu, P.K. and Agarwal, D.K. (2002) *Melampsora*. *Indian Phytopathology* 55, 248.

10 *Melampsora* Willow Rust in Chile and Northern Europe: Part of a Metapopulation?

Mauritz Ramstedt and Sergio Hurtado
Plant Pathology and Biocontrol Unit, Swedish University of Agricultural Sciences, PO Box 7035, S-75007 Uppsala, Sweden

Background

As the cultivation of willow grows globally, worldwide monitoring of pathotypes for *Melampsora* rust, the most serious disease of willow, becomes increasingly important. For safe introduction of promising new plant material from breeding work, the possible exchange of inocula for the rust between continents has to be taken into account. New clones exhibiting good resistance properties might have a rapid breakdown if new unexpected pathotypes and virulence factors appear.

In Chile, although small local willow plantations have long been used for basket making, only recently has there been increasing interest in growing willows for commercial production. This naturally leads to much larger monoclonal plantations, compared to earlier more diversified populations of willow hosts for the rust. Each growing season in Chile, rust infects both natural and cultivated stands of willows heavily (Hurtado, unpublished). Variability in virulence and aggressiveness of *Melampsora* rust is one of the main problems for diagnostics in rust control, especially for *Salix viminalis*, the species most widely grown in willow plantations.

To study the genetics underlying pathotype dynamics, molecular genetic tools are needed. Molecular markers are used increasingly to characterize the genetics of fungal plant–pathogen populations and to evaluate the levels of genetic diversity and phylogenetic relationships within and between species (Majer *et al.*, 1996). They have also been used in attempts to identify particular races or pathotypes, but this has proven to be somewhat more difficult.

Samils *et al.* (2001, 2002) have described the structure and variation of willow rust from different locations in Sweden and Northern Ireland. Complementing their study, Hurtado and Ramstedt (2002) compared the *Melampsora* rust populations at four distant locations by amplification fragment length polymorphism (AFLP) fingerprinting, as well as pathogenicity testing (Hurtado and Ramstedt, 2003). Such comparisons will facilitate our understanding of pathotype dynamics and the possible impact of long-distance migration of spores on those dynamics. Such studies are also required for improvement of management of willow rust and safer introduction of new cultivars of willows from breeding.

The Problems with Rust

For low-input tree crops such as willow, disease control by chemical means is

(eds M.H. Pei and A.R. McCracken)

difficult and economically and environmentally unrealistic. The use of resistant or tolerant clones is therefore the priority, and breeding and development of a desirable planting design become important issues. One problem encountered is that it is assumed that a willow will be in the ground for 15–25 years, while its enemy, the rust, develops very quickly. With recombination every year, with countless numbers of spores, the rust can always be one step ahead in selection and adaptation to hosts present. Rusts are also known for their great heterogeneity, with a wide range of virulence spectra, even in relatively small populations. Different pathotypes attack different willow clones/species. Pathotypes can have a narrow or broad host range, even though the European rusts on short-rotation forest (SRF) willows are mainly restricted within the host range of their respective *formae speciales* (Pei *et al.*, 1996) with a limited diversity of potential hosts (Tables 10.1 and 10.2). Chilean rusts have not been found to follow the same pattern and seem to have much higher multi-virulent properties, the same isolate being compatible on several species of willows (Table 10.3). The nature of this virulence difference is not known, reliable markers to separate the specificity for different clones or groups of clones have not yet been found. Whether there exists a genotype–pathotype correlation for *M. epitea*, and therefore a correlation between genetic diversity and pathotype diversity, remains to be investigated.

In Sweden, where breeding for willow resistance has gone on since 1986 (Gullberg, 1989), the high variability of genotype virulence in rust populations also makes evaluation of new willow clones in field trials troublesome. Field clone trials will not always be exposed to the rust of interest, since it might not be present in high enough number at the time or site of the trial. When the compatible willow clone appears in greater numbers/larger plantations, this new pathotype might be able to multiply rapidly, resulting in unexpected problems for the new clone. To have a sustainable breeding system, therefore, rust testing must also be conducted in the lab using all relevant virulence factors possible in the rust population.

Pathogenicity/Aggressiveness/Diversity

A range of rust pathotypes, divided into *formae speciales*, have been reported from the UK and Sweden (Pei *et al.*, 1996; Ramstedt *et al.*, 2002). Although the discriminating resistance mechanisms are not fully understood, different stages of interactions during germination and penetration have been described by Hurtado and Ramstedt (2000) and shown to be non-specific in several aspects.

An important question to be answered is what role the local adaptation of the pathogen and the long-distance dispersion play in development of new pathotypes, and possibly new virulence factors. Hubbes (1983) reported that approximately 20 possible species of *Melampsora* rust on willows existed worldwide, each with its own host preference and virulence pattern. *M. epitea* Thüm, the main *Melampsora* species infecting willow plantations in Europe (Royle and Ostry, 1995), is considered a collective species that was supposed to be divided into varieties of races, subspecies or *formae speciales* (Hylander *et al.*, 1953; Wilson and Henderson, 1966; Royle and Hubbes, 1992).

In SRF plantations in Europe, *M. epitea* comprises a number of pathotypes which differ in virulence and aggressiveness on *Salix* (Ramstedt, 1989, 1999; Pei *et al.* 1996; Ramstedt *et al.*, 2002). In most cases, these pathotypes are morphologically indistinguishable from each other and have the same alternate host. A pathotype is considered to be a heterogeneous group of genotypes, only having in common the same pattern of clone-specific virulence for a given set of clones. Pathotyping is therefore a means of monitoring the current state of interaction between pathogen virulence and plant resistance for a pathogen population and host-plant population.

When discussing pathotypes, it is important to remember that this classification is

Table 10.1. Pathotypes found in Sweden 1991–96.

			Test clones								
Pathotype			1	2	3	4	5	6	7	8	9
f. sp.	No.	Code	Mull	Q83	Stip	B.H.	149	Calo	Kors	Him	139
LD	1	0.0.0									
	2	0.0.4									+
LR	3	0.3.0						+	+		
	4	0.3.2					**+**	+	+		
	5	0.3.4						+	+		+
	6	0.7.0						+	+	+	
	7	0.7.2					**+**	+	+	+	
	8	0.7.4						+	+	+	+
	9	1.7.0	**+**					+	+	+	
	10	4.1.6			**+**		**+**	+			+
	11	4.3.0			**+**			+	+		
	12	4.7.0			**+**			+	+	+	
	13	4.7.2			**+**		**+**	+	+	+	
	14	4.7.4			**+**			+	+	+	+
LET	15	5.0.0	+		+						
	16	5.0.1	+		+	+					
	17	5.0.2	+		+		+				
	18	5.0.3	+		+	+	+				
	19	5.0.7	+		+	+	+				+
	20	5.1.2	+		+		+	**+**			
	21	5.1.3	+		+	+	+	**+**			
	22	5.1.6	+		+		+	**+**			+
	23	5.1.7	+		+	+	+	**+**			+
	24	5.4.3	+		+		+			**+**	
	25	6.1.0		+	+			**+**			
	26	6.1.6		+	+		+	**+**			+
	27	7.0.2	+	+	+		+				
	28	7.0.3	+	+	+	+	+				
	29	7.0.7	+	+	+	+	+				+
	30	7.1.1	+	+	+	+		**+**			
	31	7.1.3	+	+	+	+	+	**+**			
	32	7.1.5	+	+	+	+		**+**			+
	33	7.1.6	+	+	+		+	**+**			+
	34	7.1.7	+	+	+	+	+	**+**			+
	35	7.4.2	+	+	+		+			**+**	
	36	7.4.3	+	+	+	+	+			**+**	
	37	7.5.3	+	+	+	+	+	**+**		**+**	
Virulence factor			1	2	3	4	5	6	7	8	9

+, the pathotype shows virulence on the test clone.
+, indicate recombinations between the main pathotype, groups (*formae speciales*) LR (*M. larici-retusue*) and LET (*M. larici-epitea typica*). LD, *M. larici-daphnoides*.

not a measure for genetic relatedness, since virulence pattern (pathotypes) is only one minor part of the fungus genome. This means that similar pathotypes not only might be, but most probably are, different genotypes. The members of the same pathotype may have different origin and, in some cases, may not be closely related. Although those classifications sometimes do crudely reflect the relatedness, and

Table 10.2. Clones used for the pathotype coding.

No.	Clone	Species	Virulence values[a]	*Formae speciales*[b]
European willow differential				
1	Mullatin	*S. viminalis*	1	LET
2	Q 83	*S. triandra × viminalis*	2	LET
3	Stipularis	*S. aurita × viminalis × caprea*	4	LET (LR)
4	*Calodendron*	*S. caprea × cinerea × viminalis*	1	LR/LET
5	Korso	*S. burjatica*	2	LR
6	Himalayas	*S. disperma*	4	LR (LET)
7	Bowles Hybrid	*S. viminalis*	1	LET
8	78149	*S. viminalis × caprea*	2	LET
9	78139	*S. daphnoides*	4	LD/LR/LET
Chilean willow differential				
10	Chimbarongo	*S. viminalis*	1	
11	Chuyaca	*S. triandra × viminalis*	2	
12	Valdivia	*S. aurita × viminalis × caprea*	4	
13	L. de Aculeo	*S. caprea × cinerea × viminalis*	1	
14	Cascadas	*S. burjatica*	2	
15	Pte. Chanco	*S. disperma*	4	
16	Santiago	*S. viminalis*	1	
17	Llanquihue	*S. viminalis × caprea*	2	
18	San Javier	*S. daphnoides*	4	

[a]The virulence code values for a positive reaction are added together within each subgroup of test clones to give a three-digit code for virulence, e.g. **7.5.3** (1 + 2 + 4 = **7**, 1 + 4 = **5**, 1 + 2 = **3**)
[b]LET, *larici-epitea typica*; LR, *larici retusae*; LD, *larici-daphnoides*.

amounts of genetic diversity (but only concerning virulence), they do not represent amounts of general diversity. To describe the genetic diversity accurately, molecular tools are required (Samils *et al.*, 2001; Hurtado and Ramstedt, 2002). The specificity found in European willow rust is, to a large extent, expressed by grouping them into different *formae speciales* (f. sp.). The host range of each *forma specialis* is limited to certain species of willow. Within a *forma specialis*, the rust genotypes differ mainly quantitatively but not so much qualitatively in virulence properties.

To overcome the problem due to the rapid selection and adaptation of rust virulence to previously tolerant or resistant clones, mixture plantations have been established using different clones, instead of planting traditional monoclonal plantation (McCracken and Dawson, 1997, 1998). In such mixtures, as the hosts have different susceptibility patterns, the rust is expected not only to develop more slowly during the season but also to have more difficulties in selection for more aggressive 'take-all' pathotypes, such as those that can be found in monoclonal plantations. Multi-virulent types, which are required for a successful attack of all clones in a mixture, are supposed to be less aggressive than pathotypes specialized for one or a few hosts.

However, in our studies we have found that the level of multi-virulence of *Melampsora* rust does not seem to affect the level of aggressiveness (unpublished) in a negative way, thereby making the defence somewhat more vulnerable. Rust isolates, which, in laboratory tests, were compatible with a range of different clones or species (*S. viminalis*, *S. dasyclados*, *S. daphnoides*, *S. disperma* and *S. caprea*), did not give a lower disease index than the isolates with a more narrow host spectrum.

Nevertheless, the clonal mixtures have, over a number of years, performed well,

Table 10.3. Pathotypes found in Chile 1997–98 according to Swedish willow differential.

	Pathotype		Test clones									
f. sp.	No.	Code	Mull	Q83	Stip	B.H.	149	Calo	Kors	Him	139	
	1	0.0.0										s
	2	0.0.2					+					
LR	3	0.2.0							+			
	4	0.3.0						+	+			s
	5	0.3.2					+	+	+			s
	6	0.4.0								+		
X	7	1.2.0	+						+			
LET	8	1.0.0	+									
	9	1.0.2	+				+					
	10	2.0.2		+			+					
	11	1.0.4	+								+	
	12	2.2.4		+					+		+	
	13	4.3.3			+	+	+	+	+			
	14	5.0.3	+		+	+	+					s
	15	5.1.2	+		+		+	+				s
	16	5.1.3	+		+	+	+	+				s
	17	5.2.1	+		+	+			+			
	18	7.0.0	+	+	+							
	19	7.0.4	+	+	+						+	
	20	7.0.7	+	+	+	+	+				+	s
	21	7.1.3	+	+	+	+	+	+				s
	22	7.4.2	+	+	+		+			+		s
!	23	7.7.7	+	+	+	+	+	+	+	+	+	
Virulence factor			1	2	3	7	8	4	5	6	9	

+, the pathotype shows virulence against the test clone.
s, pathotypes also found in Sweden.
LR, *M. larici-retusae*; X, atypical *formae speciales*, pathogenic to one type-species from each f. sp. group; LET, *M. larici-epitae typica*; !, unexpected multivirulent f. sp., pathogenic to all species and clones of all f. sp. groups tested.

giving higher harvest and lower rust incidence (McCracken and Dawson, 1997, 1998). This indicates that the rust might have difficulties in adapting to the mixtures, or in competing with other rust pathotypes in the population. Another possibility is that, hitherto, the trials have been too small, as compared to all commercial monoclonal plantations, to affect the total rust population. Therefore, an important task in planning commercial mixtures would be to have a variation of mixture set-ups that would minimize the risk of adaptation of suitable rust pathotypes to the set-up of hosts.

Population Structure

Uncontrolled long-distance exchange of fungal inocula between countries may occur, since spores can be transported long distances by nature (wind; water; animal carriers, such as migrating birds; etc.) and humans (e.g. transportation of willow material). Within countries, long-distance inoculum exchange by wind-borne rust spores almost certainly plays a role in the local population structure of *M. epitea*, and in local as well as global evolutionary dynamics of pathotype structure.

From Chile's rust population, 64 out of 89 isolates studied, constituting 23 different pathotypes, were compatible on the European test clones (Table 10.4). Non-compatible isolates were from willow species expected to harbour other *formae speciales* of the rust, and neither were these compatible on several other Chilean clones. This was the case particularly with rust from the endogenous *S. humboltiana*, which was not infected by the rust from other willows (Table 10.5). All Chilean *S. viminalis* isolates (43), as well as rust from *S. caprea* and *S. cinerea*, however, were compatible on the European differentials (Table 10.4).

Table 10.4. The 23 pathotypes found among 89 isolates in Chile, according to the European willow differentials.

Species	No. of isolates tested	Pathotype code
Salix viminalis	2	1.0.4
	2	2.2.4
	3	4.3.3
	7	5.0.3 s
	2	5.1.2 s
	12	5.1.3 s
	2	7.0.0
	1	7.0.7 s
	3	7.0.4
	4	7.1.3 s
	1	7.4.2 s
	4	7.7.7
S. humboldtiana	12	0.0.0 s
	1	0.3.0 s
	1	0.3.2 s
	1	0.4.0
	5	1.0.2
S. caprea	1	0.2.0
	2	1.0.0
	1	1.2.0
	2	5.2.1
S. cinerea	1	1.0.0
	2	1.2.0
S. fragilis	2	0.0.0 s
	1	0.0.2
	2	2.0.2
S. purpurea	3	0.0.0 s
S. tortuosa	3	0.0.0 s
S. alba	4	0.0.0 s
S. babylonica	1	0.0.0 s
	1	0.2.0

s, pathotypes also found in Sweden.

The proposed *formae speciales* for European rust were, however, less pronounced in Chile (Table 10.3) than in Sweden (Table 10.1) (Ramstedt *et al.*, 2002), even if a few pathotypes seemed to dominate the Chilean rust collection. An important difference is that we found several isolates in Chile that were compatible on all clones in the Swedish differential: a multivirulence that has never been found in Europe's rust population (Ramstedt, 1998, 1999; Pei *et al.*, 1996, 1997).

Each virulence component detected earlier in Sweden's rust population was present in Chile's rust population. Isoenzyme analysis showed that the genetic variation of *Salix viminalis* (an introduced willow species) in Chile was low, compared to the endemic species *S. humboldtiana*. Comparison of rust populations for the two countries supports our hypothesis that intercontinental inoculum exchange can be a significant determinant of local pathotype structure, and consequently can be important for willow-resistance breeding. In Chile, larch is an introduced species and its presence is rare and dependent on human planting. Hence, larch plants in Chile can hardly be considered a natural alternate host for *M. epitea*. In fact, whether *M. epitea* has a natural alternate host in Chile is unknown, and therefore it is not known whether the rust has the opportunity to reproduce sexually. If sexual recombination is rare in Chile, the nature of successful strategies in gaining superior fitness in the pathogen might be different for the two countries. For instance, if reproduction is essentially asexual in Chile, then pathotype competition becomes important, as opposed to genotype competition. Hence, maintaining common pathotypes might require active intercontinental exchange between the two countries, at least sufficient for establishing a common basic strategy that is locally adaptable in both countries.

From this study we conclude that there is a striking difference between the Chilean and Swedish rust isolates. The Chilean isolates had greater multi-species virulence capacities than that found in the Swedish isolates. For example, certain Chilean

Table 10.5. Compatibility pattern of *Melampsora* rust isolates from different willow clones and species.

	Compatibility on willow test clones								
Rust isolated from	*humb.*	*alba*	*caprea*	*cinerea*	*fragilis*	*purpurea*	*tortuosa*	*vim.*	*vim.* hybr.
Chilean willows									
S. humboldtiana	+	–	(+)	(+)	(+)				
S. fragilis	–	–	+						
S. tortuosa	–	–	+	–	–				
S. babylonica	–	–	+						
S. alba	–	+							
S. caprea	–	–	+	+					
S. purpurea		–	–	+					
S. cinerea	–	–	+	+					
S. viminalis		–	(+)	+	+	(+)	(+)	+	+
Swedish willows									
S. viminalis	–	–	+	+	–	(+)	(+)	+	+
S. dayclados		–	–	+	+	–	–	–	–
S. daphnoides	–	–	(+)	+	–	–	–	–	+

humb., humboldtiana; vim., viminalis; hybr., hybrid.

isolates were able to attack both *S. viminalis* clones and pure *S. dasycladus* types ('Korso') (Table 10.3, Fig. 10.1), the pathotypes 1.2.0 and 7.7.7 being good examples (Table 10.3).

Perhaps Chile's rust population has simply made use of genetic heterogeneity in a different form from that of the Swedish population, that of more multi-species virulence. Thus Chilean rust pathotypes are less confined to specific species of willows (Table 10.3). Also, Chile's *Melampsora* rust population has yet to adapt to a markedly heterogeneous willow host population of natural willow stands. In Europe, however, large monoclonal willow plantations have encouraged *Melampsora* rust genotypes to become more clone-specific. However, increase of commercial plantations in Chile might change this, making a new screening of rust worthwhile in the near future.

In a comparison of rust from several countries, 10 French, 24 Swedish, 24 Northern Irish and 35 Chilean rust isolates were analysed using AFLP. The Swedish isolates were collected from Skåne in the south to Uppland in central Sweden. Eleven of the isolates were collected from the willow clone *S. viminalis* '78183', which is the most widely used *S. viminalis* clone in short-rotation forest plantations in the country. The French isolates were collected in a clone trial near Lectour in southern France during the summer of 1999. All isolates were taken from *S. viminalis* clones at one single location. The isolates from Northern Ireland were collected during the autumn of 1998 in a mixture trial at Castlearchdale with 20 *Salix* clones of 10 different species.

Unweighted pair group method with arithmetic mean (UPGMA) analysis of the data obtained from two-primer combinations (Fig. 10.2) did not group the isolates according to the geographical locations of their collection. Note, though, that the Jaccard coefficient is a conservative measure of similarity, compared to the simple matching coefficient, which can be as much as 0.4 times greater. These results are consistent with those of Samils *et al.* (2001, 2002) who found that, when Swedish populations 700 km apart were compared, no genetic differentiation was found between the geographically separated locations. Samils and co-workers also noted that genetic diversity was high within a population.

Gene Flow: Long-distance Migration

When geographical distance among pathogen populations correlates poorly with genetic distance, as appears to be the case

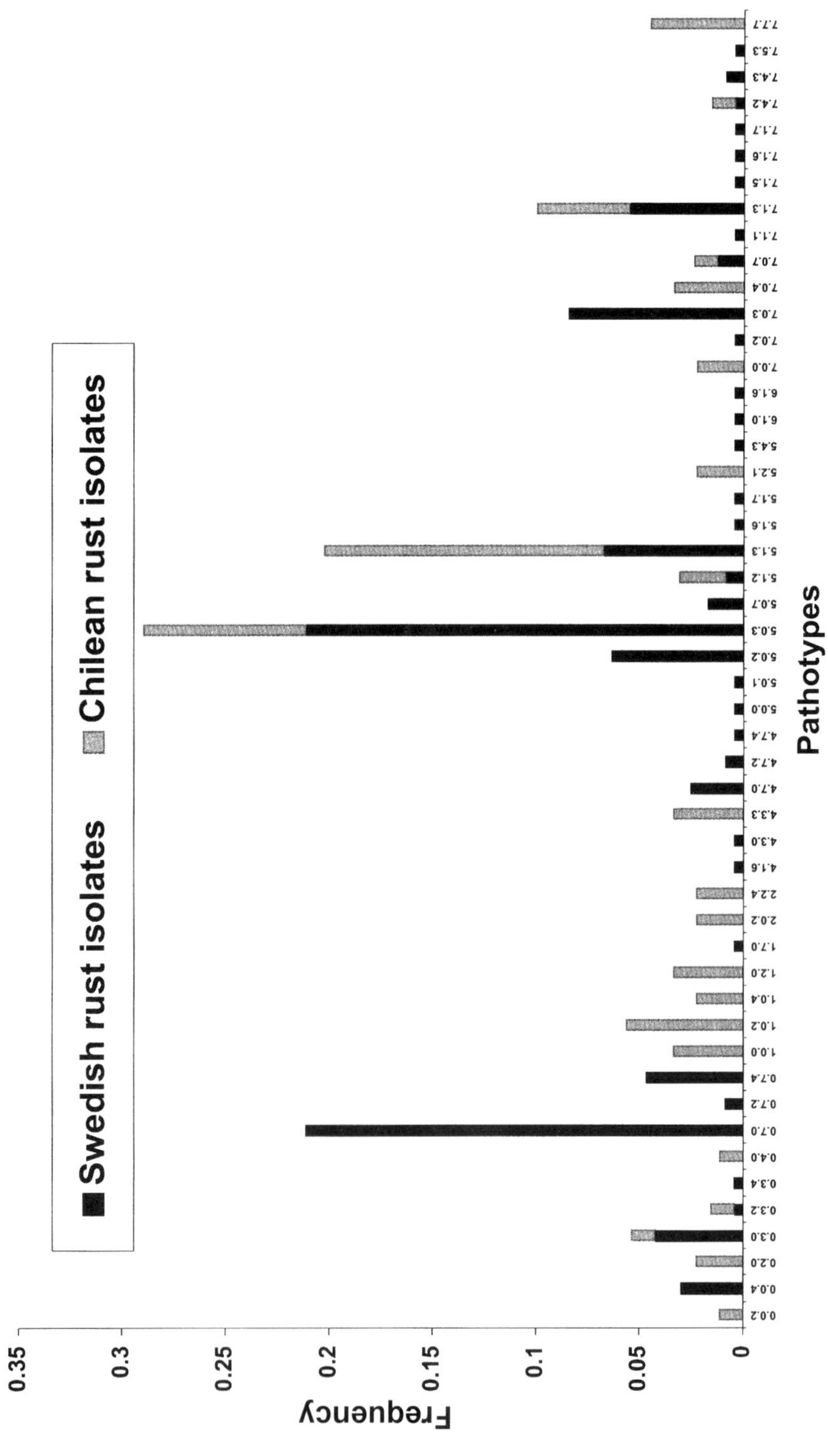

Fig. 10.1. Pathotype frequencies of *Melampsora* on willow collected from Chile and Sweden.

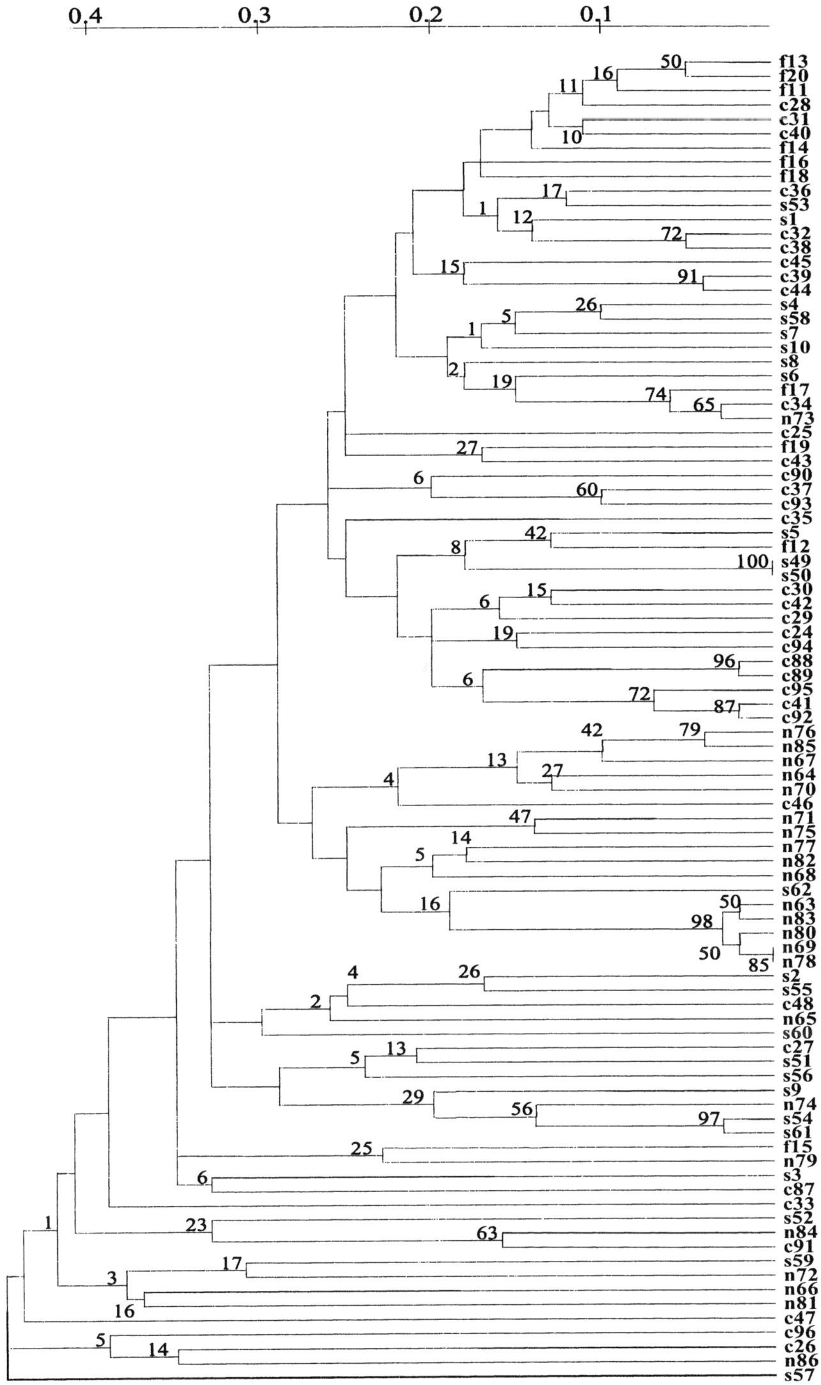

Fig. 10.2. Dendrogram generated after UPGMA using amplification fragment length polymorphism (AFLP)-based genetic distance. Numbers on the branches are bootstrap values (%) obtained from 500 replicate analyses.

for *Melampsora* rust on willows, any population sub-division that does exist can reflect changes due to the different selection pressure existing in the geographical areas (McDonald and McDermott, 1993). It would be difficult to say whether the rust isolates in our study could be grouped according to the willow hosts, because of the small numbers of samples for the species other than *S. viminalis*. It is, however, likely that the willow species present will determine the fitness for the *Melampsora* rust and affect the structure of local populations.

Gene flow between populations may be sufficient to keep populations from diverging due to random processes such as genetic drift. Thus, although the number of isolates used in this study was too limited to draw definitive conclusions, the results do suggest that country-defined, geographically distant populations have not developed separately from each other, but instead they represent part of a common population, that is, part of the same cohesive species metapopulation. Long-distance migration of rust spores can be a powerful means of maintaining such stability, and wind transportation of those spores can be an effective means of long-distance migration. Wind transportation has been reported earlier as the cause of introduction in 1978 of *M. coleosporioides* Diet., first in Asia and then Australia, and further to New Zealand the same year, by riding the trans-Tasman air current (Latch, 1980). The same wind current may have been responsible for the introduction in New Zealand from Australia of the poplar rusts *M. larici-populina* and *M. medusae*. The weather conditions in eastern Australia indicate that spores were transported at high altitudes, probably above 3000 m. Spores carried at lower levels are likely to have been washed-out by the rain (Wilkinson and Spiers, 1976).

Studying the genetic structure of the *M. epitea* populations in Swedish *S. viminalis* plantations, Samils *et al.* (2001) concluded that the genetic structure observed in *M. epitea* (i.e. high levels of both gene and genotypic variation, most of which is found within populations) is consistent with a sexually reproducing species capable of long-distance dispersal. It is also consistent with weak within-population selection. Our results affirm that *M. epitea* has high levels of gene and genotypic variation within populations. However, we also found variation between our four distant populations, suggesting that adaptation by natural selection to local hosts and local environmental conditions was strong enough to maintain the same degree of variation, despite long-distance dispersal. The long-distance dispersal, however, can explain the finding of unusual genotypes within each location, which were more similar to genotypes in a distant location. Although high rates of long-distance migration reduce variation between populations, it facilitates local adaptation by input of genetic variation. To estimate the potential for sustainable disease resistance, new resistance genes should be tested at many locations. This would not only serve as a guard against unexpected epidemic outbreaks, but would also provide clues for pathogenic variation in different pathogen populations at several locations.

Almost 40% of the 23 pathotypes found in Chile were also found in Sweden, and 25% of the 36 pathotypes found in Sweden were also found in Chile. Also, all virulence components found in Sweden were also detected in Chile, suggesting that active intercontinental exchange cannot be ruled out. From the well-documented cases of long-distance migration of rust (Wilkinson and Spiers, 1976; Latch, 1980; Aylor, 1990; Nagarajan and Singh, 1990) and especially of wind-borne rust spores, the possibility of active intercontinental exchange, at least one-way exchange, seems likely.

In addition to wind-borne spores, transfer of *Melampsora* inoculum by humans (cuttings, clothes, aeroplanes, boats) is also a probable, and very likely, way for *Melampsora* rust to spread both from Europe to Chile and from Chile to Europe. While the growing season in the southern hemisphere differs from that in the northern hemisphere, there is still a certain period when the climate conditions favourable for *Melampsora* overlap in both hemispheres (March–April or September–October). This will give the

Melampsora inoculum plenty of time to be transported by the winds present at that time of the year. According to Nagarajan and Singh (1990) the probability of viable spores reaching a target and causing infection after a 'long journey' is very low, but very real.

From these studies, we conclude that the *M. epitea* populations in Chile and Sweden may well be influenced by active intercontinental exchange of inocula, as is well documented for other pathogens (Pedgley, 1982). The importance of possible long-distance transport of rust spores deserves consideration by *Salix* breeding programmes. In addition, the consequence of unknown, unmonitored and unstudied new patterns of clone-specific virulence arriving from another region, country or continent can be devastating.

Thus, as willow cultivation becomes increasingly global and important, world-wide monitoring of the dynamics of patho-type structure in *Melampsora* populations becomes increasingly important in understanding of long-distance inoculum exchange, controlled and uncontrolled, and the effects of such exchange.

References

Aylor, D.E. (1990) The role of intermittent wind in the dispersal of fungal pathogens. *Annual Review of Phytopathology* 28, 73–92.

Gullberg, U. (1989) *Växtförädling av* Salix*, 1986–1989*. Research Notes 42, Swedish University of Agricultural Sciences, Sweden.

Hubbes, M. (1983) *A Review of the Potential Diseases of* Alnus *and* Salix *in Energy Plantations*. Ontario Tree Improvement and Forest Biomass, Toronto.

Hurtado, S. and Ramstedt, M. (2000) Compatible and incompatible reaction of *Melampsora* rust on willow leaves. *Scandinavian Journal of Forestry Research* 15, 405–409.

Hurtado, S. and Ramstedt, M. (2002) AFLP comparision of distant *Melampsora epitea* (willow rust) populations. *Mycological Research* 106, 1400–1407.

Hurtado, S. and Ramstedt, M. (2003) Comparing Chilean and Swedish rust populations: possible impact of long-distance transport of *Melampsora* rust. *Forest Pathology* 33, 69–80

Hylander, N., Jørstad, I. and Nannfeldt, J.A. (1953) Enumeratio Uredinearum Scandinavicarum. *Opera Botanica* 1, 1–102.

Latch, B.J. (1980) Weeping willow rust in New Zealand. *New Zealand Journal of Agricultural Research* 23, 535–538.

Majer, D., Mitchen, R., Lewis, B.G., Vos, P. and Oliver, R.P. (1996) The use of AFLP fingerprinting for the detection of genetic variation in fungi. *Mycological Research* 100, 1107–1111.

McCracken, A.R. and Dawson, W.M., (1997) Growing clonal mixtures of willow to reduce effect of *Melampsora epitea* var. *epitea*. *European Journal of Forest Pathology* 27, 319–329.

McCracken, A.R. and Dawson, W.M. (1998) Short rotation coppice willow in Northern Ireland since 1973: development of the use of mixtures in the control of foliar rust (*Melampsora* spp.). *European Journal of Forest Pathology* 28, 241–250.

McDonald, B.A. and McDermott, J.M. (1993) Population genetics of plant pathogenic fungi. *BioScience* 43, 311–319.

Nagarajan, S. and Singh, D.V. (1990) Long-distance dispersion of rust pathogens. *Annual Review of Phytopathology* 28, 139–153.

Pedgley, D. (1982) *Windborne Pests and Diseases. Metereology of Airborne Organisms*. Ellis Horwood, Chichester, UK.

Pei, M.H., Royle, D.J. and Hunter, T. (1996) Pathogenic specialization in *Melampsora epitea* var. *epitea* on *Salix*. *Plant Pathology* 45, 697–690.

Pei, M.H., Parker, S.R., Hunter, T. and Royle, D.J. (1997) Variations in populations of *Melampsora* willow rust and the implications for design of short rotation coppice plantations. *Aspects of Applied Biology* 49, 91–96.

Ramstedt, M. (1989) Diseases of *Salix* energy plantations in Sweden, 1988. *Proceedings of the International Energy Agency Task II Joint Workshop*, Bristol, UK, pp. 76–81.

Ramstedt, M. (1998) Diversity of the *Melampsora* rust on willows – consequences for SRF planting strategies. *Proceedings of the First IUFRO Rusts of Forest Trees WP Conference*, 2–7 August, Saariselkä, Finland. Finnish Forest Research Institute, Research papers 712, 123–130.

Ramstedt, M. (1999) Rust disease on willows – virulence variation and resistance breeding strategies. *Forest Ecology and Management* 121, 101–111.

Ramstedt, M., Hurtado, S. and Åström, B. (2002) Pathotypes of *Melampsora* rust on *Salix* in short-rotation forestry plantations. *Plant Pathology* 51, 185–190.

Royle, D.J. and Hubbes, M. (1992) Diseases and pests in energy crop plantations. *Biomass and Bioenergy* 2, 45–54.

Royle, D.J. and Ostry, M.E. (1995) Disease and pest control in the bioenergy crops poplar and willow. *Biomass and Bioenergy* 9 (1–5), 69–79.

Samils, B., Lagercrantz, U., Lascoux, M. and Gullberg, U. (2001) Genetic structure of *Melampsora epitea* populations in Swedish *Salix viminalis* plantations. *European Journal of Plant Pathology* 107, 399–409.

Samils, B., Lagercranz, U. and Gullberg, U. (2002) Genetic relationships among genetically distinct forms of *Melampsora larici-epitea* and related species based on AFLP data. *Forest Pathology* 32, 379–386.

Wilkinson A.G. and Spiers, A.G. (1976) Introduction of the poplar rusts *Melampsora larici-populini* and *M. medusae* to New Zealand and their subsequent distribution. *New Zealand Journal of Science* 19, 195–198.

Wilson, M. and Henderson, D.M. (1966) *The British Rust Fungi.* Cambridge University Press, Cambridge, UK.

11 Disease Scoring by Taking Inoculum Densities into Consideration in Leaf Disc Inoculations with Poplar and Willow Rusts

Ming Hao Pei[1] and Tom Hunter[2]
[1]*Rothamsted Research, Harpenden, Hertfordshire AL5 2JQ, UK;*
[2]*4 Wally Court Road, Chew Stoke, Bristol BS40 8XL, UK*

Disease Scoring Methods used for Poplar and Willow Rusts in Leaf Disc Inoculations

Assessment of disease is an essential part of inoculation experiments designed to characterize disease resistance/pathogen virulence. In the studies of *Melampsora* rusts on poplar and willow, the leaf disc inoculation technique has been used widely to test host resistance/pathogen virulence, because of its simplicity and efficiency. Various methods have been used to score the disease in leaf disc inoculation experiments.

With the poplar rust *M. medusae*, Prakash and Thielges (1987) used a 0–7 disease rating, i.e. 0, no macroscopic symptoms; 1, necrotic flecks only; 2, necrotic flecks and some with minute uredinia; 3, necrotic flecks with few medium-sized uredinia; 4, few, small uredinia; 5, few, large uredinia; 6, numerous, small uredinia; 7, numerous, large uredinia. In a study of the North American poplar rust, *M. occidentalis*, Hsiang and Castagner (1993) modified Prakash and Thiegle's ratings by defining the size of uredinia: < 0.4 mm diameter as small and > 0.4 mm as large; and the number of uredinia: 3–6 as few and ≥ 7 as numerous. Their inoculation experiments appeared to have used very high doses of inoculum, i.e. *c.* 7500 spores being spread over 16 mm diameter leaf discs (2.01 cm^2) using cotton wool. Newcombe *et al.* (2000, 2001) assessed resistance/susceptibility according to infection types based on the size and the presence of chlorotic or necrotic flecking, in inoculation experiments with *M.* × *columbiana*, a possible hybrid between *M. medusae* and *M. occidentalis*.

In studies of pathogenic variation in *M. epitea*, Pei *et al.* (1996) inoculated willow leaf discs at the inoculum densities 70–300 spores per 11 mm diameter (0.95 cm^2) leaf disc and, based on the number and estimated size of uredinia, scored infection types using a 0–4 scale.

In the above-mentioned cases, inoculum pressure was not clearly quantified and the criteria for disease scores were somewhat ambiguous. To date, disease scoring in leaf disc or detached leaf inoculation experiments with rust pathogens is still based largely on the assessors' experience and skills.

Inoculum Pressure and Disease Severity in Rust Fungi

In rust fungi, it is known that inoculum densities (viable spores applied on to a unit area of host tissue) can markedly influence the number of pustules/lesions produced. In wheat leaf rust *Puccinia graminis*, inoculation densities up to 2000 spores/cm^2 were linearly correlated with resulting numbers of uredinia on wheat seedlings (Peterson, 1959). In bean rust *Uromyces appendiculatus*, a linear relationship was found between the amount of inoculum (200–900 spores/cm^2) and resulting uredinia, with approximately 1 out of 10 spores producing uredinia (Schein, 1964).

Recently, quantitative relationships between inoculum density and disease variables were investigated in willow rust, *M. larici-epitea* (Pei *et al.*, 2002), and poplar rust, *M. larici-populina* (Pei *et al.*, 2003). The results from both studies suggested that rust resistance/susceptibility was expressed by the differences in both the number and the size of uredinia, and that there were close correlations between uredinial pustule area and spore production.

A practical problem in testing disease resistance/pathogen virulence involving a number of different isolates is that it is almost impossible to achieve deposition of identical numbers of spores on the leaf surface for each isolate, because each isolate has to be inoculated separately. Furthermore, viability of spores may vary considerably between isolates. Results from different inoculations would be more comparable if the inoculum densities could be accounted for in inoculation experiments.

In this chapter, we further examine the relationships between inoculum density and uredinial number/area in poplar and willow rusts, using the extensive data obtained from the quantitative inoculation experiments with *M. larici-epitea* (Pei *et al.*, 2002) and *M. larici-populina* (Pei *et al.*, 2003).

Inoculum Density and Uredinial Number/Area in Poplar and Willow Rusts

Host clones and rust isolates tested

For poplar rust, four poplar clones – *P. nigra* 'Vereecken', *P. nigra* × *P. deltoides* 'Spijk', *P. trichocarpa* 'Trichobel' and *P. trichocarpa* × *P. deltoides* × *P. deltoides* '75028' – and four *M. larici-populina* isolates having different degrees of virulence to the tested clones were used.

For willow rust, six willow clones – *S. burjatica* 'Korso', *S.* × *calodendron*, *S.* × *mollissima* 'Q83', *S.* × *stipularis*, *S. viminalis* 'Mullatin' and *S. viminalis* '78183' – and four *M. larici-epitea* isolates were used. Pathogenicity of the isolates corresponded to LET1, LET3, LET4 and LR1 (see Pei *et al.*, 1996).

Inoculation experiments

Leaf discs, 1.6 cm diameter for poplars and 1.1 cm diameter for willows, were placed on blotting-paper bridges soaked in tap water in 25 compartments of 10 × 10 cm square Petri dishes. Poplar rust spore suspensions were adjusted to seven levels (4, 8, 16, 32, 64, 128 and 256 × 10^3 spores/ml) and willow rust spore suspensions to five levels (1.5, 3, 12, 25 and 50 × 10^4 spores/ml). The spore suspensions were sprayed on to the target area (1 ml per 10 × 10 cm area) using a Humbrol air brush (Humbrol Ltd., England). Ten leaf discs were used as replicates in poplar and five discs in the willow for each host clone/rust isolate/inoculum density combination. Inoculum density for each rust isolate was determined by examining the number of germinating spores on the water agar which had been placed in the target area and then incubated at 16°C for 24 h. After inoculation, the leaf discs were incubated in a growth chamber at 16°C with 16 h per day illumination at an intensity of 80 μE/m^2/s^1.

Thirteen days after inoculation, reactions on leaf discs were recorded using a digital camera (Olympus C-2500L). Then, for each leaf disc, the spores were counted using a haemocytometer. Image analysis software (SIGMASCAN Pro 5.0; SPSS Inc.) was used to count uredinial numbers and to measure the diameters of erumpent uredinia. Further details of the inoculations with the poplar rust were described in Pei *et al.* (2003) and that with the willow rust in Pei *et al.* (2002).

Data analyses

Uredinial pustule area was calculated using the total number and average diameter of the uredinia on the leaf disc. A square root transformation was used to transform the data. A simple linear regression model ($y = ax + b$) was used to examine the relationships between inoculum density (x) and uredinial number/area (y). In the equation, a is the slope factor and b the intercept. A measure of the strength of the linear relationship was given by the percentage variance accounted for, calculated as:

% variance accounted for (VAF) = 100 × (total mean square – residual mean square)/total mean square.

A series of models were fitted, i.e. a single overall line, parallel lines (a common slope but different intercepts for the different clone/isolate combinations), separate lines with different slopes but with a common intercept (intercept = 0) and finally separate lines with different slopes as well as intercepts. The analyses were carried out using GENSTAT 5 Release 4.2 (Genstat Committee, 2000).

Poplar rust

Inoculum densities of *M. larici-populina* were estimated up to 680 viable spores/cm^2. Compared to a single overall line, the relationship between inoculum density and uredinial number was better expressed by the model for parallel lines for different clone/isolate combinations (%VAF increased from 48.3 to 73.1, $P < 0.001$) (Table 11.1). Allowing the slopes and intercepts to change independently for each combination further improved the fit ($P < 0.001$; %VAF = 79.1) compared to the parallel lines model. The model with a common intercept (intercept = 0) but individual slopes gave significant increase in fit compared to a single overall line having intercept = 0 (%VAF increased from 48.3 to 77.8, $P < 0.001$).

Table 11.1. Percentage variance accounted for (%VAF) for different models fitted to the plots between inoculum density and uredinial number/area of *M. larici-populina* and *M. larici-epitea*.

	Poplar		Willow (inoculum densities up to 740 spore/cm^2)		Willow (inoculum densities up to 400 spore/cm^2)	
Model	Uredinial number	Uredinial area	Uredinial number	Uredinial area	Uredinial number	Uredinial area
Single overall line	48.3	43.6	55.6	36.1	41.3	36.1
Parallel lines	73.1	69.4	66.8	58.6	65.3	63.1
Lines with independent slopes and intercepts	79.1	75.2	68.4	61.9	70.2	70.2
Single overall line having intercept = 0	48.3	43.7	43.9	20.3	33.0	33.4
Individual lines with common intercepts (= 0)	77.8	73.6	58.3	48.7	52.4	64.3

Between inoculum density and uredinial pustule area, parallel lines gave a better correlation than the single overall line (%VAF increased from 43.6 to 69.4, $P < 0.001$). The model having individual slopes and intercepts gave slight but significant improvement compared to the parallel lines model (%VAF = 75.2, $P < 0.001$). The model with a common intercept (intercept = 0) but having individual slopes gave significant increase in fit compared to a single overall line (%VAF increased from 43.7 to 73.6, $P < 0.001$).

Willow rust

Inoculum densities of *M. larici-epitea* were estimated up to 740 viable spores/cm^2. Between inoculum density and uredinial number, the model for parallel lines for different clone/isolate combinations showed a significant increase in fit compared to a single overall line ($P < 0.001$; %VAF increased from 55.6 to 66.8). Allowing the slopes and intercepts to change independently for each combination also improved the fit (%VAF = 68.4, $P < 0.001$) compared to the parallel lines model. The model with a common intercept (intercept = 0) gave significant increase in fit compared to a single overall line (%VAF increased from 43.9 to 58.3, $P < 0.001$).

Between inoculum density and uredinial pustule area, parallel lines gave better fit than a single overall line (%VAF increased from 36.1 to 58.6, $P < 0.001$). The model having individual slopes and intercepts gave slight but significant improvement compared to the parallel lines model (%VAF = 61.9, $P < 0.001$). The model with a common intercept (intercept = 0) gave significant increase in fit compared to a single overall line having intercept = 0 (%VAF increased from 20.3 to 48.7, $P < 0.001$).

When analyses were done with the data restricted to inoculum densities < 400 spores/cm^2, the model for parallel lines for different clone/isolate combinations showed a significant increase in fit compared to a single overall line ($P < 0.001$; %VAF increased from 41.3 to 65.3) between inoculum density and uredinial number. Allowing the slopes and intercepts to change independently for each combination improved the correlation (%VAF = 70.2, $P < 0.001$) compared to the parallel lines model. The model with a common intercept (intercept = 0) gave significant increase in fit compared to a single overall line having intercept = 0 (%VAF increased from 33.0 to 52.4, $P < 0.001$).

For the data obtained at the inoculum densities < 400 spores/cm^2, parallel lines gave a better fit than single overall line (%VAF increased from 36.1 to 63.1, $P < 0.001$) between inoculum density and uredinial pustule area. The model having individual slopes and intercepts gave significant improvement compared to the parallel lines model (%VAF = 70.2, $P < 0.001$). The model with a common intercept (intercept = 0) gave significant increase in fit compared to a single overall line having intercept = 0 (%VAF increased from 33.4 to 64.3, $P < 0.001$).

Disease Scoring by Taking Inoculum Densities into Consideration

Based on the present results, we propose that inoculum densities should be taken into account when disease is scored in leaf disc or detached leaf inoculations. This can be done conveniently by using the linear relationships between inoculum density and uredinial number/area. In our analyses, parallel lines and individual lines having different slopes and intercepts gave somewhat better correlations between inoculum density and uredinial number/area compared with the individual lines with a common intercept (intercept = 0). However, because of its simplicity and the biological implications that no uredinia develop without inoculum, individual lines with a common intercept may be reasonably adequate to weigh the differences in disease due to the differences in the amount of inoculum applied. As rust resistance/susceptibility was expressed by the differences in both the number and size of uredinia (Pei *et al.*, 2002, 2003), the use of uredinial pustule area data would be more meaningful in disease scoring because the uredinial pustule area is calculated based on both the number and the size of uredinia.

A possible way of scoring disease is to use a highly susceptible clone/isolate combination as a reference and then calculate the slope based on its inoculum density and uredinial pustule area. With other combinations, slopes can also be calculated according to respective inoculum densities and uredinial pustule area measurements. Then, the slopes can be categorized into different scales and the disease scores can be obtained accordingly. Otherwise, percentage figures may be given to indicate the disease severity in relation to the highly susceptible reference clone.

As an example, the data for the uredinial pustule area and inoculum density of four *M. larici-populina* isolates on two poplar clones were plotted in Fig. 11.1. *Populus nigra* 'Vereecken' was infected by all the four isolates, *P. nigra* × *P. deltoides* 'Spijk' by isolates BE and BO. For the model having individual lines with a common intercept, the slope factor for the most susceptible clone/isolate combination BE/ 'Spijk' was 0.175. If 0–4 scale disease ratings are used, given the assumption that $y = 0.175x$ represents the mid-point for the scale 4 rating, then the mid-point for scale 3 is 0.125, scale 2 is 0.75 and scale 1 is 0.25. Lines (dashed in Fig. 11.1) can be drawn arbitrarily as $y = 0.15x$ to indicate the border between the scales 3 and 4, $y = 0.10x$ between 2 and 3 and $y = 0.05x$ between 1 and 2. Then disease scores for various clone/ isolate combinations can be obtained by plotting their uredinial pustule area against respective inoculum densities. It should be noted that, at very low inoculum densities, small variations in uredinial pustule area data may give very different disease scores. Therefore, care should be taken to avoid applying extremely low or excessively high number of spores in inoculation experiments. It appears that the proposed method of disease scoring may be well suited for inoculum densities between 100 and 400 spores/cm^2 for both poplar and willow rusts.

Apart from inoculum density, disease scores can be influenced by temperature, light, the age and growing conditions of plants. Therefore, these conditions should be standardized in the preparation of inoculation experiments. Consideration should also be given to the fact that viability of spores may vary and spore concentration

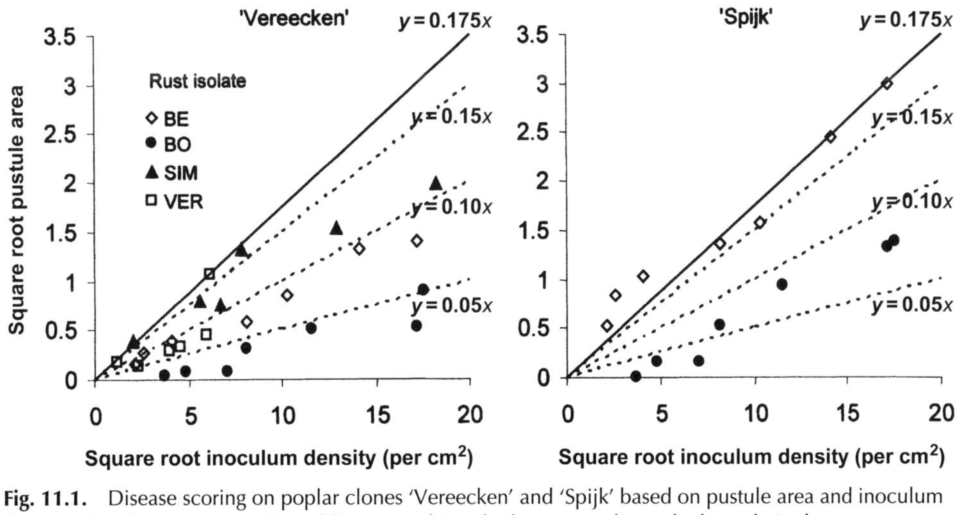

Fig. 11.1. Disease scoring on poplar clones 'Vereecken' and 'Spijk' based on pustule area and inoculum density of *M. larici-populina*. Dotted lines are arbitrarily drawn to indicate the boundaries between scores when 0–4 scale rating is used.

may not necessarily reflect the number of viable propagules deposited on a host surface. It is desirable to use, wherever possible, fresh spores for inoculation and to quantify inoculum pressure according to both the density and viability of the spores.

Concluding Remarks

We examined the relationships between inoculum density and uredinial number/area using the data obtained from quantitative inoculations with poplar rust *M. larici-populina* and willow rust *M. larici-epitea*, using leaf discs. Simple regression models having parallel lines, individual lines with different slopes and intercepts, and individual lines with a common intercept were fitted for different host clone/rust isolate combinations. For poplar rust, all the models fitted well, % variance accounted for being 69–79. In willow rust, reasonable linear relationships between inoculum density and uredinial number/area were maintained at the inoculum densities < 400 spores/cm^2 (% variance accounted for being 52–65). A method was proposed to score disease in leaf disc inoculations using slope factors which were calculated based on the uredinial pustule area and inoculum density.

The proposed method has several attractive attributes. First, disease scores based on uredinial pustule area can be indicative of the extent of spore production because uredinial pustule area is closely correlated to the number of spores produced. Spore production is one of the most important driving variables in the development of disease epidemics and is also one of the most accurate and least subjective ways of assessing the growth of pathogens and the susceptibility of hosts (Johnson and Taylor, 1976). In *M. larici-populina*, there was a close correlation between the number of spores produced and uredinial pustule area (Spearman rank correlation 0.94) (Pei *et al.*, 2003). In *M. larici-epitea*, spore yield was better correlated with uredinial pustule area (accounting for 61.2% variance) than to the number of uredinia (42.7% variance). Secondly, the size of host leaf area would not affect the slope factor. For example, the slope factor obtained based on uredinial pustule area and inoculum density on a 2 cm^2 leaf disc would be the same as that obtained on a 10 cm^2 leaf segment. Thirdly, the data obtained between different inoculation experiments can be made more comparable by including known host clone/rust isolate combinations in both inoculation experiments. The proposed method would also be useful in studies of disease resistance/pathogen virulence in other plant/rust systems.

Acknowledgements

This study was funded by the Department for Environment, Food and Rural Affairs (DEFRA), UK, and the European Union. Rothamsted Research receives grant-aided support from the Biotechnology and Biological Sciences Research Council of the UK.

References

Genstat Committee (2000) GenStat 5 for Windows, Release 4.2, 5th edition. Numerical Algorithms Group, Oxford, UK.

Hsiang, T. and Castagner, G.A. (1993) Variation in *Melampsora occidentalis* rust on poplars in the Pacific Northwest. *Canadian Journal of Plant Pathology* 15, 175–181.

Johnson, R. and Taylor, A.J. (1976) Spore yield of pathogens in investigation of the race-specificity of host resistance. *Annual Review of Phytopathology* 14, 97–119.

Newcombe, G., Stirling, B., McDonald, S. and Bradshaw, H.D. (2000) *Melampsora* × *columbiana*, a natural hybrid of *M. medusae* and *M. occidentalis*. *Mycological Research* 104, 261–274.

Newcombe, G., Stirling, B. and Bradshaw, H.D. (2001) Abundant pathogenic variation in the new hybrid rust *Melampsora* × *columbiana* on hybrid poplar. *Phytopathology* 91, 981–985.

Pei, M.H., Royle, D.J. and Hunter, T. (1996) Pathogenic specialization of *Melampsora epitea* var. *epitea* on *Salix*. *Plant Pathology* 45, 679–690.

Pei, M.H., Ruiz, C., Hunter, T., Arnold, G.M. and Bayon, C. (2002) Quantitative relationships between inoculum of *Melampsora larici-epitea* and corresponding disease on *Salix*. *Plant Pathology* 51, 443–454.

Pei, M.H., Ruiz, C., Harris, J. and Hunter, T. (2003) Quantitative inoculations of poplars with *Melampsora larici-populina*. *European Journal of Plant Pathology* 109, 269–276.

Peterson, L.J. (1959) Reactions between inoculum density and infection of wheat by uredospores of *Puccinia graminis* var. *tritici*. *Phytopathology* 49, 607–614.

Prakash, C.S. and Thielges, B.A. (1987) Pathogenic variation in *Melampsora medusae* leaf rust of poplars. *Euphytica* 36, 563–570.

Schein, R.D. (1964) Design, performance, and use of a quantitative inoculater. *Phytopathology* 54, 509–513.

12 Interactions Between Poplar Clones and *Melampsora* Populations and their Implications for Breeding for Durable Resistance

Jean Pinon and Pascal Frey

UR Pathologie Forestière, INRA, F-54280 Champenoux, France

Background

In the 1950s, bacterial canker (*Xanthomonas populi* (Ridé) Ridé and Ridé) was a major concern in the northern part of Europe in the area used for poplar cultivation. During the 1960s, the major foliar disease on poplar in Europe was *Marssonina brunnea* (Ell. and Ev.) Magn., following its introduction from North America. In the past, rust caused problems sporadically, only in favourable weather conditions and following infections by *Discosporium populeum* (Sacc.) Sutton, which resulted in decline. Three *Melampsora* species are pathogenic on cultivated poplars in Europe: *M. larici-populina* Kleb., *M. allii-populina* Kleb. and *M. medusae* Thuem. The latter is confined to south-western France, Spain and Portugal. To date, no significant damage caused by *M. medusae* has been observed, even though susceptible cultivars are widely planted (Pinon, 1986). *M. allii-populina* was frequent in western France (Pinon, 1991) and its frequency increased in Belgium (M. Steenackers, personal communication) due to the cultivation of 'Beaupré' which is relatively susceptible to *M. allii-populina*. Since 1997 the frequency of this species has decreased significantly with 'Beaupré' becoming severely infected by new pathotypes of *M. larici-populina*. This latter species is now a major concern all over Europe.

Variability within a given *Melampsora* species is mainly defined by the presence of pathotypes (or physiological races), this second level of variability becoming increasingly important. The first evidence of pathotypes was mentioned for *M. larici-populina* by Van Vloten (1949), although the occurrence of different pathotypes in the field did not become obvious until the 1980s. At present, eight virulences are known within *M. larici-populina*, with potentially 256 pathotypes (Table 12.1). During the past 20 years, many cultivars have lost their complete resistance to *M. larici-populina*, due to new virulences and development of new pathotypes. Virulences were also described within *M. allii-populina* (Frey and Pinon, 1997). To date, no major change in clone behaviour has been observed in relation to the pathotypes of *M. allii-populina*. This is probably due to the fact that no breeder took susceptibility to this species into account and therefore no selection was made for complete resistance.

It has been demonstrated that populations of *M. larici-populina* are not

Table 12.1. Virulences known within *M. larici-populina*.

Virulence number	Year of description	Initial location	Clone used to detect virulence
1	1982	Belgium, France	'Ogy'
2	1986	France	'Aurora'
3	1949	The Netherlands	'Brabantica'
4	1974	France	'Unal'
5	1982	France, Belgium?	'Rap'
6	1994	France	'87B12'
7	1994	Belgium, France	'Beaupré'
8	1997	Belgium, France	'Hoogvorst'

homogeneous and can evolve very quickly in a given location. During the same growing season, populations are diverse, according to sites, and a new virulence can spread easily, with recombinations giving rise to new pathotypes. Host population is the main selection force exerted on rust populations (Pinon and Frey, 1997) and the fungus can adapt efficiently to the host. Climatic aspects do not seem to play a major role in population evolution.

The main challenge for poplar breeding and cultivation, therefore, is incorporation of durable resistance to rusts that are composed of diverse and changing populations.

Virulences and Pathotypes within *Melampsora* on Poplar

Melampsora larici-populina

Variability within M. larici-populina

The variability of pathogenicity within *M. larici-populina* is detected at two levels: virulence and aggressiveness. Recognition by Van Vloten of the first pathotypes in 1949 was not considered by breeders for several decades. In 1974 rust infections were detected on poplar cultivars such as 'Unal' and 'Hunnegem', which had been selected for complete resistance (V. Steenackers, personal communication), indicating a new virulence in the fungus. Steenackers (1982) observed breakdown of complete resistance on 'Isières', 'Ogy', 'Spijk' and 'Rap' in Belgium in 1982 and suggested that this was caused by a new pathotype. This was confirmed in France after experimental inoculations in the laboratory (Pinon and Bachacou, 1984). Studying pathotype populations, Pinon and Bachacou (1984) found that most isolates pathogenic to 'Unal' were also pathogenic to 'Rap', but with some exceptions, and they therefore proposed the existence of two distinctly different virulences. In 1986, some completely resistant cultivars from Italy ('Cima', 'Luisa Avanzo') and the Netherlands ('Hees', 'Ellert') became infected at INRA, Nancy in France, and artificial inoculations confirmed the existence of a new virulence (Pinon and Peulon, 1989). Some *Populus deltoides* 'Marsh' clones studied at INRA, Orléans were completely resistant to the above-mentioned virulences but showed some infections in the nursery. Isolates collected on these clones (87B11, 87B12, 87B29 and 87B37) expressed a new virulence (Pinon and Lefèvre, 1994). In Belgium in 1994 (Steenackers *et al.*, 1994), and in France shortly afterwards (Pinon, 1995), another breakdown of complete resistance was detected and subsequent rust outbreaks brought about a major economical impact. Many Belgian ('Beaupré', 'Boelare', 'Gibecq', 'Ghoy', 'Primo') and Dutch ('Barn', 'Donk', 'Dorskamp', 'Flevo') clones lost their complete resistance. According to the extent of symptoms seen in Belgium on adult trees in 1994, it is probable that this new virulence had appeared 1 or 2 years previously. The eighth virulence was found, again in Belgium and in France in 1997, when the complete resistance of newly released cultivars 'Hoogvorst' and 'Hazendans' was broken down.

For simplicity, pathotypes were gathered into groups, with 'E' meaning Europe (Table 12.2): E1 pathotypes are older pathotypes which have never exhibited virulences 1, 2, 5, 6, 7 or 8. Most frequently they have one virulence (4) or two (3, 4). E2 pathotypes have at least virulence 1, E3 virulence 2, E4 virulence 7 and E5 pathotypes virulence 8. For scientific studies in pathology and tree breeding, it is recommended that isolates are described by their composition in virulences (pathotypes) according to the number allocated in Table 12.1.

After the introduction of *M. larici-populina* to Australia and its migration to New Zealand, pathotypes were also described in these countries. It was possible to compare European pathotypes with the ones from New Zealand where European cultivars were included in pathological studies. Pathotypes (NZ-1 and NZ-2) described in New Zealand (Latch and Wilkinson, 1980) have the same host range as pathotypes described in Europe as E1 and E3. NZ-2 was probably introduced from Europe but was detected in New Zealand before its description in France.

Complete (or pathotype-specific) resistance is known in cultivars of *P.* × *euramericana* Guinier (syn. *P.* × *canadensis* Moench) and *P.* × *interamericana* Brockh. Both hybrids have *P. deltoides* as a parent, which is the probable source of pathotype-specific resistance (Dowkiw, 2003). As pointed out by Newcombe (1998), *P. deltoides* has not evolved with *M. larici-populina* but is the source of complete (exapted) resistance to this rust agent. It seems that breeders have, in the past, mainly used this source of resistance for their hybrids. Frey and Pinon inoculated a few *P. trichocarpa* Torr. and Gray clones with different pathotypes, but none exhibited pathotype-specific resistance. In 1975, we assessed many *P. trichocarpa* clones in Orléans and all those assessed were found to be infected. Among *P. nigra* L., 46 clones of French origin ('Gard', 'Durance' and 'Drôme rivers') were inoculated with 36 isolates of *M. larici-populina* from the same origin.

Table 12.2. Groups of pathotypes of *M. larici-populina* responsible for infection on the most common cultivars.

Groups of pathotypes	Corresponding needed virulence	Type of cultivars
		Differential (pathotype-specific resistance)
E2	1	'Isières', 'Ogy', 'Rap', 'Spijk'
E3	2	'Altichiero', 'Bellotto', 'Luisa Avanzo', 'Aurora', 'Carpaccio', 'Cima', 'Enza', 'Ellert', 'Hees', 'Grimminge', 'Olona', 'Orba', 'Tiepolo', 'Veronese'
E4	7	'A73', 'Barn', 'Beaupré', 'Boelare', 'Donk', 'Dorskamp', 'Flevo', 'Gaver', 'Ghoy', 'Gibecq', 'H-402', 'H-490', 'Kopecky', 'Muur', 'Oglio', 'Oudenberg', 'Pannonia', 'Pegaso', 'Primo', 'Soligo', 'Tasman', 'Vesten'
E5	8	'Hazendans', 'Helix', 'Hoogvorst'
		Universal cultivars (no pathotype-specific resistance)
All	No virulence needed among 1, 2, 6, 7, 8	'Adda', 'Adige', 'Agathe F', 'Alcinde', 'Arno', 'Bellini', 'Bietigheim', 'BL Costanzo', 'Blanc du Poitou', 'Blom', 'Boccalari', 'Branagesi', 'Brenta', 'Büchig', 'Cappa Bigliona', 'Carolin', 'Columbia river', 'Culasso', 'Dora', 'Dvina', 'Eridano', 'Eugenei', 'Faux-Gaver', 'Florence Biondi', 'Fritzi Pauley', 'Gattoni', 'Gelrica', 'Gomol', 'Guardi', 'Guariento', 'Heimburger', 'Hunnegem', 'I-45/51', 'I-154', 'I-214', 'I-455', 'I 74–76', 'Isonzo', 'Italica', 'Koster', 'Lambro', 'Lena', 'Lima', 'Lux', 'Mella', 'Mellone Carlo' (= MC), 'Muhle Larsen', 'Neva', 'Onda', 'Ongina', 'Panaro', 'Pourtet', 'Raspalje', 'Rintheim', 'Robusta', 'Rochester', 'Scott Pauley', 'Serchio', 'Sesia', 'Sile', 'Stura', 'Tannenhoft', 'Tardif de Champagne', 'Taro', 'Ticino', 'Timavo', 'Trebbia', 'Trichobel', 'Triplo', 'Unal', 'Veneziano'

These isolates were all pathogenic to all the clones, although various levels of general resistance were observed on the same clones following natural infections in the nursery. Pathotype-specific resistance in *P. trichocarpa* and *P. nigra* cannot be ruled out, although it seems infrequent. The frequency of pathotype-specific resistance in *P. deltoides* is unknown, but it appears quite easy to find and it is likely to have been favoured by breeders.

The second level of variability is aggressiveness. The first evidence for variability in aggressiveness between isolates belonging to the same pathotype was found among 1, 3, 4, 5 and 7 pathotypes collected in 1998 on 'Beaupré' at Moy-de-l'Aisne (Aisne, France). On 'Beaupré', under constant temperature (20°C) in the laboratory, the latent period of these isolates was 7 days, compared to 8 days required by the older isolates of the same pathotype. Variation in infection (described as number of uredinia per unit of leaf area) was also recorded, although this characteristic was most sensitive to experimental conditions and was less reliable in distinguishing isolates. Dowkiw (2003) considered sporulation. While most isolates produced small uredinia on selected clones, he found that some isolates were able to develop significantly larger uredinia on the same clones. Finally, all three quantitative parameters used to quantify isolate aggressiveness seemed to vary between isolates, including those within the same pathotype. Probably aggressiveness is favoured each year through sexual reproduction of the fungus on larch and is then selected as an advantage.

A frequently asked question in plant pathology is, 'What is the cost for the fungus of unnecessary virulences?' In order to describe rust populations, rust samples have been collected from 'Robusta', which is, up to the present, susceptible to all known pathotypes (i.e. absence of pathotype-specific resistance). Since the detection of the first isolates with virulence 7, 10 years ago, the most frequent isolates are still 1, 3, 4, 5 and 7, despite the fact that only virulence 7 is necessary to infect the common cultivar 'Beaupré'. Since 1994, new pathotypes bearing virulence 7 have been detected each year. Without taking virulence 8 into account, 20 pathotypes were found and, when virulence 8 is included, 25 are known altogether. Some are less complex, combining two virulences (4, 7), while others are more complex (7 virulences, 1, 2, 3, 4, 5, 6 and 7 and 1, 2, 3, 4, 5, 7 and 8). Miot (1999) conducted several laboratory experiments to determine the effect of the accumulation of unnecessary virulences on fungal fitness. In each experiment, one simple and one complex isolate were mixed and inoculated on compatible cultivars. Many cycles were successively performed and the proportion of the two pathotypes was determined. As a result, a significant change in the proportion of the isolates was found, but there was no clear tendency suggesting that unnecessary virulences had a cost for the fungus. The results were typically dependent on the choice of the isolates. This is in agreement with previous observations that the original pathotypes, with five virulences, were still in the majority, even 10 years after their first appearances. It was concluded that this fungus is able to accumulate virulences, thus increasing its host range, without losing aggressiveness or fitness.

Interactions between poplar clones and pathotypes of M. larici-populina

Clone susceptibility to *Melampsora* species has been reviewed previously (Lemoine and Pinon, 1978; Pinon, 1992; Pinon and Valadon, 1997). Furthermore, many poplar clones have been inoculated in the laboratory to establish their susceptibility to *M. larici-populina* pathotypes. Only those clones that are commercially available will be cited, with no reference to experimental clones (Table 12.2). A new case was found with clone 'A4A'. This clone was initially free of rust when introduced in France. From 1998 some infection was detected on this clone and the level of infection has increased year by year. Pathotypes infecting this clone belong to the E4 group but were of a new type, combining virulences 2 and 7. This new combination is likely to have appeared during sexual recombination on larch. In addition, a few experimental clones were also infected, only by such pathotypes. Of the clones tested, only 'A79'' has remained completely resistant.

Some cultivars showed quantitative interactions with pathotypes. For example, 'Blanc du Poitou' was almost resistant to E3 isolates, but somewhat more susceptible to E4 (without damage). During the epidemic on 'Beaupré', it became evident that another interamerican cultivar, 'Raspalje'', performed better against rust than had been previously recorded. This clone was compatible with all known pathotypes. However, when inoculated in the laboratory separately with E1 and E4 isolates, it appeared to be less susceptible to E4 than to E1 (longer latency and less infection). The same response was observed with the clone 'Unal' (Fig. 12.1). Consequently, 'Raspalje' has become more popular, but there is a risk that its increasing cultivation could exert a selection pressure on the rust populations, with an increase in the proportion of the more pathogenic E1 pathotypes. The risk will also increase with the concomitant decrease of 'Beaupré' cultivation leading to fewer E4 pathotypes.

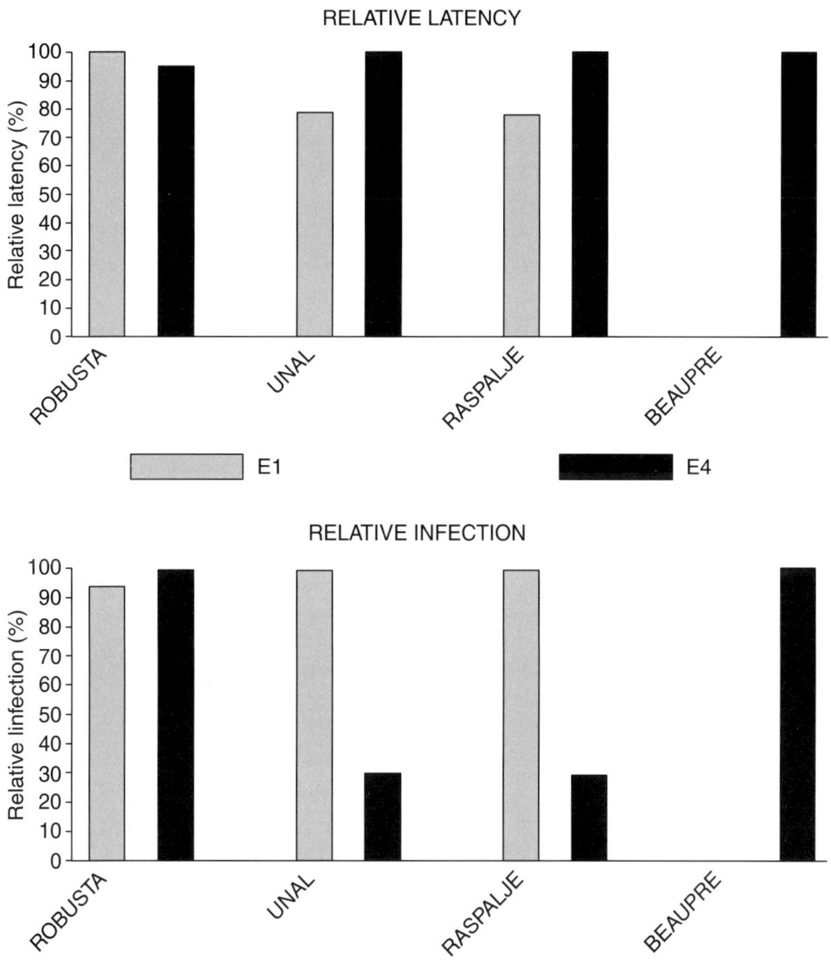

Fig. 12.1. Quantitative interactions between clones and *M. larici-populina* isolates of E1 (pathotype 3–4) and E4 (pathotype 3–4–7) groups. Data (latency and infection, i.e. number of uredinia) expressed as relative values from the most pathogenic isolate on each cultivar.

Cultivars in which complete resistance was broken, were still able to exhibit various levels of general resistance. For instance, with E4 infection, 'Beaupré' may have been badly damaged, while 'Soligo' remained quite healthy. The general resistance was usually poor within *P.* × *interamericana* but better (and often sufficient in practice) within *P.* × *euramericana*. Some exceptions may exist, e.g. the level of susceptibility of 'Ghoy' is relatively high and this clone must only be used where the risk of rust is low.

Melampsora allii-populina

Formae speciales were described within *M. allii-populina* according to the host range of alternate hosts. Viennot-Bourgin (1937) described one *forma specialis* (f. sp. *allii-populina*) which infected *Arum* spp. and *Allium* spp., while another (f. sp. *muscaridis populina*) was pathogenic to *Muscari comosum* and *Allium sphaerocephalum*. On the latter, he described (under a light microscope) thin spinules at the apex of the urediniospores that were not observed on recent samples using a scanning microscope (Frey and Pinon, 1997). Dupias (1965) described a third *forma specialis* (f. sp. *typica*) pathogenic on all known alternate hosts. No experimental data were collected to confirm the validity of these *formae speciales* and it would be necessary to conduct genetic analysis involving isolates from single pustules.

Magnani (1966) reported a possible variability in pathogenicity to poplars within *M. allii-populina*, and indicated the presence of pathotypes. However, his experimental data could also be interpreted as being due to differences in aggressiveness, potentially attributed to inoculum quality. Fourteen isolates from poplar (on various clones and locations) and 28 on alternate hosts (*Allium* spp., *Arum italicum* and *Muscari comosum*) were collected and inoculated on to a wide range of poplar clones (Frey and Pinon, 1997). Many clones were infected by all of the isolates (i.e. universal clones) while others were infected only by some isolates (i.e. differential clones) (Table 12.3).

According to the pathogenicity of isolates originating from poplars and from alternate hosts, a first list of differential clones is proposed (Table 12.3). 'Altichiero' may be added to this list, although further pathotype studies are needed to confirm that the virulence necessary to infect this clone is distinct from the present list of eight virulences. This list is not necessarily complete, because a selected set of clones was used to inoculate the isolates and, also, our collection of isolates may not represent all the variability of the pathogen.

As isolates from different alternate hosts were tested, their pathogenicity to poplar clones was compared, in relation to their original alternate host. No link was found, and it is hypothesized that genes for pathogenicity to poplar and to alternate hosts are completely different (Frey and Pinon, 1997). The differential cultivars for *M. allii-populina* belong to either pure species (i.e. *P. trichocarpa* for 'Fritzi Pauley') or hybrids (*P.* × *euramericana* and *P.* × *interamericana*). This is the first time that a balsam poplar was found as a source of complete resistance towards a rust agent. Only one clone of *P. nigra* was included ('Italica') and, therefore, it cannot be concluded that there is an absence of complete resistance in this species. In the future, it will be necessary to collect more *M. allii-populina* isolates on *P. nigra* and to inoculate them on to a wider range of *P. nigra* clones to determine whether complete resistance exists in this host species that co-evolved with the rust pathogen.

Table 12.3. Reaction of poplar clones to isolates of *M. allii-populina*.

Universal clones	Differential clones
'Aurora', 'Beaupré', 'Bellini', 'Bellotto', 'Boelare', 'Büchig', 'Carpaccio', 'Fritzi Pauley', 'Ghoy', 'I-154', 'I-214', 'Isières', 'Italica', 'Primo', 'Raspalje', 'Rintheim', 'Robusta', 'Tiepolo', 'Unal', 'Véronèse'	'Altichiero', 'Carpaccio', 'Cima', 'Fritzi Pauley', 'I-154', 'Isières', 'Luisa Avanzo', 'NL 2842', 'Rap', 'Spijk'

Work was carried out on *M. allii-populina* populations. For example, in 1993 populations were collected in south-eastern France (where wild *P. nigra* is predominant) and in a nursery of poplar cultivars (Guémené-Penfao, Loire-Atlantique) in western France (Fig. 12.2). Seven virulences were examined and it appeared that the frequencies of the virulences varied according to sites. To the present time 42 different pathotypes of *M. allii-populina* have been detected in France. Some isolates were simple, being pathogenic on cultivars such as 'Beaupré' but without any of the virulences listed above. Others were more complex, combining up to 7 virulences (all except virulence 1). The most frequent pathotype in 1990 in Fontainebleau and Nancy was 2–4–5–8, which was collected from *Muscari comosum* and *Allium niveale*.

Unlike *M. larici-populina*, no cultivar with defeated complete resistance to *M. allii-populina* has been selected up to the present. The effect of poplar populations on pathogen populations is therefore not documented. In addition, no studies were conducted to examine quantitative interactions between poplar clones and isolates or pathotypes of the rust. No information is yet available on the variability of the fungus for aggressiveness. Laboratory inoculations indicated that the infection efficiency of *M. allii-populina* was lower than that of *M. larici-populina*, with approximately twice the number of urediospores needed to get the same level of infection.

Melampsora medusae

In Europe, *M. medusae* was introduced a long time ago and first described in Spain by Fragoso (1925) and then in south-western France by Dupias (1965). It was identified on 'I-214' leaves received from Portugal in the 1980s. Little attention was given to *M. medusae*, for two probable reasons. First, no real damage or epidemic was observed (for unclear reasons), despite the fact that some cultivars popular in France were found to be susceptible to *M. medusae* in New Zealand (Pinon, 1986). Secondly, its distribution was restricted to south-western

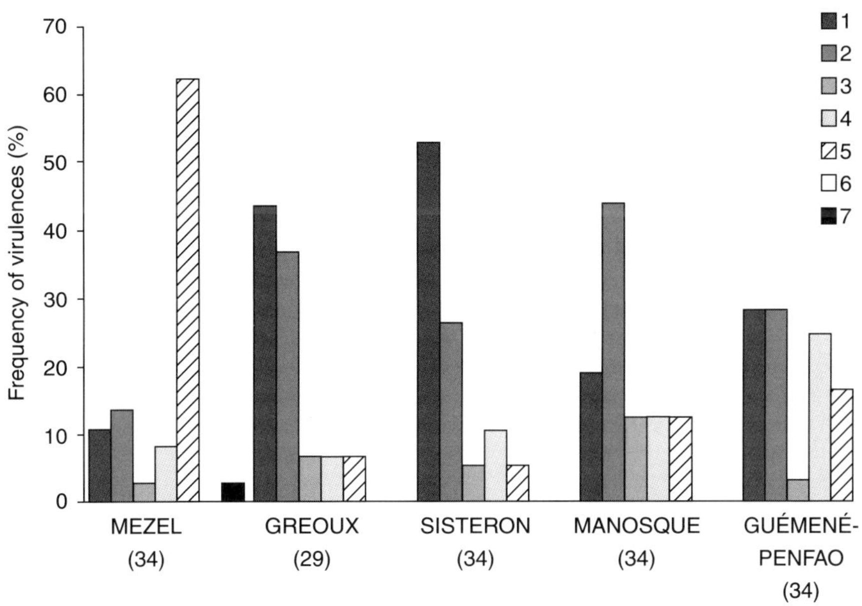

Fig. 12.2. Frequency of seven virulences in five populations of *M. allii-populina* in 1993. South-eastern France: Mézel, Gréoux, Sisteron, Manosque; western France: Guémené-Penfao. In brackets: number of isolates studied.

Europe. Evidence from surveys carried out in France, showed that the infected area may be defined as a triangle between Bordeaux, Toulouse and Bayonne. The rust studied fitted the description of *M. medusae* f. sp. *deltoidae* defined by Shain (1988).

During the 1980s, there was also one case of infection in the western part of France (Guémené-Penfao) on 'Rap'. In 1999 infections were found in Nancy (C. Husson, 1999, personal communication) that allowed some recent clones to be added to the host list. In the following year, both in Guémené-Penfao and in Nancy, *M. medusae* was no longer detectable. It seemed that this fungus might sometimes move from south-western France over long distances, during the summer, and initiate very small and discrete infections far from its established locations. However, in both cases, it seemed unable to overwinter. This fungus is a quarantine pathogen in the EU and the French Plant Protection Service inspects poplar plants in nurseries. Between 1993 and 1997, *M. medusae* was detected in some commercial nurseries in south-western France. Since 1998, it has not been detected in France (J. Tournut, Nancy, 2003, personal communication). Therefore, only a list of clones which have been found to be infected with *M. medusae* may be drawn up. Slight infections were found on some cultivars in Bordeaux between 1988 and 1991 (Table 12.4). In most cases, these clones were also infected by *M. allii-populina* and *M. larici-populina*. Van Kraayenoord (1984) published information on the level of susceptibility of many clones.

Some variability in pathogenicity was described within *M. medusae* in North America (Prakash and Thielges, 1987; Newcombe *et al.*, 2000) and Australasia (Singh and Heather, 1982). Interactions between poplar clones and rust isolates were often quantitative and sometimes clearly qualitative. Temperature may affect the expression of poplar clone × *M. medusae* isolate interactions (Singh and Heather, 1982; Prakash and Thielges, 1989). Although some clones were described as possessing complete resistance and isolates as having matching virulence, no official list of differential clones was proposed. Also, poplar clones used in each study were different and they were often not widely grown cultivars (experimental clones). Comparing data from these two geographical areas, it was found that some clones could act as differential clones, possibly in the case of 'Eugenei', 'I-154' and 'I-455' (Pinon, 1992).

Impact of the Breakdown of Complete Resistance

Until the present, only the breakdown of resistance by *M. larici-populina* affected

Table 12.4. Clones found infected by *M. medusae* in France.

Natural infection in nursery		
Bordeaux (1988–1991)	Nancy (1999)	Laboratory inoculation
'Alabama', 'Aladin', 'Albane', 'Altichiero', 'Angulata de Chautagne', 'Barn', 'Batard de Haute Rive', 'Beaupré', 'Blanc du Poitou', 'Carolin', 'Carpaccio', 'Cima', 'Culasso', 'Donk', 'Estoup', 'Florence Biondi', 'Gaver', 'Ghoy', 'Gibecq', 'Guariento', 'Hees', 'Heidemij', 'Hoogvorst', 'I-214', 'I 45–51', 'I-455', 'I-488', 'Isières', 'Luisa Avanzo', 'Lux', 'Olin', 'Rap', 'Raspalje', 'Robusta', 'Spijk', 'Unal', 'Veneziano', 'Veronese', 'Virginiana', 'Virginie de Frignicourt'	'Hazendans', 'Hoogvorst', 'NL 2233', 'NL 3261', 'NL 4040'	72501[a], 041.3.402[a], 71077–308[a], 71093R1[a], 71097R2[a], 72142R4[a], 72156R5[a], 92519R9[a], 95545R1[a], 96546R1[a], 87B12[b], 'Aurora', 'Beaupré', 'Brabantica', 'Flevo', 'Hoogvorst', 'I 45-51', 'Italica', 'Ogy', 'Pourtet', 'Rap', 'Robusta', 'Sarrazin', 'Simonii', 'Unal', 'Vereecken'

[a]Native *Populus nigra* clones.
[b]*P. deltoides* from INRA, Orléans.

poplar cultivation in Europe. In France, at least, the breakdown of complete resistance by pathotypes from group E2 (virulence 1 and 5) had no significant effect. The reason for this was that the corresponding clones were not cultivated commercially. Pathotypes with virulence 2 (E3 group) induced significant damage on 'Luisa Avanzo', grown mainly in south-western France. In this area, although larch is infrequent, infections were serious and some stands had to be sprayed with fungicide. Damage due to rust was sometimes severe, although the overall effect was limited because this clone was grown in a relatively small area. Some clones susceptible to virulence 2 were assessed each year for rust infection in Nancy. More than 15 years after the outbreak of this group of pathotypes, infection is still heavy in the nursery (Fig. 12.3).

The most important damage was that due to the outbreak of virulence 7 (E4 pathotypes) because this overcame the complete resistance of a popular cultivar, 'Beaupré'. Virulence 7 was detected for the first time in northern France in 1994. In 1997 a serious outbreak occurred in this region. Some stands planted close to larch were defoliated early in July. Before the occurrence of E4 pathotypes, 'Beaupré' was infected by *M. allii-populina* without damage, except in some short-rotation coppices. Virulence 7 was detected in Nancy on larch in spring 1996. Infection increased steadily on 'Beaupré' in the nursery, while on 'Robusta' it fluctuated, according to the climate or assessment date (Fig. 12.4). The frequency of virulence 7 (E4 group) increased during the same period. In 2002, 68% of the isolates on 'Robusta' had virulence 7 and some cases of mortality were observed in relatively young stands of 'Beaupré'. Damage due to *M. larici-populina* is important. According to Gastine *et al.* (2003) the current annual increment of 'Beaupré' was reduced by 20–30% in 1998 and by 50–60% in 2000.

More recently, virulence 8 was discovered (E5 group). This virulence is necessary to infect some recent cultivars such as 'Hoogvorst' and 'Hazendans'. Isolates with virulence 8 were found just at the time when these cultivars were registered in Belgium, and so laboratory tests could be conducted in France in order to forecast their level of susceptibility. 'Hoogvorst' was found to be as susceptible to E5 pathotypes as 'Beaupré' was to E4 pathotypes (Fig. 12.5). Even if infection was somewhat less on 'Hazendans', poplar growers were advised against planting these two clones. E5 pathotypes therefore did not have these clones to build up their population and had difficulty spreading in France where their frequency

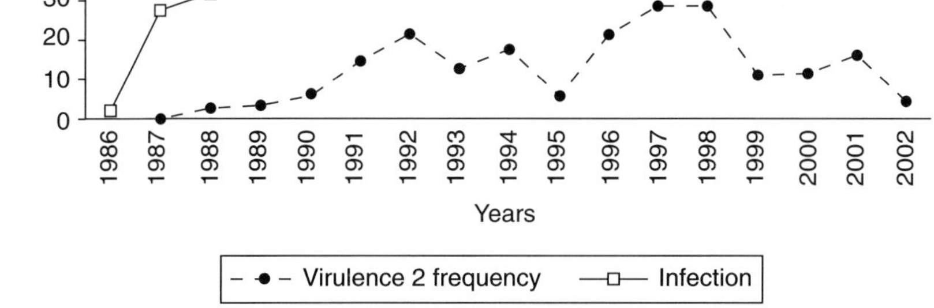

Fig. 12.3. Relationship between the infection level of cultivars susceptible to virulence 2 and virulence 2 frequency in Nancy.

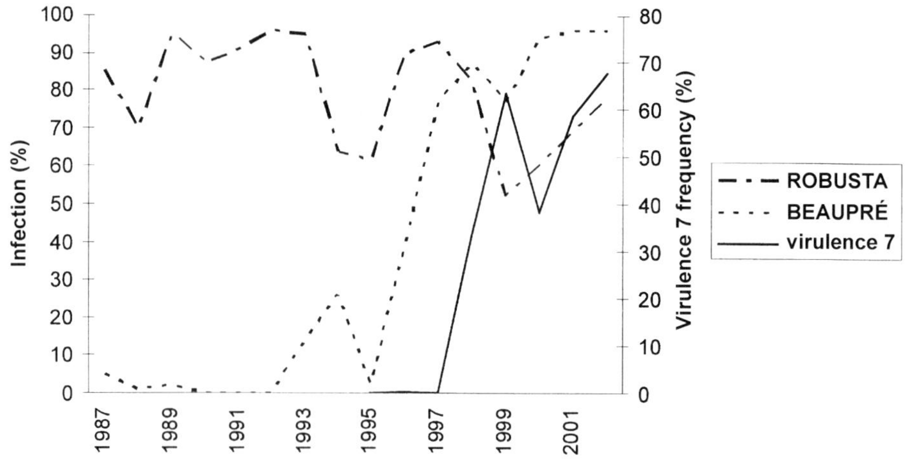

Fig. 12.4. Relationship between infection and virulence 7 frequency in Nancy.

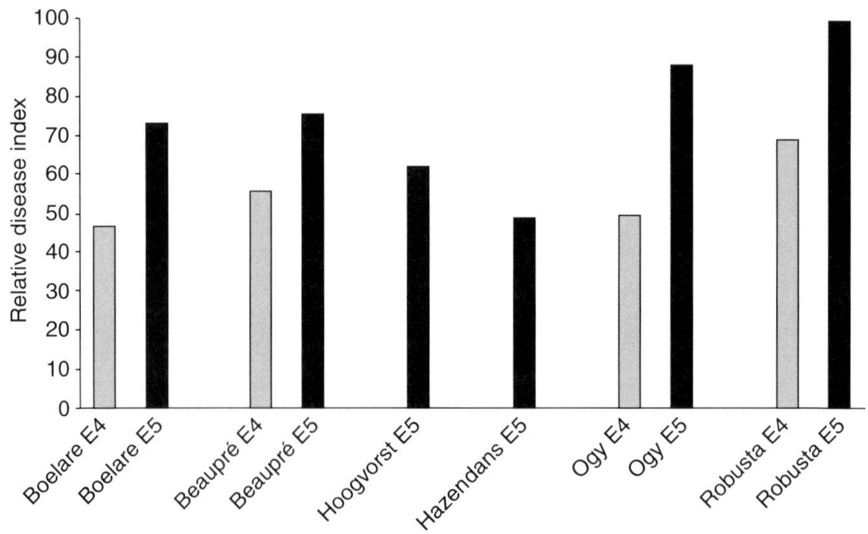

Fig. 12.5. Comparison of the aggressiveness of E4 and E5 pathotypes on five cultivars. Aggressiveness is expressed as a relative disease index (number of uredinia divided by latency) adjusted to 100 for 'Robusta' inoculated with E5.

remains low. The area planted with these two clones in France is low and their health remains good, in contrast to recent mortalities described in Belgium on 'Hoogvorst'. Comparing clone populations and frequencies of the relevant virulences, it became obvious that poplar rust fits perfectly with Van der Plank's theory on the effect of clonal popularity on infection (Van der Plank, 1968).

Adaptation of Poplar Cultivation on the Evolution of Pathotype Populations

E3 pathotypes affected mainly two clones, planted generally in south-western France, 'Luisa Avanzo' and 'Cima'. These two clones were a minority in poplar plantations in France. Their maximum nursery plant production was around 4–6% between 1987 and 1992 (Fig. 12.6). Later their proportion

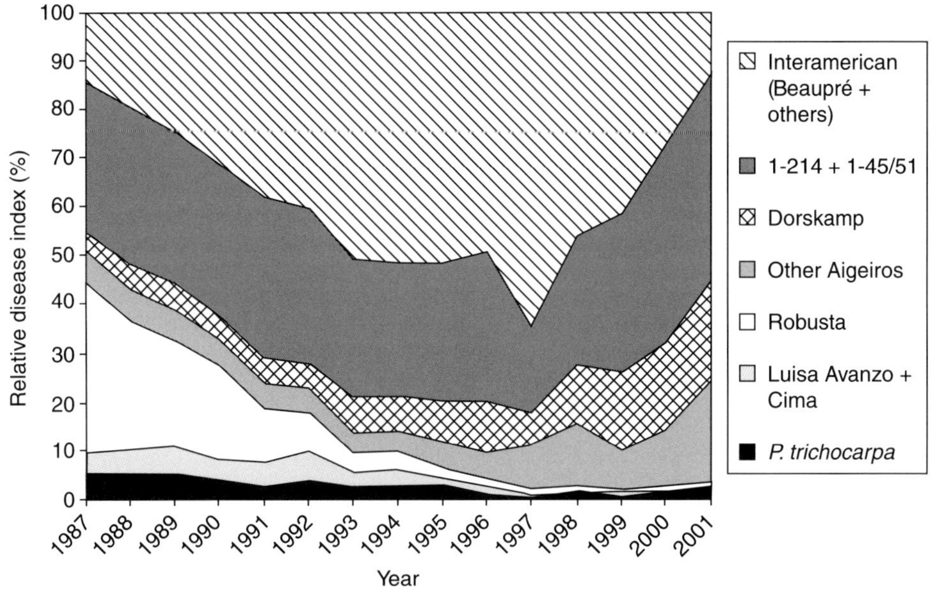

Fig. 12.6. Changes in poplar plant production in France according to the Ministère de l'Agriculture, de l'Alimentation, de la Pêche et des Affaires Rurales (1987–2001).

in new plantations decreased to as low as 0.08% during the winter of 2001–2002. This decrease in popularity was due to rust susceptibility, as well as to declines relating to flooding during the growing season.

Virulence 7 was first detected, at very low frequency, in 1994. Advice was given to growers to reduce the proportion of 'Beaupré' and 'Boelare' in new plantations. During the winter of 1996–97, 'Beaupré' represented 57% of the total number of poplar plants purchased (Fig. 12.6). In addition, this clone was especially popular in the northern half of France, where 80% of new plantations were planted with 'Beaupré'. Damage became evident in northern France in 1997 and, from the winter of 1996–97 until the winter of 2001–2002, the general demand for this clone decreased sharply, and in winter 2001–2002, 'Beaupré' contributed only 2.75% of the plantations (Fig. 12.6). Such a boom and bust situation had been described previously by Van der Plank (1968) for agricultural crops.

In the meantime, new interest emerged in older cultivars, most of which ('I 214', 'Dorskamp', 'I-45/51', 'Flevo') belonged to *P.* × *euramericana*. Hence the performance of 'I-214' and ''I-45/51' against rust was not modified by the new virulences of *M. larici-populina*. 'Dorskamp' and 'Flevo' had originally been completely resistant and became infected with the emergence of virulence 7. Nevertheless, they possessed a good level of general resistance and some new clones such as 'Triplo' and 'Koster' have also been produced. There has been renewed interest in one clone of *P.* × *interamericana* ('Raspalje'). In the past, it had been relatively susceptible to rust, and so 'Beaupré' was preferred because of its health and fast growth. When virulence 7 developed, 'Beaupré' became badly infected while 'Raspalje' appeared healthier, especially when planted close to 'Beaupré'. 'Raspalje' was less susceptible to E4 pathotypes than to E1 pathotypes (Fig. 12.1). Because 'Beaupré' is still common, in many places E4 pathotypes are the most frequent, especially in northern France.

Testing Clones Considering Fungal Diversity

Methods used for rust resistance testing must be reliable. One risk is to declare a clone as resistant when it is not, either because the pathotypes capable of causing infection were not present in the test or present only at a very low frequency. Three levels of variability must be taken into account: species, virulences and aggressiveness. Although complete resistance was easily defeated, susceptible clones became acceptable if it was demonstrated that there was a high level of tolerance, i.e. obvious infection without serious suppressive effect on growth.

Melampsora allii-populina

In the past, *M. allii-populina* occurred mixed with *M. larici-populina*, especially in western France. Rust was assessed in a nursery located in this area (Guémené-Penfao) and samples were taken from plants for further species identification under the microscope. This was not completely satisfactory because the proportion of infection due to each *Melampsora* species differed according to clone, and it would have been too time-consuming to determine the precise proportions. In 1999 a clonal trial was established in the nursery with an alternate host (leek). Leaves bearing telia of *M. allii-populina*, collected in the previous year, were placed near to the leeks. However, host alternation was largely unsuccessful. In 2000, the clonal trial was inoculated in early June in a different way. All larch plants present in the nursery were sprayed with a fungicide in order to prevent their infection by *M. larici-populina*. Therefore the nursery could only be infected by *M. larici-populina* later in the season by inocula coming from outside. Plants of ''Beaupré' (susceptible to *M. allii-populina*) were grown in a glasshouse at INRA, Nancy, inoculated in the laboratory with several isolates of *M. allii-populina* (covering the whole range of known virulences) and transported to Guémené-Penfao nursery on 6 June, just before sporulation. On 19 June all clones were observed and around 60% showed first symptoms. Epidemics started and infection was well established by mid-July. For the first time it was possible to compare poplar clones for their susceptibility to *M. allii-populina*. The reactions of clones ranged from free of rust to severely infected. In the future, those clones with defeated resistance need to be tested under sufficient inoculum pressure of the relevant virulence so that they fully express their susceptibility.

Melampsora larici-populina

Because of the high risk of infection and damage posed by *M. larici-populina*, a sophisticated strategy has been developed (Pinon and Berthelot, 2004). Most of the time, clones are evaluated in the nursery with a controlled inoculum, so as to ensure that all virulences are present. Larch plants are grown in the trial and in spring poplar leaves bearing telia are placed on the ground under the larch. These infected leaves had been collected during the previous autumn from differential clones (Table 12.1). Samples were also collected from 'Robusta' at the time of each assessment, in order to quantify the prevalent pathotype population. For most clones, laboratory tests were conducted to determine which virulence(s) was (were) required to infect clones being tested. Symptoms are also assessed on control clones (including those of Table 12.1) of known susceptibility. Results were interpreted by taking the following into account: the level of symptoms on each clone, the pathotype population and the virulences required for infection of clones. This strategy has proved successful, but a complementary approach has also been developed for clones with intermediate reactions to the rust, as it is difficult to define the acceptable threshold for such clones. Recently, shoot growth on some clones was compared between fungicide-sprayed and unsprayed control plants. It appeared that the rust impact was much greater on *P.* × *interamericana* than on *P.* × *euramericana*. Within a hybrid type, clones with the same level of infection did not suffer from rust to the same extent.

Therefore, not only clones with fewer rust symptoms, but also those exhibiting a fair level of tolerance, may be selected. After a genetic analysis, Dowkiw (2003) showed that the link between resistance (expressed by symptoms) and tolerance (growth under infection) is weak and consequently, these two characters may have to be selected independently. He suggested that a major gene for partial resistance is inherited from the *P. trichocarpa* parent.

When a new virulence appears, breaking down the complete resistance of some clones, nursery tests do not allow a pertinent evaluation of such clones, because the frequency of the new virulence is too low. Several years are needed to increase the new virulence frequency to such a level that allows the clones to express their true level of resistance. This was evident in the popular clones such as 'Beaupré' when E4 appeared, and in the recently released cultivars such as 'Hoogvorst' and 'Hazendans' when E5 was detected. Poplar nursery staff, poplar growers and officers in charge of forest policy need to be able to make an early prognosis and, through laboratory tests, the future of such cultivars can be forecast. Tests were conducted on discs cut from leaves taken from poplar cuttings grown in the greenhouse. 'Robusta', which is susceptible to all pathotypes, was used as a control and latency and infection (the number of uredinia per disk) were measured. Figure 12.5 shows the high level of susceptibility of 'Beaupré', 'Boelare' and 'Ogy' to E4 and an even more serious infection by E5. With such early tests it was possible to forecast possible heavy infections on 'Hazendans' and 'Hoogvorst' by E5. In 2003 those cultivars were badly damaged in Belgium (J. Van Slycken, Belgium, 2003, personal communication). A method, based on a particle counter, was also developed in cooperation with Husson (Husson, unpublished data) to estimate the level of sporulation. Sporulation depends on the clone being tested and, to some extent, on the isolate or pathotype. 'Beaupré' and 'Boelare' supported a high level of sporulation of the pathogen and the level increased progressively after the appearance of uredinia. Sporulation was less intense on 'Robusta' and this is one of the reasons why this clone is less damaged by rust than 'Beaupré'.

Nursery tests are preferred because they are less time-consuming and can take into account the effect of plant physiology and phenology on rust development. Nevertheless, when a new virulence does appear, an early prognosis can be obtained on a limited number of clones in laboratory tests.

Mixing Clones and *M. larici-populina* Populations

Poplar cultivation is generally based on monoclonal stands. Mixing cultivars is a potential approach to rust control. This approach is well documented for willow rust (Chapters 16 and 17, this volume). These authors obtained positive effects on health and growth from willow mixtures. Miot *et al.* (1999) studied a mixture of three poplar clones whose complete resistance was defeated by different virulences of *M. larici-populina* (*P.* × *jackii* 'Aurora', *P.* × *interamericana* 'Boelare' and 'Rap'). Each clone was inoculated with an isolate avirulent on the two other clones. Over 3 years, the benefit of the mixture was small on infection and often none on growth. The experiment was located in a glade in a beech and oak forest with little poplar cultivation in the surrounding areas. Nevertheless, natural infection occurred during the second and third years due to 1–3–4–5–7 isolates. Such isolates were virulent on 2 out of 3 mixture components. As discussed previously, *M. larici-populina* is able to cumulate virulences and maintain a fair level of aggressiveness and fitness. This feature may reduce the interest of poplar clonal mixtures and, even more, such mixtures may favour complex and aggressive isolates.

Discussion

The breakdown of complete resistance to *M. larici-populina*, the low level of general resistance in many *P.* × *interamericana* cultivars and the popularity of these cultivars had a major impact on poplar cultivation in Europe. The population of the pathogen changed rapidly, with new virulent isolates spreading easily on the poplar clones with defeated resistance. In any given location, the frequency of such isolates increased quickly. In addition, higher aggressiveness (shorter latency and higher sporulation) was later detected. When a new virulence (virulence 7) was detected in complex isolates, the incubation period was longer on clones with defeated resistance (e.g. 'Beaupré') than on universal cultivars such as 'Robusta'. Later, isolates with reduced incubation periods were detected. At present, the pathogen population is much more diverse in cultivation areas, made up of complex isolates (combining many virulences) with a high level of aggressiveness. Dowkiw (2003) also found longer incubation periods with complex isolates having the recent virulence 8.

It is too early to know if cultivating more cultivars with general resistance will reverse the present situation towards populations with fewer and simpler pathotypes. Until now, population studies were mainly devoted to virulences and more recently to fungal genotypes. A key question to be asked is about the variability in aggressiveness, for which some evidence is available. Again, it is not known if the cultivation of cultivars with higher levels of general resistance will exert some selection pressure on the fungus for more aggressiveness.

The high level of adaptability of this pathogen to its host is clearly demonstrated and the choice of clones offering diversity is a prerequisite for sustainable poplar cultivation. Poplar growers were quite often obliged to spray in order to control the disease, and new plantations were made up of clones with more general resistance, but less vigour. Increase of diversity was advocated, but this was sometimes difficult to achieve as many clones were genetically related. Some new clones with more durable resistance were proposed for cultivation, but often information on their characters at adult age (adaptation to ecological conditions, wood quality and plasticity) was incomplete or missing.

The breakdown of resistance also had a marked influence on poplar breeders, who were obliged to select new clones and change their strategy, choosing alternative parents and initiating new crosses. Numerous vigorous cultivars were discarded. Interest in complete resistance (not necessarily durable) was questioned. Dowkiw *et al.* (2003) demonstrated recently that there is a positive side-effect of complete (but defeated) resistance. This resistance is linked with general resistance towards some virulences. They suggested that this general resistance implicates a major gene from *P. trichocarpa*. Dowkiw (2003) also proved that progenies with a given *P. trichocarpa* father can exhibit various level of resistance according to the *P. deltoides* mother parent. It means that some associations of parents are more interesting than others. Also some backcrosses of *P.* × *interamericana* towards *P. deltoides* carried out by the Instituut voor Bosbouw en Wildbeheer (IBW) in Geraardsbergen (Belgium) produced hybrids with a sufficient level of resistance. Thus it seems too early to discard definitively, complete resistance. Recently it was found, among *P. deltoides* clones under study at INRA Orléans, that some individuals were without this complete resistance but had a good level of general resistance. These clones may be used as parents for creating new *P.* × *interamericana* clones. New hybrid combinations can also be made. For instance, IBW recently submitted hybrids between *P. trichocarpa* and *P. maximowiczii*. These clones proved to exhibit only general resistance in laboratory tests and the first symptom assessments in the nursery suggested a good level of resistance.

Basic knowledge is still needed in defining clone management in space and time. This is essential because host populations exert an effective and rapid selection pressure on the pathogen population, which affects the health of poplars, under a boomerang effect. The ability of the fungus to develop complex isolates with high aggressiveness and fitness makes things more complicated. The diversification of resistance sources and the deployment of rust tolerance are important areas for future investigation.

Acknowledgements

The authors thank Claude Husson for his collaboration, especially in the development of a method for sporulation assessment.

References

Dowkiw, A. (2003) Analyse génétique de la résistance et de la tolérance de peupliers hybrides *Populus deltoides* × *Populus trichocarpa* aux rouilles foliaires à *Melampsora larici-populina*. PhD thesis, The University of Orléans, France.

Dowkiw, A., Husson, C., Frey, P., Pinon, J. and Bastien, C. (2003) Partial resistance to *Melampsora larici-populina* leaf rust in hybrid poplars: genetic variability in inoculated excised leaf disk bioassay and relationship with complete resistance. *Phytopathology* 93, 421–427.

Dupias, G. (1965) Les rouilles des peupliers dans les Pyrénées et le Bassin sous-pyrénéen. *Bulletin de la Société Mycologique de France* 81, 188–196.

Fragoso, R.G. (1925) *Flora Iberica. II. Uredinales*. Museo Nacional de Ciencias Naturales, Madrid, pp. 198–199.

Frey, P. and Pinon, J. (1997) Variability in pathogenicity in *Melampsora allii-populina* expressed on poplar cutivars. *European Journal of Forest Pathology* 27, 397–407.

Gastine, F., Berthelot, A., Bouvet, A., Servant, H. and Roy, B. (2003) La protection phytosanitaire du cultivar 'Beaupré' est-elle efficace? *Informations Forêt* 2, 6 pp.

Latch, B.J. and Wilkinson, A.G. (1980) New poplar clones help distinguish races of *Melampsora larici-populina* Kleb. in New Zealand. *Australasian Plant Pathology* 9, 112–113.

Lemoine, M. and Pinon, J. (1978) Différences clonales de sensibilité des peupliers aux rouilles à *Melampsora larici-populina* et *M. allii-populina*. *Revue Forestière Française* 30, 181–185.

Magnani, G. (1966) Investigations on the possibility of physiological specialization in *Melampsora allii-populina*. *Pubblicazioni del Centro di Sperimentazione Agricola e Forestale Roma* 8, 127–133.

Miot, S. (1999) Rôle de la variabilité de *Melampsora larici-populina* agent de la rouille des peupliers et de la structure de la population hôte sur l'évolution des populations de l'agent pathogène. PhD thesis, The University Henri Poincaré, Nancy, France.

Miot, S., Frey, P. and Pinon, J. (1999) Varietal mixture of poplar clones: effects on infection by *Melampsora larici-populina* and on plant growth. *European Journal of Forest Pathology* 29, 411–423.

Newcombe, G. (1998) A review of exapted resistance to disease of *Populus*. *European Journal of Forest Pathology* 28, 209–216.

Newcombe, G., Stirling, B., McDonald, S. and Bradshaw, H.D. Jr (2000) *Melampsora columbiana*, a natural hybrid of *M. medusae* and *M. occidentalis*. *Mycological Research* 104, 261–274.

Pinon, J. (1986) Situation de *Melampsora medusae* en Europe. *Bulletin OEPP* 16, 547–551.

Pinon, J. (1991) Eléments de répartition des rouilles des peupliers cultivés en France. *Comptes-rendus de l'Académie d'agriculture de France* 77, 109–115.

Pinon, J. (1992) Variability in the genus *Populus* in sensitivity to *Melampsora* rusts. *Silvae Genetica* 41, 25–34.

Pinon, J. (1995) Présence en France d'une nouvelle race de *Melampsora larici-populina*, agent de la rouille foliaire des peupliers cultivés. *Revue Forestière Française* 47, 230–234.

Pinon, J. and Bachacou, J. (1984) Existence de deux groupes d'isolats différant par leur pouvoir pathogène chez *Melampsora larici-populina* Kleb. *Comptes-rendus de l'Académie d'agriculture de France* 70, 114–122.

Pinon, J. and Berthelot, A. (2004) *La prise en compte des problèmes sanitaires par le GIS Peuplier*. Les Cahiers du DSF, 1–2003/2004 (La santé des Forêts [France] en 2002). Min. Agri. Alim. Pêche Aff. Rur. (DGFAR), Paris, pp. 84–87.

Pinon, J. and Frey, P. (1997) Structure of *Melampsora larici-populina* populations on wild and cultivated poplar. *European Journal of Plant Pathology* 103, 159–173.

Pinon, J. and Lefèvre, F. (1994) A new virulence found among isolates of *Melampsora larici-populina*. Réunion de la Commission Internationale du Peuplier, Izmit (Turkey), 3–7 October.

Pinon, J. and Peulon, V. (1989) Mise en évidence d'une troisième race physiologique de *Melampsora larici-populina* Kleb en Europe. *Cryptogamie Mycologie* 10, 95–106.

Pinon, J. and Valadon, A. (1997) Comportement des cultivars de peupliers commercialisables dans l'Union européenne vis-à-vis de quelques parasites majeurs. *Annales des Sciences Forestières* 54, 19–38.

Prakash, C.S. and Thielges, B.A. (1987) Pathogenic variation in *Melampsora medusae* leaf rust of poplars. *Euphytica* 36, 563–570.

Prakash, C.S. and Thielges, B.A. (1989) Interaction of geographic isolates of *Melampsora medusae* and *Populus*: effect of temperature. *Canadian Journal of Botany* 67, 486–490.

Shain, L. (1988) Evidence for *formae speciales* in poplar leaf rust fungus *Melampsora medusae*. *Mycologia* 80, 729–732.

Singh, S. and Heather, W.A. (1982) Temperature sensitivity of race–cultivar interactions in *Melampsora medusae* Thum and *Populus* spp. *European Journal of Forest Pathology* 12, 123–127.

Steenackers, V. (1982) Nouvelle race physiologique de *Melampsora larici-populina* en Belgique (provisional communication). FAO–CIP 22nd session du groupe de travail maladies Casale–Monferrato (Italy), September 1982.

Steenackers, M.; Steenackers, V. and Delporte, T. (1994) A new physiological race of *Melampsora larici-populina* in Belgium. FAO–CIP, Réunion du groupe de travail sur les maladies, Izmit (Turkey), 3–7 October.

Van der Plank, J.E. (1968) *Disease Resistance in Plants*. Academic Press, New York.

Van Kraayenoord, C.W.S. (1984) *National Report on Activities Related to Poplar and Willow*. International Poplar Commission, Ottawa.

Van Vloten, H. (1949) Kruisingsproeven met rassen van *Melampsora larici-populina* Klebahn. *Tijdschrift over Plantenziekten* 55, 196–209.

Viennot-Bourgin, G. (1937) Contribution à l'étude de la flore cryptogamique du Bassin de la Seine (11ième note). Deux urédinées nouvelles. *Revue de Pathologie et d'Entomologie Agricole de France* 24, 78–85.

13 Transgenic Hybrid Aspen with Altered Defensive Chemistry: a Model System to Study the Chemical Basis of Resistance?

Johanna Witzell[1], Marlene Karlsson[1], Marisa Rodriguez-Buey[2], Mikaela Torp[1] and Gunnar Wingsle[1]
[1]*Umeå Plant Science Center, Department of Forest Genetics and Plant Physiology, Swedish University of Agricultural Sciences, SE-901 83 Umeå, Sweden;*
[2]*Umeå Plant Science Center, Department of Plant Physiology, Umeå University, S-901 87 Umeå, Sweden*

Phenolic Compounds of *Salicaceae* Plants

Plants of the *Salicaceae* family (poplars and willows) are rich in phenolic secondary metabolites, including low molecular weight phenolic glycosides (salicylates) and high molecular weight polyphenols (flavonoids and condensed tannins) (Lindroth *et al.*, 1987, 2001; Nichols-Orians *et al.*, 1993; Ruuhola, 2001). During recent decades, there has been considerable accumulation of evidence corroborating their versatile role as mediators of interactions between *Salicaceae* plants and their insect and mammalian herbivores (Pasteels *et al.*, 1983; Tahvanainen *et al.*, 1985; Kolehmainen *et al.*, 1994; Ruuhola *et al.*, 2001). Phenolic metabolites are also recognized as antimicrobial agents and signal compounds in many plant–pathogen interactions (for reviews see Nicholson and Hammerschmidt, 1992; Dixon and Paiva, 1995). However, there are only a few reports considering their functions in the interactions between pathogens and *Salicaceae* plants (Scaysbrook *et al.*, 1992; Hakulinen, 1998). Thus, their possible importance for the resistance of willows and poplars to pathogens has remained poorly understood. In willows, *Melampsora* rust infection has been shown to induce the accumulation of certain phenolic metabolites (Hakulinen, 1998; Hakulinen *et al.*, 1999; Hakulinen and Julkunen-Tiitto, 2000), but the molecular mechanisms behind this induction, or its possible consequences to the invading fungi, have not been studied.

MYB Genes Regulate Phenylpropanoid Metabolism in Plants

The biosynthesis of phenolic compounds involves different metabolic pathways. Aromatic amino acids, such as phenylalanine, which serve as precursors in the synthesis of phenolics, are products of the shikimic acid pathway. Enzymatic deamination of phenylalanine yields *trans*-cinnamic acid, which can be further metabolized to *para*-coumaric acid. The subsequent additions of hydroxyl groups and other substituents yield a variety of derivatives, phenylpropanoids, which are precursors of more complex phenolic substances, such as lignin (Ruuhola, 2001). The regulation of the enzymes of phenylpropanoid metabolism is likely to be complex, given the high structural diversity of phenolics, as well as the great variation in phenolic levels due to plant inherent factors (genotype, ontogeny) and environmental factors (Coleman *et al.*, 1992; Herms and Mattson, 1992). At the transcriptional level, different branches of the phenylpropanoid metabolism have been shown to be regulated by DNA-binding proteins encoded by the MYB-domain genes (Grotewold *et al.*, 1994; Sablowski *et al.*, 1994; Moyano *et al.*, 1996; Tamagnone *et al.*, 1998). Members of the MYB family have been found in nearly all eukaryotes, and in plants they comprise the largest family of transcription factors. In addition to regulation of the phenylpropanoid pathway, they have been shown to influence hormone responses (Urao *et al.*, 1993), cell shape (Noda *et al.*, 1994), trichome differentiation (Oppenheimer *et al.*, 1991) and plant defence responses (Yang and Klessig, 1996).

Hybrid Aspen Carrying Antisense Constructs of MYB Genes Show Altered Phenotype and Chemical Deviations

Recently, two MYB genes, *PttMYB46* and *PttMYB76*, were isolated and cloned from hybrid aspen (Karlsson, 2003). These genes were selected from hybrid aspen expressed sequence tag (EST) library from the Swedish Centre for Tree Functional Genomics project database (PopulusDB) and their cDNA cloned. A polymerase chain reaction (PCR) fragment carrying *Hin*dIII and *Bam*HI restriction sites was generated from the 3′ non-conserved part of the genes and cloned in antisense orientation in a pJIT60 vector between the double cauliflower mosaic virus (CaMV) 35S promoter and the CaMV 35S terminator (for more details, see Karlsson, 2003). Cuttings of hybrid aspen (*Populus tremula* L. × *P. tremuloides* Michx., clone T89) were transformed with *Agrobacterium tumefaciens* strains carrying the gene of interest and regenerated as described previously (Nilsson *et al.*, 1992). Several lines carrying the antisense construct of *PttMYB46* or *PttMYB76* were recovered, multiplied as cuttings and rooted. Wild type (WT) plants (clone T89) were multiplied in the same way. The plants were grown in a climate chamber illuminated by Osram HQI-TS 400 W lamps giving a minimum quantum flux density of 150 μE m^{-2} s^{-1}, with 16 h photoperiod and 20°C/16°C day/night temperature cycles.

The transformed lines showed altered phenotypes with a stunted appearance. More detailed studies confirmed that the transgenic lines had reduced number and length of internodes. In addition, some lines showed increased trichome frequency and anatomical alterations in the stem tissues. Such notable morphological and anatomical deviations from the WT plants suggested underlying metabolic and chemical differences (Karlsson, 2003). The lines 46 III, 46 IV, 76 III and 76 V, which showed a clearly deviating phenotype from the WT, were chosen for chemical analyses. The bark of the transgenic 76 V plants was found to have a markedly decreased carbohydrate/lignin ratio, suggesting enhanced allocation of carbon to phenylpropanoid metabolism in the bark of these plants (Karlsson, 2003). This inspired us to compare the levels of individual phenolic compounds in WT and transgenic plants. We used liquid chromatography (HPLC) to measure the concentrations of methanol-extractable phenolics in the leaf tissues of the WT and transgenic plants. From 6-week-old plants ($n = 5$ for WT and each line), the first full-grown leaf was collected, air-dried at room temperature and stored in a desiccator until extracted and analysed as described by Karlsson (2003).

The analyses showed that the phenolic pool of aspen leaves was dominated by salicin (2-*O*-β-D-glucoside of salicyl alcohol) and its derivatives (salicortin and tremulacin). In addition, chlorogenic acid and cinnamic acid derivatives and some flavonoids (quercetin and kaempferol derivatives) were detected. Clear-cut qualitative differences in phenolic profiles were not detected between the three genotypes, whereas quantitative differences were apparent. Compared to the WT plants, transgenic lines had elevated levels of phenolic acids and reduced levels of total salicylates in their full-grown leaves (Fig. 13.1).

Transgenic Plants with Altered Phenolic Levels: a Model System to Study the Chemical Basis of Resistance?

Phenolic compounds are frequently implicated as potential defensive chemicals against pathogens in *Salicaceae* plants (Bradshaw *et al.*, 1991; Scaysbrook *et al.*, 1992; Nicolescu *et al.*, 1996). So far, however, the possible links between the phenolic metabolism and pathogen resistance in *Salicaceae* plants have remained unclear. In several cases, the analyses of different aspects of phenolic metabolism did not include testing of the actual resistance. In most of the studies addressing the question, the degree of resistance/susceptibility has been correlated with the chemical composition or concentrations of a clone or a selection of clones. However, this approach is not without drawbacks and may not lead to profound understanding of the role of secondary chemicals in plant–pathogen interactions, or to identification of reliable biochemical markers for disease resistance/susceptibility in *Salicaceae* plants. As compared with traditional studies with different clones, the transformed hybrid aspen lines with altered phenolic profiles provide a much narrower genetic background for variation in resistance. This should facilitate identification of the chemical characters that are critical for resistance or susceptibility.

As a first attempt to explore this field, we performed experiments where we inoculated the detached leaves of the WT and different transgenic lines with conidia from *Pollacia radiosa*, the anamorph of *Venturia tremulae* (*Ascomycotina*). This fungus is the causal agent of aspen and poplar leaf and twig blight, with the characteristic 'shepherd's crook' symptoms. The leaves were detached from young trees grown in the same conditions as above, surface-sterilized and maintained on MS/agar medium. Conidia suspension or water (control) was spread on the leaves. During a 14-day incubation period, the development of the disease was monitored by measurements of the necrotic area on leaves (Fig. 13.2). The results of the inoculation experiments indicated varying responses to infection among the studied aspen genotypes. Compared to WT, the line 76 III did not show marked differences in sensitivity to infection. However, on leaves of the line 76 V infection seemed to spread faster than on WT leaves (Fig. 13.2A). The lines carrying the antisense construct of *PttMYB46* seemed to be sensitive to the infection. We found that in the line 46 IV the first symptoms appeared 2 days earlier than in WT. Moreover, although the infection on leaves of line 46 III started at

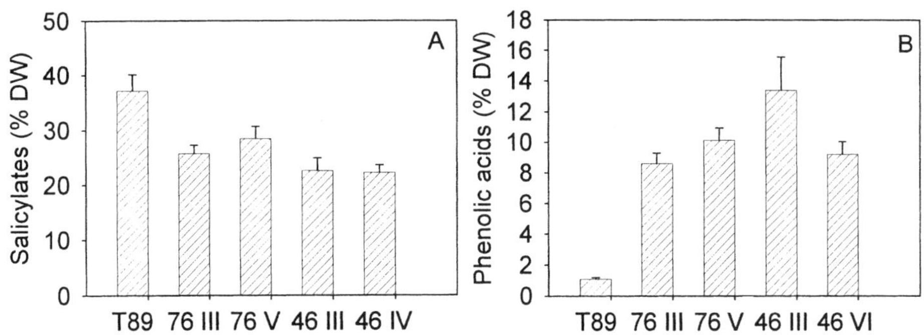

Fig. 13.1. Relative concentrations of (A) salicylates (sum amount of salicin, salicortin and tremulacin) and (B) phenolic acids (sum amount of three chlorogenic acid and two cinnamic acid derivatives) in hybrid aspen (clone T89 = wild type) and transgenic plants.

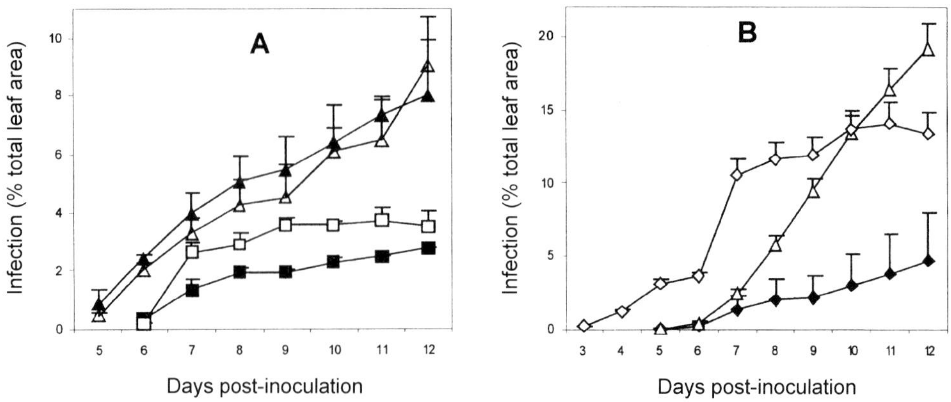

Fig. 13.2. Infected necrotic area expressed as the percentage of the total leaf area from (A) hybrid aspen (▲, ■) and its transgenic lines 76 III (△) and 76 V (■), carrying an antisense construct of *PttMYB76* gene; and (B) hybrid aspen (◆) and its transgenic lines 46 III (△) and 46 IV (◇), carrying an antisense construct of *PttMYB46* gene. Shown are the means of two different experiments using at least three different leaves from two individuals for every line in every experiment. Vertical bars represent standard error of the mean.

the same time as on WT leaves (i.e. 5 days after the inoculation) it spread faster than on WT leaves (Fig. 13.2B). The results suggest that in our study system the analysed phenolic compounds did not participate in the defence against the pathogen infection, or the observed changes in the phenolic pool did not compromise the plant response to the pathogen. It is also possible that the pathogen was able to use the higher amount of phenolic compounds as a carbon source.

At present, we are testing the sensitivity of the transgenic plants against infections by other fungal pathogens that use different strategies for infection than the studied ascomycete fungus. Among these are leaf rusts (*Melampsora* sp., *Basidiomycotina*) and stem-infecting *Neofabraea populi* (*Ascomycotina*). Detailed studies on the resistance of the hybrid aspen plants carrying antisense constructs of MYB genes against pathogens and other natural enemies, such as insect herbivores, should provide us with novel insights into the molecular and chemical basis of resistance in plants. The amenability of hybrid aspen to modern molecular techniques (e.g. microarrays) and the increasing knowledge of its genetics open completely new possibilities for this type of investigation.

Conclusions

The phenylpropanoid pathway produces a variety of phenolic secondary metabolites, which may function as signal compounds and defensive chemicals in plant–pathogen interactions. Members of the MYB family of transcription factors have been recognized as regulators of different branches of phenylpropanoid metabolism. Recently, two MYB genes, *PttMYB46* and *PttMYB76*, were isolated and cloned from hybrid aspen. The morphological and anatomical deviations of transgenic plants from the wild-type plants suggested underlying chemical differences, which could be linked to phenylpropanoid metabolism. We studied the levels of methanol-soluble phenolic compounds in four transgenic hybrid aspen lines, each carrying an antisense construct of one of these genes. Marked differences in the levels of individual phenolic compounds were found between wild-type and transgenic lines. In general, the level of phenolic acids was raised and the level of salicylates (salicin and its derivatives, salicortin and tremulacin) was reduced in the transgenic plants. These transgenic lines may provide a novel model system for investigations of the role of phenolics in interactions between poplars and their natural enemies, such as *Melampsora* rusts.

References

Bradshaw, H.D. Jr, Parsons, T.J. and Gordon, M.P. (1991) Wound-responsive gene expression in poplars. *Forest Ecology and Management* 43, 211–224.

Coleman, J.S., Jones, C.G. and Krishik, V.A. (1992) Phytocentric and exploiter perspectives of phytopathology. *Advances in Plant Pathology* 8, 149–195.

Dixon, R.A. and Paiva, N.L. (1995) Stress-induced phenylpropanoid metabolism. *Plant Cell* 7, 1085–1097.

Grotewold, E., Drummond, B.J., Bowen, B. and Peterson, T. (1994) The myb-homologous P gene controls phlobaphene pigmentation in maize floral organs by directly activating a flavonoid biosynthetic gene subset. *Cell* 76, 543–553.

Hakulinen, J. (1998) Nitrogen-induced reduction in leaf phenolic level is not accompanied by increased rust frequency in a compatible willow (*Salix myrsinifolia*)–*Melampsora* rust interaction. *Physiologia Plantarum* 102, 101–110.

Hakulinen, J. and Julkunen-Tiitto, R. (2000) Variation in leaf phenolics of field-cultivated willow (*Salix* sp.) clones in relation to occurrence of *Melampsora* rust. *Forest Pathology* 30, 29–41.

Hakulinen, J., Sorjonen, S. and Julkunen-Tiitto, R. (1999) Leaf phenolics of willow clones differing in resistance to *Melampsora* rust. *Physiologia Plantarum* 105, 662–669.

Herms, D.M. and Mattson, W.J. (1992) The dilemma of plants, to grow or to defend. *Quarterly Review of Biology* 67, 283–335.

Karlsson, M. (2003) Molecular factors involved in the formation of secondary vascular tissues and lignification in higher plants. Studies of CuZn-SOD and members of MYB and zinc-finger transcription factors families. PhD thesis, Swedish University of Agricultural Sciences, Sweden. *Silvestria* 280.

Kolehmainen, J., Roininen, H., Julkunen-Tiitto, R. and Tahvanainen, J. (1994) Importance of phenolic glycosides in host selection of shoot galling sawfly *Euura amerinae*, on *Salix pentandra*. *Journal of Chemical Ecology* 20, 2455–2466.

Lindroth, R.L., Hsia, M.T.S. and Schriber, M.J. (1987) Characterization of phenolic glycosides from quaking aspen. *Biochemical Systematics and Ecology* 15, 677–680.

Lindroth, R.L., Roth, S. and Nordheim, E.V. (2001) Genotypic variation in response of quaking aspen (*Populus tremuloides*) to atmospheric CO_2 enrichment. *Oecologia* 126, 371–379.

Moyano, E., Martinez-Garcia, J.F. and Martin, C. (1996) Apparent redundancy in myb gene function provides gearing for the control of flavonoid biosynthesis in antirrhinum flowers. *Plant Cell* 8, 1519–1532.

Nicholson, R.L. and Hammerschmidt, R. (1992) Phenolic compounds and their role in disease resistance. *Annual Review of Phytopathology* 30, 369–389.

Nichols-Orians, C.M., Fritz, R.S. and Clausen, T.P. (1993) The genetic basis for variation in the concentration of phenolic glycosides in *Salix sericea*, clonal variation and sex-biased differences. *Biochemical Systematics and Ecology* 21, 535–542.

Nicolescu, C., Sandre, C. and Jouanin, L. and Chriqui, D. (1996) Genetic engineering of phenolic metabolism in poplar in relation with resistance against pathogens. *Acta Botanica Gallica* 143, 539–546.

Nilsson, O., Alder, T., Sitbon, F., Little, C.H.A., Chalupa, V., Sandberg, G. and Olsson, O. (1992) Spatial pattern of cauliflower mosaic virus 35S promoter–luciferase expression in transgenic hybrid aspen trees monitored by enzymatic assay and non-destructive imaging. *Transgenic Research* 1, 209–220.

Noda, K.I., Glover, B.J., Linstead, P. and Martin, C. (1994) Flower colour intensity depends on specialized cell shape controlled by a Myb-related transcription factor. *Nature* 369, 661–664.

Oppenheimer, D.G., Herman, P.L., Sivakumaran, S., Esch, J. and Marks, M.D. (1991) A myb gene required for leaf trichome differentiation in *Arabidopsis* is expressed in stipules. *Cell* 67, 483–494.

Pasteels, J.M., Rowell-Rahier, M., Braekman, J.C. and Dupont, A. (1983) Salicin from host plants as precursor of salicylaldehyde in defensive secretion of *Chrysomelinea* larvae. *Physiological Entomology* 8, 307–314.

Ruuhola, T. (2001) Dynamics of salicylates in willows and its relation to herbivory. PhD thesis, University of Joensuu, Finland. *Dissertations in Biology* 8.

Ruuhola, T., Sipura, M., Nousiainen, O. and Tahvanainen, J. (2001) Systemic induction of Salicylates in *Salix myrsinifolia* (Salisb.) *Annals of Botany* 88, 483–497.

Sablowski, R.W.M., Moyano, E., Schuch, W., Martin, C. and Bevan, M. (1994) A flower-specific Myb protein activates transcription of phenylpropanoid biosynthetic genes. *EMBO Journal* 13, 128–137.

Scaysbrook, T., Greenaway, W. and Whatley, F.R. (1992) Relation of 'antimicrobial' compounds present in poplar bud exudates to disease resistance by poplars. *Zeitschrift für Naturforschung* C 47, 197–200.

Tahvanainen, J., Helle, E., Julkunen-Tiitto, R. and Lavola, A. (1985) Phenolic compounds of willow bark as deterrents against feeding by mountain hare. *Oecologia* 65, 319–323.

Tamagnone, L., Merida, A., Parr, A., Mackay, S., Culianez-Macia, F.A. and Martin, C. (1998) The AmMYB308 and AmMYB330 transcription factors from Antirrhinum regulate phenylpropanoid and lignin biosynthesis in transgenic tobacco. *Plant Cell* 10, 135–154.

Urao, T., Yamaguchi, S., Urao, S. and Shinozaki, K. (1993) An Arabidopsis myb homolog is induced by dehydration stress and its gene product binds to the conserved MYB recognition sequence. *Plant Cell* 5, 1529–1539.

Yang, Y. and Klessig, D.F. (1996) Isolation and characterization of a tobacco mosaic virus-inducible myb oncogene homolog from tobacco. *Proceedings of the National Academy of Sciences USA* 93, 14972–14977.

14 Basidiospore-derived Penetration by Species of *Cronartium* and *Melampsora*: an Outline

Alessandro Ragazzi[1], Nicola Longo[2], Biancamaria Naldini[2], Salvatore Moricca[2] and Irene Dellavalle[3]

[1]Dipartimento di Biotecnologie Agrarie, Sezione di Patologia Vegetale, Universita di Firenze, Piazzale delle Cascine 28, 50144 Frienze, Italy; [2]Dipartimento di Biologia Vegetale, Universita di Firenze, Via la Pira 4, 50121 Firenze, Italy; [3]CNR, Istituto per la Protezione delle Piante, Area della Ricerca del CNR di Firenze, Via Madonna del Piano, 50019, Sesto Fiorentino, Italy

Background

Penetration into host tissue is the very first step of infection by fungal pathogens. Research conducted so far with rust fungi has shown that there are essentially two types of penetration: (i) *direct* penetration, when the rust germ tube invades directly through an epidermal cell wall; and (ii) *indirect* penetration, when it first enters a stoma and then invades the cell (Fig. 14.1).

The most generally accepted theory regarding penetration is that direct penetration is associated with the germ tubes of basidiospores (monokaryotic stage), while indirect penetration is associated with the germ tubes of aeciospores and urediniospores (dikaryotic stage). With direct penetration the (basidiospore) germ tube develops an intra-epidermal infection structure (vesicle and primary infection hypha), while with indirect penetration the (aeciospore/urediniospore) germ tube forms a substomatal infection structure (Littlefield and Heath, 1979; Bushnell and Roelfs, 1984; Hoch and Staples, 1991; Mendgen *et al.*, 1996; Epstein and Nicholson, 1997; Mendgen, 1997).

Patton and Johnson (1970) reported that penetration on the needles of *Pinus strobus* L. by the basidiospores of *Cronartium ribicola* J.C. Fisher ex Rabenh. was initiated indirectly through the stomatal openings and followed by the formation of a substomatal vesicle. However, later studies with many rust fungi confirmed the direct penetration by the basidiospores (Miller *et al.*, 1980; Gray *et al.*, 1983; Gold and Mendgen, 1984, 1991; Hopkin *et al.*, 1988; Longo *et al.*, 1988, 1991, 1994, 1997; Morin *et al.*, 1992) and the researchers considered the indirect penetration of *P. strobus* by *C. ribicola* (Patton and Johnson, 1970) and of *Pinus banksiana* Lamb. by *Cronartium comandrae* Pk. (Bergdhal and French, 1985) as anomalies to be explained.

Gold and Mendgen (1984) reported two other cases of indirect penetration from basidiospores: *Crysomyxa abietis* on *Picea abies* (Grill *et al.*, 1978) and *Coleosporium* spp. on *Pinus* spp. (Bauer, 1983). They considered that direct penetration from

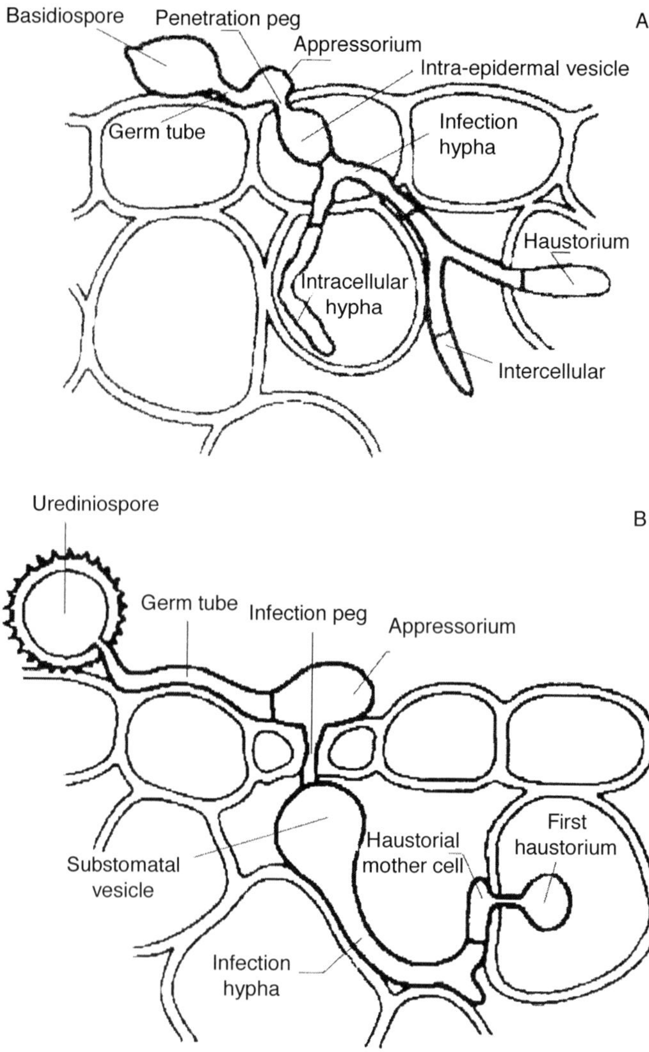

Fig. 14.1. Diagrammatic representation of the monokaryotic (A) and dikaryotic (B) infection process of a rust fungus. (A) Direct penetration; (B) indirect penetration. (Modified from Mendgen, 1997.)

basidiospores was typical of rusts on angiosperms, while indirect penetration from basidiospores was typical on gymnosperms. Penetration of gymnosperm leaves was indirect because of the high level of cutinization and the thick epidermal cell wall. On the other hand, Longo *et al.* (1988, 1991, 1997) found that basidiospore germ tubes of *Melampsora pinitorqua* (Br.) Rostr. and *M. larici-tremulae* Kleb. penetrate into *Pinus* spp. and *Larix decidua* Mill. directly, regardless of any organ they infect. Similar results were obtained with *Cronartium quercuum* (Berk.) Miyabe ex Shirai f. sp. *fusiforme* (Cumm.) Burds. and Snow on *Pinus elliottii* Engelm. var. *elliottii* (Miller *et al.*, 1980) and with *Endocronartium harknessii* (J.P. Moore) Y. Hiratsuka on *Pinus contorta* Dougl. var. *latifolia* Engelm. (Hopkin *et al.*, 1988).

From these findings it can be concluded that the same organ of a gymnosperm can be penetrated either directly or indirectly by a basidiospore germ tube. The manner of penetration depends on the rust species and probably on the tree organ each rust usually infects in nature. Thus *M. pinitorqua*, *C. quercuum* f. sp. *fusiforme* and *E. harknessii* penetrate directly into growing shoots, where there are few or no stomata, while *C. ribicola* and *C. comandrae* penetrate indirectly into the needles, which have many stomata.

Mode of Penetration of *Cronartium* spp.

Cronartium flaccidum (Alb. et Schw.) Wint. is a macrocyclic, heteroecious rust that forms spermogonia and aecia on some species of pine, and uredia and telia on some angiosperms such as *Vincetoxicum hirundinaria* Medicus, *Peonia* spp. and *Gentiana* spp. Basidiospores of *C. flaccidum* infect the primary and secondary needles of pine (Wilson and Henderson, 1966). Its haploid mycelium, as it colonizes the mesophyll, forms stomata in the tissue it invades, giving rise to yellowish spots that are the first symptom of rust infection (Raddi, 1976; Ragazzi and Moriondo, 1979, 1980).

Ragazzi *et al.* (1987) and Ragazzi and Dellavalle Fedi (1992) examined the germ tubes of *C. flaccidum* on the needles of *Pinus pinaster* Ait. and *Pinus nigra* Arn. subsp. *laricio* (Poiret) and concluded that, like *C. ribicola* (Patton and Johnson, 1970), its basidiospore germ tubes entered indirectly through the stoma, and not directly through the epidermal cell wall. However, these authors also stressed that the behaviour of the germ tube after it had invaded the substomatal antechamber was still unclear (Fig. 14.2).

Longo *et al.* (2000) examined the penetration and early colonization of *P. pinea* needles by *C. flaccidum* and found that, on the needle surface, the germ tubes were generally single, not branched, fairly uniform in diameter, and grew in length without directing themselves towards a stoma (Figs 14.3, 14.4). Penetration into a stomatal antechamber was therefore a random event, and indeed, of the many germ tubes of varying size observed, only a few entered stomata. In some cases a germ tube appeared initially to lodge in a stomatal antechamber, but then grew out of it again, continuing its growth without penetrating the stomatal opening. Several germ tubes may enter the same chamber, but usually only one continues to grow, remaining more or less attached to the walls of the subsidiary cells that are located above the guard cells, until it crosses the stomatal opening (Fig. 14.5).

C. flaccidum penetrates into the stomatal opening by the apex of the germ tube, without forming appressoria. At the stomatal opening, germ tube diameter decreases by about half; it becomes flattened and ribbon-like, and takes up only a small, and sometimes peripheral, part of the opening (Figs 14.5 and 14.6). Within the stoma the germ tube is more flattened when closer to the side of the stomatal opening, less so when near centre. Preferential points of germ tube passage through the stomatal opening were not noted. The ultrastructure of the germ tube at the stomatal opening is typical of a vegetative hypha, with its cytoplasm rich in organelles and granules of glycogen. Once a germ tube passes through the stomatal opening and enters the substomatal chamber, the hypha derived directly from this germ tube becomes an infection structure. The term 'infection structure' is used to refer to the hypha that forms inside the substomatal chamber before it initiates intracellular colonization (Longo *et al.*, 2000) (Figs 14.5 and 14.6).

The lack of an appressorium-like structure had already been noted in *C. flaccidum* by Ragazzi and Dellavalle Fedi (1992), in *C. ribicola* by Patton and Johnson (1970) and in *C. comandrae* by Bergdahl and French (1985). The mode of penetration of the three species is similar at their monokaryotic stage. With *C. flaccidum* the germ tube narrows at the point where it passes through the stomatal opening and remains distinctly flattened thereafter. Similar observations were made with *C. ribicola* (Patton and Johnson, 1970) and with *C. comandrae* (Bergdahl and French, 1985).

In *E. harknessii* on *P. contorta* (Hopkin *et al.*, 1988) and *C. quercuum* f. sp. *fusiforme* on *P. elliottii* var. *elliottii* (Miller *et al.*, 1980), basidiospore germ tubes in some cases invaded the stomatal antechamber apparently in a manner of indirect penetration, but then immediately penetrated the anticlinal wall of one of the subsidiary cells surrounding the stoma. There it formed an intracellular structure like that produced by direct penetration through the epidermal cell wall.

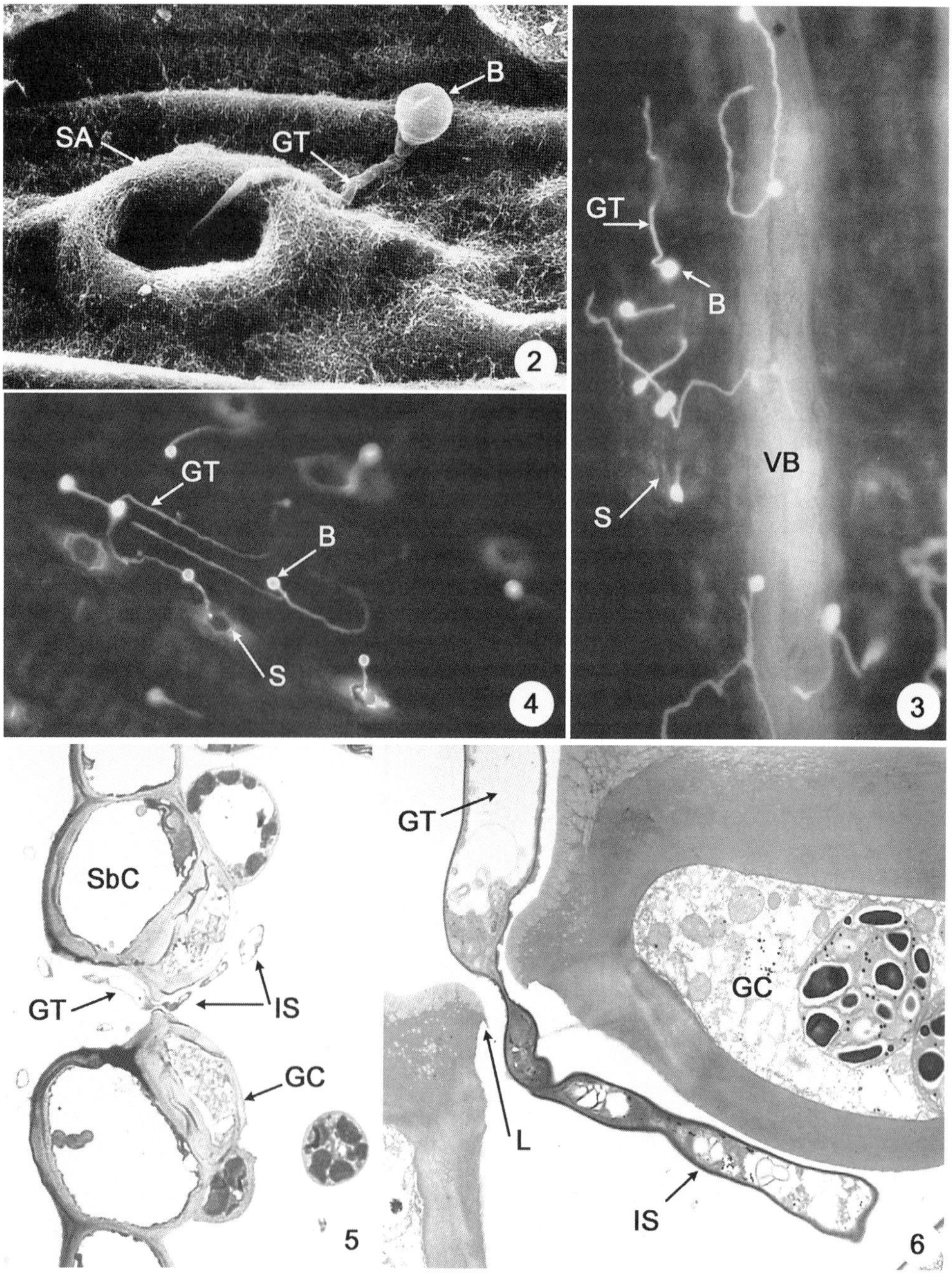

Figs 14.2–6.

The appressorium of the *Uredinales* (Littlefield and Heath, 1979; Harder and Chong, 1984; Hoch and Staples, 1991) is highly differentiated at the dikaryotic infection stage and is separated from the germ tube by a septum. At the monokaryotic stage, the appressorium is generally present but not well differentiated, being merely a swelling of the apex of the germ tube and not normally separated from it by a septum (Miller *et al.*, 1980; Gray *et al.*, 1983; Gold and Mendgen, 1984; Hopkin *et al.*, 1988; Longo *et al.*, 1991, 1994; Morin *et al.*, 1992).

At the dikaryotic stage the penetration peg is described as a thin protuberance growing from the innermost layer of the appressorium wall, which advances, growing apically into the stomatal opening (Littlefield and Heath, 1979).

The penetration peg of the monokaryotic stage also grows directly out of the appressorium but it does not differentiate like that of the dikaryotic stage (Gold and Mendgen, 1984; Mendgen, 1997). The apex which penetrates into the cuticle of the epidermal cell is very thin. It enlarges in the cell wall and, as it grows into the host cell, is sheathed by the cell's plasma membrane.

As far as the basidiospore-derived prepenetration is concerned, *C. flaccidum* differs from most other observed rusts in that its penetration is always indirect (Longo *et al.*, 2000). The same type of penetration is also seen with *C. ribicola* (Patton and Johnson, 1970) and with *C. comandrae* (Bergdahl and French, 1985). In contrast, with *E. harknessii* (Hopkin *et al.*, 1988) or *C. quercuum* f. sp. *fusiforme* (Miller *et al.*, 1980), penetration in some cases is initiated indirectly into the stomatal antechamber, but then it becomes direct, through the anticlinal walls of the subsidiary cells of the stoma.

Mode of Penetration of *Melampsora* spp.

Melampsora pulcherrima (Bub.) Maire, the causal agent of Mediterranean white poplar (*Populus alba* L.) rust, forms its spermogonia and aecia on *Mercurialis annua* L. (Magnani, 1961; Moriondo *et al.*, 1989).

Naldini *et al.* (1993) showed that haploid infection *M. pulcherrima* begins on the host leaves and only thereafter spreads to the stem and lateral branches. Its infection process has been described by Longo *et al.* (1994). The germ tubes generally measure 5–6 μm in length, but may go up to 20 μm in some cases, making them long enough to reach any point of the periclinal wall of an epidermal cell, or the anticlinal walls of two contiguous cells. However, the penetration is always directly through the wall of an epidermal cell. Direct penetration also occurs through the walls of the stomatal subsidiary cells, and sometimes through the wall of the guard cells (Fig. 14.7). The fully developed, intracellular infection structure is a more or less round vesicle which is attached to the

Fig. 14.2. Scanning electron micrograph of *Cronartium flaccidum* on a needle of *Pinus pinaster*. The germ tube (GT) of a basidiospore (B) is going into the stomatal antechamber (SA). ×1000. (Ragazzi and Dellavalle Fedi, 1992).

Fig. 14.3. Epifluorescence optics of *Cronartium flaccidum* on a cotyledon of *Pinus pinea*. The basidiospore germ tubes (GT) show a random behaviour without moving directly towards the stomata (S). B, Basidiospore; VB, vascular bundle. ×220. (Longo *et al.*, 2000.)

Fig. 14.4. Epifluorescence optics of *Cronartium flaccidum* on a secondary needle of *Pinus pinea*. The basidiospore germ tubes (GT) show a random behaviour without moving directly towards the stomata (S). ×180. (Longo *et al.*, 2000.)

Fig. 14.5. Light microscopy (thin transverse section) of *Cronartium flaccidum* on a cotyledon of *Pinus pinea*. Stomatal indirect penetration carried out by a basidiospore germ tube (GT) without appressorium formation. SbC, subsidiary cell; IS, infection structure; GC, guard cell. ×920. (Longo *et al.*, 2000.)

Fig. 14.6. Transmission electron micrograph of *Cronartium flaccidum* on a cotyledon of *Pinus pinea*: the same penetration as Fig. 14.5. Below the stromatal aperture, in the substomatal chamber, the hypha from the germ tube (GT) becomes the 'substomatal infection structure' (IS), which appears not to be diametrically sectioned. GC, Guard cell; L, lip. ×6800. (Longo *et al.*, 2000.)

epidermal cell wall by a narrow neck and, at the distal end, by a hypha-like structure (infection hypha) consisting of two parts, one wide, and one narrow and long. These parts are divided by septa. The infection structure enlarges within the penetrated epidermal cell. In many cases it runs along the cell wall and then turns back to its penetration point (Figs 14.7 and 14.8). The fully developed infection structure generally has some branches that develop in succession from the distal end of the vesicle, near the septum, then from its wider part, and finally from its terminal, narrower end. These branches extend towards the anticlinal wall of the cell, growing into contiguous epidermal cells, or towards the lower periclinal wall of the epidermal cell that was invaded, from which they grow out. The intracellular branches that form terminal structures (haustoria) in contiguous epidermal cells are more numerous in infections on the lower-leaf epidermis. These intracellular branches become infection hyphae that grow out through the lower periclinal walls of the epidermal cells, and form intercellular hyphae in the spongy parenchyma. These hyphae produce haustoria in the spongy parenchyma cells. The intracellular branches in the contiguous epidermal cells are less numerous in infections on the upper-leaf epidermis, where numerous intracellular branches appear to grow out of the epidermal cells and form haustoria in the palisade cells. The still young haustoria produced in the epidermal cells or in the cells of the spongy and palisade parenchyma, although partially developed into the host cells, are similar in shape to infection hyphae, with a larger proximal part, and a distinct and thinner distal part. These intracellular structures have not been seen to grow out of the host cells.

Some of the most important features relating to the basidiospore germ tube penetration and the infection structures of *M. pulcherrima* on leaves of *M. annua* are as follows:

1. Germ tubes are short and random-growing, with a swollen apex that is an undifferentiated appressorium. The germ tube and the appressorium adhere to the epidermal cell wall by means of extracellular material.

2. Penetration is direct: a thin penetration peg comes out of the appressorium and grows through the epidermal cell wall, producing an intracellular infection structure consisting of a vesicle and a septate hypha. The branches of this septate hypha grow out of the epidermal cells and produce both intercellular hyphae in the mesophyll and, directly, hypha-like haustoria in the contiguous epidermal cells and the underlying mesophyll.

3. A matrix can be found between the plasma membrane of the host cell ensheathing the invading fungus and the fungal wall. The plasma membrane encloses all the intracellular fungal structures (infection hypha, branches of the infection hypha, haustoria) except the intra-epidermal vesicle.

Collars with a host cell wall-like structure encircle the neck of the intra-epidermal vesicle in the penetration pore, and the intracellularly growing rust structures (Longo *et al.*, 1994).

The formation of appressoria and the direct germ tube penetration at any point in the epidermal cell wall, including the stomatal subsidiary and guard cells, do not occur frequently. Many researchers described that penetration through the epidermis is more likely to occur near the place where the cell walls come together, towards which the germ tube directs itself in growth (Gold and Mendgen, 1984; Hopkin *et al.*, 1988; Morin *et al.*, 1992).

Differences in the behaviour of germ tubes between different species can be ascribed to characteristics of the epidermal tissue of the host. Even in direct penetration from basidiospores (monokaryotic), there are some plant–fungus interactions based on contact stimuli and chemical signals (Desprez-Loustau and Le Menn, 1989; Gold and Mendgen, 1991), which are generally typical of indirect penetration by dikaryotic urediniospores into the stomata (Hoch and Staples, 1991).

The basidiospore appressorium which is not separated from the germ tube by septa, appearing as a slightly enlarged apex of the germ tube, as in *M. pulcherrima*, is commonly described for basidiospore germlings (Gold and Mendgen, 1984; Mims and Richardson, 1989; Morin *et al.*, 1992). This type of structure is very different from the differentiated appressorium produced during urediniospore germination which gives rise to indirect penetration (Staples and Macho, 1984).

In *M. pulcherrima*, the extracellular material can be seen around the appressorium and around most of the basidiospore germ tube (Fig. 14.8). This is thought to be an exudate that characteristically accompanies basidiospore germination (Gold and Mendgen, 1991) and is generally associated with the fungal structures involved in direct penetration (Nicholson and Epstein, 1991). It not only protects the appressorium but also serves as a reserve for enzymatic penetration (Gold and Mendgen, 1984). Extracellular material probably plays the same enzymatic role in the case of *M. pulcherrima*, the penetration peg of which produces a pore with a sharply defined edge, which might be the result of enzyme activity (Longo *et al.*, 1994).

In other rusts, the basidiospore penetration peg is very thin between the epicuticular wax and the cuticle, and generally enlarges where it passes through the cell wall (Metzler, 1982; Gray *et al.*, 1983; Gold and Mendgen, 1984; Longo *et al.*, 1991), while in *M. pulcherrima* it is of relatively uniform diameter (Fig. 14.8).

In conclusion, ultrastructural studies (Longo *et al.*, 1994) of *M. pulcherrima* on the leaves of *M. annua* confirmed the behaviour and type of direct penetration typical of basidiospore germ tubes.

Melampsora pinitorqua Rostr. is the causal agent of twist rust of pine shoots, and *M. larici-tremulae* Kleb. is the agent of rust of larch needles. They are morphologically very similar. They both have uredial and telial stages on the same range of poplar species, *P. tremula* L., *P. alba* L. and *P. canescens* Sm. and are considered to be *formae speciales* of a single species, *M. populnea* (Pers.) Karst. (Boerema and Verhoeven, 1972). More recently, it has been hypothesized that speciation is taking place in the *M. populnea* complex, and that the above-mentioned entities are evolving towards two new species, reproductively isolated from their sympatric progenitor (Longo *et al.*, 1997, 2002).

The basidiospore germ tubes of these two rusts on their respective aecial hosts (*M. pinitorqua* on *Pinus* spp. and *M. larici-tremulae* on *Larix decidua*) generally measure 5–10 μm, rarely more. Germ tube growth on the host proceeds in a random manner. The apex of the basidiospore germ tube forms an appressorium that appears to be the undifferentiated, slightly enlarged apex of the germ tube itself (Fig. 14.9). When *P. sylvestris* is inoculated with *M. larici-tremulae*, in contrast, most basidiospore germ tubes are longer (50 μm and more), twisted, with irregular swellings and branchings, and without apparent appressoria. In this particular host–parasite interaction, secondary basidiospores are also frequently observed (Longo *et al.*, 1997).

The morphology of the direct type of penetration and of the intra-epidermal infection structures of both rusts on their respective hosts and of *M. larici-tremulae* on *P. sylvestris* is similar. In the host epidermis, the penetration peg derived from the appressorium is narrow at the point where it penetrates the cuticle, and more expanded in the cell wall. The intracellular infection structure derived from the penetrating peg looks somewhat like an ovoid, vacuolated vesicle, linked to the epidermal cell wall by a narrow neck, while its distal part becomes elongated, forming the infection hypha (Figs 14.10–14.12). In the epidermal cell this infection hypha is generally straight. Its ramifications, branching off in succession from the proximal part towards the distal part, grow intercellularly into the first layers of the underlying parenchyma, where they form both intercellular and intracellular hyphae (Fig. 14.10). According to Longo *et al.* (1988, 1991) the intracellular hyphae and the haustoria are characteristic of the first and last stages of rust mycelium development in the host cell, respectively.

The general morphology of the first infection stages of the two rust agents in

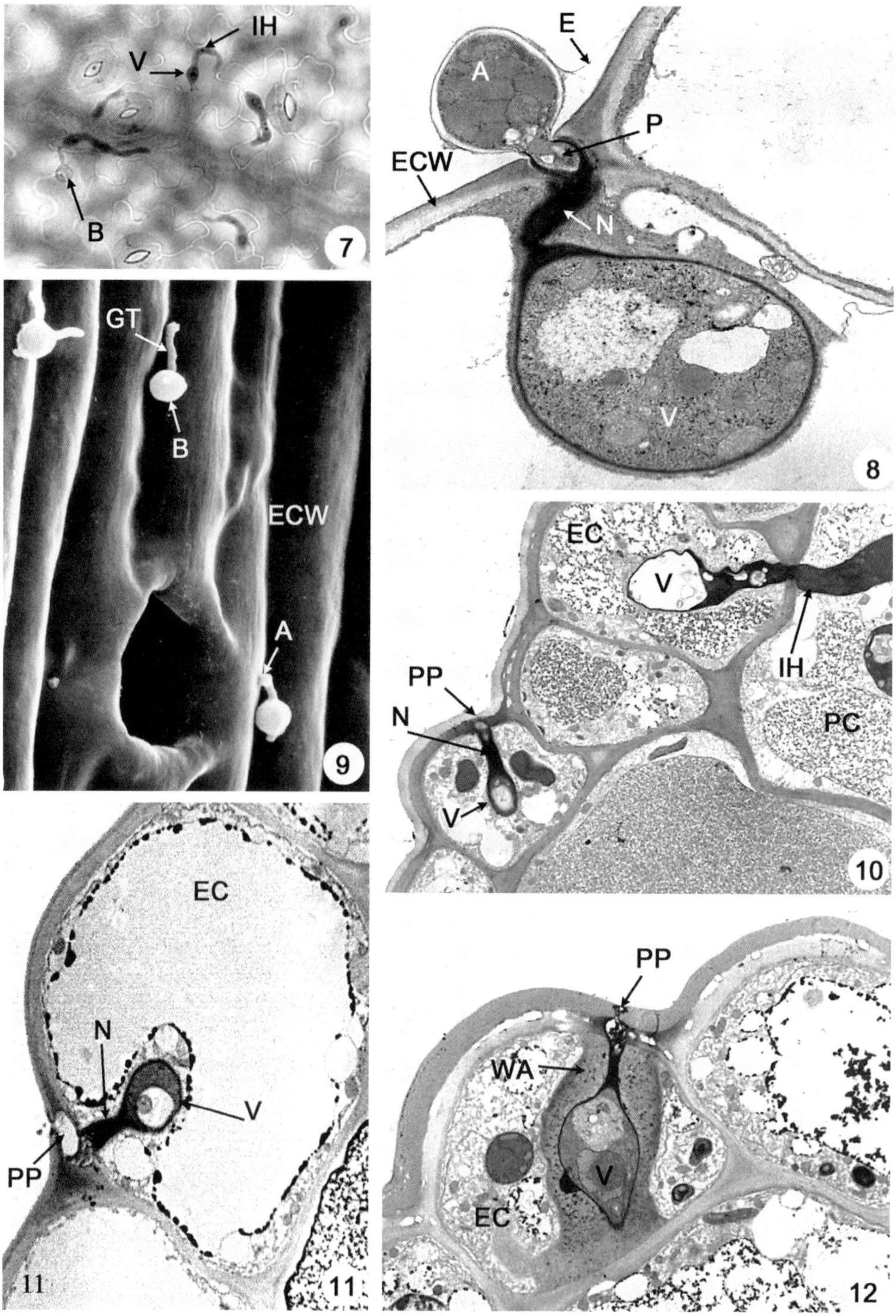
IH
V
B
7
E
A
P
ECW
N
V
8
GT
B
ECW
A
9
EC
V
IH
PP
N
V
PC
10
EC
N
V
PP
11
11
PP
WA
V
EC
12

Figs 14.7–12.

their respective hosts is typical of a rust infection from basidiospores (Gold and Mendgen, 1991), as has already been reported for *M. pinitorqua* by Longo *et al.* (1991) and, more recently, by Morin *et al.* (1992) and Longo *et al.* (1994).

Concluding Remarks

In this chapter, we have reviewed the literature on direct and indirect penetration at the monokaryotic and the dikaryotic stage of the *Uredinales*. It reveals that the penetrating processes and the structural elements of these two types are fundamentally different.

Penetration at the dikaryotic stage

Indirect penetration of the host by the dikaryotic stage of the fungus has the following characteristics: the germ tubes direct themselves purposely towards the stomata; the appressorium is separated from the germ tube by a septum; and the stomatal opening is entered by a penetration peg (Longo *et al.*, 2000) (Fig. 14.1B).

The stages of the dikaryotic infection process are mediated by a series of physical and chemical stimuli that are released by the host surfaces (Hoch and Staples, 1991; Mendgen *et al.*, 1996; Epstein and Nicholson, 1997), particularly by the stomata (Terhune *et al.*, 1991). These stimuli, which condition the behaviour of the fungal structures on the host surface, are more effective when the fungus is more strongly adhered to host surface. Adhesion is secured by mucilaginous exudates produced by the fungus and containing numerous lytic enzymes, especially cutinases, important factors in ensuring the adhesion of the rust structures and causing them to differentiate after adhesion (Mendgen *et al.*, 1996; Howard, 1997; Mendgen, 1997). More specifically, according to Mendgen *et al.* (1996) and Epstein and Nicholson (1997), each enzyme changes the host cuticle from hydrophobic to hydrophilic, like the fungal surface itself, and in this way promotes adhesion. The hydrophobic nature of the host surface therefore seems to condition the entire pre-penetration process by the fungus. In indirect penetration, the adhesion of the appressorium is very important since it must remain correctly placed on the guard cells of the stoma (Epstein and Nicholson, 1997).

Fig. 14.7. *Melampsora pulcherrima* on a leaf of *Mercurialis annua*. Intracellular infection structures from direct penetrations of basidiospore (B) germlings into cells of lower epidermis. The roundish naked vesicles (V) and developing infection hyphae (IH) are visible. ×310. (Longo *et al.*, 1994.)

Fig. 14.8. Transmission electron micrograph of *Melampsora pulcherrima* on a leaf *Mercurialis annua*. Penetration peg (P) through the epidermal cell wall and the naked vesicle (V) of an intracellular infection structure. The appressorium (A) is visible on the epidermal cell wall (ECW). E, Exudate; N, vesicle neck. ×8900. (Longo *et al.*, 1994.)

Fig. 14.9. Scanning electron micrograph of *Melampsora pinitorqua* on a primary needle of *Pinus pinea*. Germinated basidiospores (B) with short germ tubes (GT): a basidiospore germ tube shows the appressorium (A) at the penetration point on the epidermal cell wall (ECW). ×1000. (Longo *et al.*, 1991.)

Fig. 14.10. Transmission electron micrograph of *Melampsora pinitorqua* on a shoot of *Pinus sylvestris*. Two infection structures in two epidermal cells (EC). Penetration point (PP) through the epidermal cell wall and naked vesicle (V) (left structure); a vacuolated vesicle (V) and the transcellularly growing infection hypha (IH) from epidermal into underlying parenchyma cell (PC) (right structure). N, vesicle neck. ×2800. (Longo *et al.*, 1997.)

Fig. 14.11. Transmission electron micrograph of *Melampsora larici-tremulae* on a needle of *Larix decidua*. Penetration point (PP) through the epidermal cell wall and the naked vesicle (V) of an infection structure. EC, Epidermal cell; N, vesicle neck. ×4000. (Longo *et al.*, 1997.)

Fig. 14.12. Transmission electron micrograph of *Melampsora larici-tremulae* on a shoot of *Pinus sylvestris*. Penetration point (PP) through the epidermal cell wall and the naked vesicle (V) of an infection structure. The 'cell wall-like apposition' (WA) in the epidermal cell (EC) completely encases the infection structure. ×3350. (Longo *et al.*, 1997.)

The characteristics of the dikaryotic stage after penetration into the stoma are: the substomatal vesicle is separated from the penetration peg by a septum; and the infection hypha has a distal septum that separates it from the haustorial mother cell, which is terminal and gives rise to the first haustorium (Longo *et al.*, 2000) (Fig. 14.1B).

In the substomatal chamber all the signals that regulate the interaction between fungus and host are present along the entire structure of the rust fungus, from the vesicle to the first haustorium which, in the dikaryotic stage, is the first intracellular structure (Mendgen *et al.*, 1988; Heath, 1989, 1995).

Penetration at the monokaryotic stage

Direct penetration of the host surface by the monokaryotic stage of the fungus has the following characteristics: the germ tubes grow indiscriminately, moving in an undirected manner over the leaves; the appressorium is poorly differentiated; the penetration peg penetrates the epidermal cell wall (Longo *et al.*, 2000) (Fig. 14.1A).

The signals between the host surface and the fungus during pre-penetration at the monokaryotic stage are not as well known as at the dikaryotic stage (Mendgen, 1997). Some authors have postulated the occurrence at the monokaryotic stage of some signals that condition the structures formed by the rust (Desprez-Lousteau and Le Menn, 1989; Gold and Mendgen, 1991; Longo *et al.*, 1994). It has been observed that the germ tubes of some rusts show preferential growth lines along the anticlinal walls of the epidermal cells (Gold and Mendgen, 1984; Hopkin *et al.*, 1988; Morin *et al.*, 1992). On the other hand, a close adhesion of the fungus to the host surface is thought to be important, especially between appressorium and the host epidermis (Epstein and Nicholson, 1997). According to Gold and Mendgen (1991), the mucilaginous matrix commonly present around the external structures of the rust is not only useful to favour rust adhesion, but above all serves as an enzyme reservoir to assist penetration. It should be borne in mind that in direct penetration the turgor pressure inside the appressorium plays a crucial role, and mechanical action combined with enzymatic action then renders the cell wall penetrable (Mendgen and Deising, 1993; Howard, 1997).

The characteristics of the monokaryotic stage after direct penetration are: (i) an intra-epidermal vesicle that is not separated from the penetration peg by a septum, but is separated by a septum from the infection hypha; and (ii) an infection hypha which forms intracellular hyphae and haustoria, and also intercellular hyphae that then give rise to haustoria in the parenchyma cells (Longo *et al.*, 2000) (Fig. 14.1A).

Signals that regulate the host–parasite interaction after penetration can be found at the monokaryotic stage too, but unlike the dikaryotic stage, all such signals occur at the epidermal cells, where penetration takes place (Heath, 1989, 1995).

The infection process at the dikaryotic stage of the *Uredinales* produces a series of distinct and well-differentiated structures from both a morphological and functional point of view, and each of these structures reacts in a specific way with its host. This indicates that infection at the dikaryotic stage of the fungus is evolutionarily a more complex process than infection at the monokaryotic stage (Gold and Mendgen, 1991; Mendgen and Deising, 1993; Heath, 1995; Mendgen *et al.*, 1996; Mendgen, 1997).

On the whole, the first stages of the monokaryotic infection process of *C. flaccidum* investigated by Longo *et al.* (2000) (Fig. 14.13) bear a resemblance to the infection process of rusts at the dikaryotic stage. *C. flaccidum* also penetrates its host indirectly and forms a substomatal infection structure and a first haustorium in a parenchyma cell. However, this resemblance is a matter of behaviour only, since the morphological and functional characteristics that emerge with this type of infection process are essentially those of the monokaryotic stage. And this is particularly true as the *C. flaccidum* germ tubes on the host surface do not grow in a specific direction; it is only by chance that they grow into a stomatal

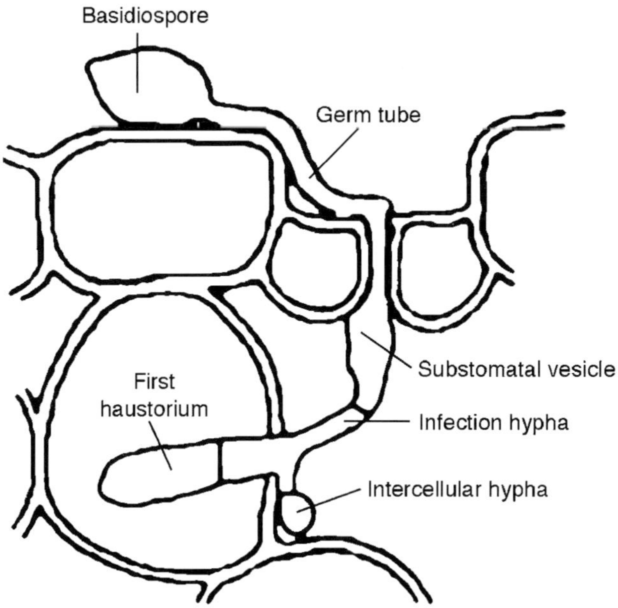

Fig. 14.13. Diagrammatic representation of the monokaryotic infection process of *Cronartium flaccidum*. Indirect penetration (Longo *et al.*, 2000).

opening. When they do so, they do not differentiate an appressorium or a penetration peg; the infection structures in the substomatal chamber have only one septum between the vesicle and the infection hypha. All haustoria, including those in the subsidiary cells of the stoma, are typical monokaryotic structures, especially the first, which, on account of its particular shape, acts even more like a hypha. Moreover, the distinctive features of the first haustorium are also seen in the monokaryotic intracellular structures of *M. pinitorqua* (Longo *et al.*, 1988, 1991) which, deriving directly from the infection hypha in the epidermal cell, colonize both the epidermal cells and the contiguous parenchyma cells. The first haustoria of *C. flaccidum* can therefore be considered an extension of the substomatal infection hypha, the behaviour of which they maintain until they become covered with a haustorial membrane and a matrix and assume the new function of exchanging substances between the host and the fungus. The same thing happens with the intracellular infection hypha, which in the case of direct, monokaryotic penetration derives from an intra-epidermal vesicle: this hypha, unlike the vesicle, is covered with a matrix, indicating its haustorial function (Longo *et al.*, 2000).

The penetration process of *C. flaccidum* at the monokaryotic stage differs from both the monokaryotic penetration of most other rust fungi and from the penetration of a dikaryotic stage. It is initiated by the germ tube without the formation of an appressorium, and hence without a true penetration peg. Since the purpose of the appressorium is to cause the fungus to adhere to the host, the lack of this specialized structure has a biological significance that becomes clear considering that *C. flaccidum* does not penetrate the intact epidermis of the host directly (as is characteristic of monokaryotic penetration), nor indirectly through a stoma (typical dikaryotic penetration) with a germ tube that grows purposely towards a stoma (Longo *et al.*, 2000).

It is interesting to consider the similar behaviour of *C. flaccidum* and rusts at the dikaryotic stage in regard to the growth towards a cutinized surface. Indeed, for the dikaryotic penetration process, the hydrophobic nature of the cutinized host-cell walls, which extend from the outer surface of the epidermis to the substomatal chamber due to the continuity of the cuticle, seems to stimulate the penetration process (Mendgen *et al.*, 1996; Epstein and Nicholson, 1997) and the formation of intercellular infection structures in the substomatal chamber (Terhune *et al.*, 1991). As for the intracellular colonization, the hydrophobic nature of the cutinized inner tangential wall of the epidermal cells seems to prevent the differentiation of the haustorial mother cells and hence the formation of haustoria, as demonstrated in *Uromyces appendiculatus* on *Phaseolus vulgaris* by Terhune *et al.* (1991).

In regard to *C. flaccidum* at the monokaryotic stage, it is not known what signals are exchanged between the host cell walls and the fungal pre-penetration and infection structures, nor what is the role of the enzymes in the exchange of these signals. It has, however, been noted that *C. flaccidum* at this stage, like the rusts at the dikaryotic stage, does not enter through the epidermal cell walls, but produces its first intracellular structures in the parenchyma (Longo *et al.*, 2000). It does not produce any such structures in the epidermal cells. Cutin is scarce or absent in parenchyma cell walls and the epidermal cells are impregnated with cutin.

As a final comment, it can be said that the basidiospore germ tubes of *C. flaccidum* in nature infect organs such as the secondary needles of *Pinus* spp., that are strongly cutinized but also densely covered with stomata. Since these germ tubes are unable to exert a mechanical action (which, in monokaryotic penetration, is always combined with enzymatic activity), in order to succeed in colonizing the secondary needles they changed their mode of penetration from direct, characteristic of the *Uredinales* at their monokaryotic stage, to indirect. The penetration structures of this rust at the monokaryotic stage nevertheless still retain the morphological and functional characteristics of the typical monokaryotic structures, even if their behaviour is reminiscent of dikaryotic structures (Longo *et al.*, 2000).

It is concluded that with *C. flaccidum* at the monokaryotic stage it is the nuclear set that determines the morphology and functions of the structures involved in the infection process; this holds true even though the cytological and histological characteristics of the host organ, which this rust fungus is adapted to infect in nature, condition its mode of penetration.

Acknowledgements

We are grateful to Gabriele Tani, Department of Plant Biology, University of Florence, for skilful assistance with the preparation of photographic plates. Appreciation is also expressed to Mrs E. Bruno, Department of Agricultural Biotechnology, University of Florence, for technical assistance with the scanning electron microscopy.

References

Bauer, R. (1983) Experimentell-ontogenetische und karyologische Untersuchungen an *Uredinales*. Doctoral dissertation, Universität Tübingen, Tübingen, Germany.

Bergdahl, D.R. and French, D.W. (1985) Penetration of the primary tissues of *Pinus banksiana* by *Cronartium comandrae*. In: Barrows-Broaddus, J. and Powers, H.R. Jr (eds) *Proceedings of the IUFRO 'Rusts of hard Pines' Conference*, October 1984, University of Georgia, Athens, pp. 179–192.

Boerema, G.H. and Verhoeven, A.A. (1972) Check-list for scientific names of common parasitic fungi. Series la: fungi on trees and shrubs. *Netherland Journal of Plant Pathology* 78 (suppl. 1).

Bushnell, W.R. and Roelfs, A.P. (eds) (1984) *The Cereal Rusts*, Vol. I. *Origins, Specificity, Structure and Physiology*. Academic Press, Orlando, Florida.

Desprez-Loustau, M.L. and Le Menn, R. (1989) Epicuticular waxes and *Melampsora pinitorqua* Rostr. pre-infection behaviour on maritime pine shoots. A scanning electron microscope study. *European Journal of Forest Pathology* 19, 178–188.

Epstein, L. and Nicholson, R. (1997) Adhesion of spores and hyphae to plant surfaces. In: Carroll, G.C. and Tudzynski, P. (eds) *The Mycota*, Vol. V (part A). Springer, Berlin, pp. 11–25.

Gold, R.E. and Mendgen, K. (1984) Cytology of basidiospore germination, penetration and early colonization of *Phaseolus vulgaris* by *Uromyces appendiculatus* var. *appendiculatus*. *Canadian Journal of Botany* 62, 1989–2002.

Gold, R.E. and Mendgen, K. (1991) Rust basidiospore germlings and disease initiation. In: Cole, G.T. and Hoch, H.C. (eds) *The Fungal Spore and Disease Initiation in Plants and Animals*. Plenum Press, New York, pp. 67–99.

Gray, D.J., Amerson, H.V. and Van Dyke, C.G. (1983) Ultrastructure of the infection and early colonization of *Pinus taeda* by *Cronartium quercuum* f. sp. *fusiforme*. *Mycologia* 75, 117–130.

Grill, D., Lackner, E. and Scharner, M. (1978) Untersuchungen an mit *Chrysomyxa abietis* befallenen Fichtennadeln. *Phyton* 19, 71–82.

Harder, D.E. and Chong, J. (1984) Structure and physiology of haustoria. In: Bushnell, W.R. and Roelfs, A.P. (eds) *The Cereal Rusts*, Vol. I. Academic Press, Orlando, Florida, pp. 431–476.

Heath, M.C. (1989) A comparison of fungal growth and plant responses in cowpea and bean cultivars inoculated with urediospores or basidiospores of the cowpea rust fungus. *Physiological and Molecular Plant Pathology* 34, 415–426.

Heath, M.C. (1995) Signal exchange between higher plants and rust fungi. *Canadian Journal of Botany* 73, 616–623.

Hoch, H.C. and Staples, R.C. (1991) Signaling for infection structure formation in fungi. In: Cole, G.T. and Hoch, H.C. (eds) *The Fungal Spore and Disease Initiation in Plants and Animals*. Plenum Press, New York, pp. 25–46.

Hopkin, A.A., Reid, J., Hiratsuka Y. and Allen, E. (1988) Initial infection and early colonization of *Pinus contorta* by *Endocronartium harknessii* (western gall rust). *Canadian Journal of Plant Pathology* 10, 221–227.

Howard, R.J. (1997) Breaching the outer barriers. Cuticle and cell wall penetration. In: Carroll, G.C. and Tudzynski, P. (eds) *The Mycota*, Vol. V (part A). Springer, Berlin, pp. 43–60.

Littlefield, L.J. and Heath, M.C. (1979) *Ultrastructure of the Rust Fungi*. Academic Press, New York.

Longo, N., Naldini Longo, B., Tani, G. and Drovandi, F. (1988) Osservazioni sull'infezione basidiosporica di *Melampsora pinitorqua* Rostr. (*Uredinales*) in alcuni ospiti del gen. *Pinus*. *Giornale Botanico Italiano* 122 (suppl. 1), 155.

Longo, N., Naldini Longo, B., Tani, G. and Drovandi, F. (1991) Penetration and early colonization in basidiospore-derived infection of *Melampsora pinitorqua* Rostr. on *Pinus*. Structural and ultrastructural observations. In: Y. Hiratsuka, J.K. Samoil, P.V. Blenis, P.E. Crane and B.L. Laishley (eds) *Proceedings of IUFRO 'Rusts of Pine' Conference, September 1989, Banff, Canada*. Information Report NOR-X-317, Northern Forestry Centre, Forestry Canada, Edmonton, Canada, pp. 120–127.

Longo, N., Naldini, B., Drovandi, F., Gonnelli, T. and Tani, G. (1994) Penetration and early colonization in basidiospore-derived infection of *Melampsora pulcherrima* (Bub.) Maire on *Mercurialis annua* L. *Caryologia* 47, 207–222.

Longo, N., Naldini, B., Paolillo, A., Drovandi, F., Tani, G. and Gonnelli, T. (1997) Morphological aspects of early host–parasite interactions in infections of *Melampsora pinitorqua* and *Melampsora larici-tremulae* on *Pinus sylvestris*. Implications in the taxonomical relationship of the two rust fungi. *Caryologia* 50, 35–57.

Longo, N., Poggiolesi, S., Naldini, B. and Tani, G. (2000) Penetration and early colonization in basidiospore-derived infection on needles of *Pinus pinea* L. by *Cronartium flaccidum* (Alb. et Schw.) Wint. *Caryologia* 53, 9–29.

Longo, N., Naldini, B., Drovandi, F. and Gonnelli, T. (2002) Signs of resistance to *Melampsora larici-tremulae* on species of *Pinus* hosts of *Melampsora pinitorqua*: implications regarding the taxonomic relationship between the two rust fungi. *Phytopathologia Mediterranea* 41, 212–219.

Magnani, G. (1961) Ruggine mediterranea del pioppo bianco da *Melampsora pulcherrima* (Bub.) Maire. *Pubblicazioni Ente Nazionale Cellulosa e Carta* 5, 221–236.

Mendgen, K. (1997) The *Uredinales*. In: Carroll, G.C. and Tudzynski, P. (eds) *The Mycota*, Vol. V (part B). Springer, Berlin, pp. 79–85.

Mendgen, K. and Deising, H. (1993) Infection structures of fungal plant pathogens – a cytological and physiological evaluation. *New Phytologist* 124, 193–213.

Mendgen, K., Schneider, A., Sterk, M. and Fink, W. (1988) The differentiation of infection structures as a result of recognition events between some biotrophic parasites and their hosts. *Journal of Phytopathology* 123, 259–279.

Mendgen, K., Hahn, M. and Deising, H. (1996) Morphogenesis and mechanism of penetration by plant pathogenic fungi. *Annual Review of Phytopathology* 34, 367–386.

Metzler von, B. (1982) Untersuchungen an Heterobasidiomyceten. 23: Basidiosporenkeimung und infektionsvorgang beim Birnengittrrost. *Phytopathologische Zeitschrift* 103, 126–138.

Miller, T., Patton, R.F. and Powers, H.R. Jr (1980) Mode of infection and early colonization of slash pine seedlings by *Cronartium quercuum* f. sp. *fusiforme*. *Phytopathology* 70, 1206–1208.

Mims, C.W. and Richardson, E.A. (1989) Ultrastructure of appressorium development by basidiospore germlings of the rust fungus,

Gymnosporangium juniperi-virginianae. *Protoplasma* 148, 111–119.

Morin, L., Brown, J.F. and Auld, B.A. (1992) Teliospore germination, basidiospore formation and the infection process of *Puccinia xanthii* on *Xanthium occidentale*. *Mycological Research* 96, 661–669.

Moriondo, F., Naldini Longo, B., Longo, N., Drovandi, F. and Gonnelli, T. (1989) Some observations on the life-cycle of *Melampsora pulcherrima* (Bub.) Maire. *Phytopathologia Mediterranea* 28, 46–52.

Naldini, B., Longo, N., Drovandi, F. and Gonnelli, T. (1993) *Melampsora pulcherrima* (Bub.) Maire on *Mercurialis annua* L.: some epidemiological experiments. *Phytopathologia Mediterranea* 32, 129–134.

Nicholson, R.L. and Epstein, L. (1991) Adhesion of fungi to the plant surface: prerequisite for pathogenesis. In: Cole, G.T. and Hoch, H.C. (eds) *The Fungal Spore and Disease Initiation in Plants and Animals*. Plenum Press, New York, pp. 3–23.

Patton, R.F. and Johnson, D.W. (1970) Mode of penetration of needles of eastern white pine by *Cronartium ribicola*. *Phytopathology* 60, 977–982.

Raddi, P. (1976) Indagine sui tipi di macchie fogliari e sul loro significato in discendenze di *Pinus pinaster* infette da *Cronartium flaccidum*. *Rivista di Patologia Vegetale* 12, 91–98.

Ragazzi, A. and Dellavalle Fedi, I. (1992) Penetration of *Cronartium flaccidum* into pine needles. *European Journal of Forest Pathology* 22, 278–283.

Ragazzi, A. and Moriondo, F. (1979) Suscettibilità del *Pinus sylvestris* L. e del *Pinus nigra* Arn. alla ruggine vescicolosa *Cronartium flaccidum* (Alb. et Schw.) Wint. *L' Italia Forestale e Montana* 34, 121–129.

Ragazzi, A. and Moriondo, F. (1980) Results of *Cronartium flaccidum* (Alb. et Schw.) Wint. inoculations on eight-year-old plants of *Pinus pinea* L. In: Powers, H.R., Grasso, V. and Raddi, P. (eds) Proceedings of the IUFRO 'Rusts of hard pines' Meeting, September 1979, Florence, Italy. *Phytopathologia Mediterrana* 19, 51–56.

Ragazzi, A., Fagnani, A. and Dellavalle Fedi, I. (1987) Telial and basidiospore stages of *Cronartium flaccidum*: light and scanning electron microscopy observations. *Phytopathologia Mediterranea* 26, 81–84.

Staples, R.C. and Macho, V. (1984) Germination of urediospores and differentiation of infection structures. In: Bushnell, W.R. and Roelfs, A.P. (eds) *The Cereal Rusts*, Vol. I. *Origins, Specificity, Structure and Physiology*. Academic Press, Orlando, Florida, pp. 255–283.

Terhune, B.T., Allen, E.A., Hoch, H.C., Wergin, W.P. and Erbe, E.F. (1991) Stomatal ontogeny and morphology in *Phaseolus vulgaris* in relation to infection structure initiation by *Uromyces appendiculatus*. *Canadian Journal of Botany* 69, 477–484.

Wilson, M. and Henderson, D.M. (1966) *British Rust Fungi*. Cambridge University Press, Cambridge, UK.

15 Host Diversity, Epidemic Progression and Pathogen Evolution

Chris C. Mundt
Department of Botany and Plant Pathology, 2082 Cordley Hall, Oregon State University, Corvallis, OR 97331-2902, USA

Introduction

The simultaneous culture of plant genotypes differing in their reactions to pathogen genotypes is one of several (Mundt, 1994b) strategies for disease control and management of host plant resistance. Multiline cultivars (mixtures of lines bred for phenotypic uniformity of agronomic traits) and cultivar mixtures (mixtures of agronomically compatible, cultivated varieties, with no additional breeding for phenotypic uniformity) are being used increasingly in commercial production, and have been the subject of recent reviews (Garrett and Mundt, 1999; Finckh *et al.*, 2000; Mundt, 2002). It has been emphasized that randomly chosen mixture components will not necessarily provide adequate disease control. Instead, mixtures must be relevant, or functional, against the specific pathogen population in question (Schmidt, 1978; Mundt and Browning, 1985b; Wolfe, 1985). Effects of mixtures on disease levels as compared to their component pure stands can range from a disease increase to nearly complete disease control (Smithson and Lenne, 1996). This review will discuss factors impacting the level of disease control in mixtures, the effects of host diversity on pathogen evolution, and comment briefly on the potential to use mixtures for control of *Melampsora* rust of willow.

Specialized Foliar Pathogens of Small Grains

The majority of studies regarding disease control through use of host mixtures have been with rusts (*Puccinia* spp.) and powdery mildews (*Blumeria graminis* f. spp.) of small grains, or with rice (*Oryza sativa*) blast, caused by *Magnaporthe grisea*. Several mechanisms have been postulated to explain the reduction in severity of such diseases caused by mixtures (Browning and Frey, 1969; Chin and Wolfe, 1984b; Mundt and Browning, 1985b; Garrett and Mundt, 1999; Koizumi, 2001). Inoculum dilution caused by the increased distance between plants of the same genotype is often the most important mechanism (Burdon and Chilvers, 1977; Chin and Wolfe, 1984b; Wolfe, 1985). However, other mechanisms can be important as well. For example, induced resistance is predicted to be important by simulation models (Lannou *et al.*, 1995) and accounted for about 30% of the total reduction of wheat (*Triticum aesivum*) yellow rust (caused by *Puccinia striiformis*) (Calonnec *et al.*, 1996). In another wheat yellow rust study, cultivars resistant to all inoculated races compensated for susceptible cultivars through increased tillering, which sometimes substantially increased the proportion of healthy host tissue in mixtures (Finckh and Mundt, 1992a,b). Approximately 25% of the total powdery mildew reduction observed in a barley (*Hordeum vulgare*) cultivar mixture was attributed to disruptive selection caused by quantitative adaptation of the pathogen to the genetic background of the different cultivars in the mixture (Wolfe *et al.*, 1981). A highly successful mixture approach for control of rice blast involves row mixtures of a tall, susceptible cultivar with a shorter, more resistant one (Zhu *et al.*, 2000). Height differences between cultivars resulted in reduced leaf wetness, which may have reduced pathogen infection (Y. Zhu, Yunnan Agricultural University, personal communication).

Mixture Composition

There must usually be a correspondence between the resistance genes incorporated in the mixture and the avirulence genes present in the target pathogen population if diversity is to be functional. This can be accomplished in either formal or informal ways (Mundt, 2002), but often involves attempts to minimize the percentage of the host population that a given race will be virulent against. If there is functional host/pathogen diversity present, disease severity will decrease with decreasing frequency of a host genotype in mixture. This has been demonstrated numerous times with rusts and powdery mildews of small grains (Browning and Frey, 1969; Mundt and Browning, 1985b; Wolfe, 1985), and with rice blast (Mundt, 1994b; Koizumi, 2001). Leonard (1969) predicted that the infection rate of the pathogen will decline logarithmically with the proportion of a genotype in the mixture, and this relationship has been demonstrated in the field for several small grain diseases (Leonard, 1969; Burdon and Chilvers, 1977; Luthra and Rao, 1979; Elliot *et al.*, 1980; Koizumi and Kato, 1987). Similarly, field evidence suggests that disease levels may decline when an increasing number of host genotypes is added to equiproportional mixtures (Mundt, 1994a; Newton *et al.*, 1997).

From a disease control standpoint, one would expect diminishing returns for decreasing the frequency of a host genotype in a mixture or to increasing the number of host genotypes in a mixture (Leonard, 1969; Mundt, 1990). On the other hand, incorporating a larger number of genotypes in a mixture could be useful in countering pathogen evolution towards complex virulence (Marshall, 1989). In practical applications of mixtures, considerations such as agronomic performance of potential mixture components, agronomic compatibility among mixture components, and characteristics of the seed-handling infrastructure may dominate over other considerations. Multiline cultivars have included as many as 8–12 lines (Browning and Frey, 1981; Allan *et al.*, 1983). In contrast, cultivar mixtures often include only the most agronomically competitive genotypes and, thus, usually contain only 2–5 components.

Epidemic Intensity

The effect of epidemic intensity on mixture efficacy can vary. Theoretically, one would expect the difference in disease levels between pure stands and mixtures to increase with increasing number of pathogen generations (Leonard, 1969). The level of initial infection, the generation time of the pathogen, and the rate of epidemic development will, in turn, determine the number of pathogen generations that occur in a given epidemic. In the field, disease levels initially diverge over time, and then converge as the host's carrying capacity for disease is approached (e.g. see Wolfe and Barrett, 1980). The rate of approach to carrying capacity will greatly impact the effectiveness of mixtures for disease control (Mundt, 1990), and a fast approach can substantially reduce mixture effectiveness.

Large amounts of external inoculum can greatly reduce the ability of mixtures to control disease. For example, potato mixtures provided better late blight control in areas with little or no outside inoculum, than in an area where external spore showers are heavy and frequent (Garrett and Mundt, 2000; Garrett *et al.*, 2001). Thus, high disease pressure caused by fast epidemics and large amounts of outside inoculum may reduce mixture efficacy, while a large number of pathogen generations may both increase disease pressure and increase mixture effectiveness.

Spatial Scale

Mixtures may provide greater disease control in production-scale situations than is observed in small-scale experimental plots (Wolfe, 1985; Mundt, 1994b, 2001), for several reasons. Unnaturally high inoculation rates and the effects of interplot interference can reduce observed levels of disease control in experiments, and mixtures are especially sensitive to such effects (Wolfe, 1985; Mundt, 1994b, 2002). Mixtures reduce the velocity of spread of disease from an initial focus (Van den Bosch *et al.*, 1990), and the nature of this expansion is such that the velocities of pure stands versus mixtures may diverge as epidemics expand in time and space (Mundt, 2001, 2002). For spatially uniform epidemics, mixtures may reduce inoculum levels regionally, resulting in increased levels of disease control at larger spatial scales (Mundt *et al.*, 2002).

Some evidence exists to support the view that mixtures perform better at larger spatial scale. Field observations, though lacking replication or controls, have sometimes suggested that mixtures provide greater levels of disease control in commercial production than in small experimental field plots (Wolfe and Barrett, 1980; Browning and Frey, 1981; Wolfe, 1985; Mundt, 1994b). In the early 1980s, the use of barley cultivar mixtures to control powdery mildew in the German Democratic Republic expanded rapidly to eventually incorporate 92% (360,000 ha) of the barley area. This large-scale mixture adoption was associated with a sharp decline in both mildew levels and fungicide use (Wolfe, 1992; Wolfe *et al.*, 1992). In China, mixtures of sticky rice cultivars highly susceptible to blast were mixed with non-glutinous cultivars that were more resistant. A total of 812 and 3342 ha of contiguous rice fields were grown to mixtures in 1998 and 1999, respectively. Observation plots of non-diversified rice indicated that the mixtures reduced panicle blast severity by an average of 94% on the susceptible sticky rice cultivars (Zhu *et al.*, 2000). This practice expanded to 550,000 ha in 2003 (Y. Zhu, Yunnan Agricultural University, personal communication).

Autoinfection and Host Geometry

The impact of mixtures on disease levels is expected to decline with increasing proportions of autoinfection (Barrett, 1980; Mundt *et al.*, 1986; Goleniewski and Newton, 1994), i.e. infections 'in which the donor (infector) host individual is the same as the recipient (infected) host individual' (Robinson, 1976). Although the degree of autoinfection can be influenced by pathogen characteristics such as dispersal and lesion expansion (Garrett and Mundt, 1999; Mundt, 2002), this section will focus on the effects of host geometry. Everything else being constant, autoinfection would be expected to increase with increasing genotype unit area (GUA), i.e. the ground area occupied by an independent, genetically homogeneous unit of host tissue (Mundt and Browning, 1985a). The GUA can be approximated as the ground area occupied by an individual plant in a random mixture of plants, and the area occupied by a row and field for row mixtures and interfield diversification, respectively.

Computer simulations and field studies have shown that mixture efficacy will decrease when the degree of aggregation of plant genotypes is changed to alter GUA (summarized in Mundt, 2002). However,

GUA interacts with other spatial variables to determine disease levels in mixtures. Computer simulations (Mundt *et al.*, 1986) and field studies with rust pathogens (Mundt and Browning, 1985b; Mundt and Leonard, 1985, 1986) showed that GUA had little influence on mixture efficacy if initial inoculum was distributed in a single focus, rather than uniformly over plots. Later theoretical work suggested that the number of host units may be more important than their size, and that mixtures of large genotype units can provide substantial disease control, even when inoculum is distributed uniformly, provided that the total number of host units is sufficiently large (Mundt and Brophy, 1988). However, a field test of this concept provided inconclusive results (Mundt *et al.*, 1996).

Spatial manipulations may demonstrate effects of GUA when compared under constant conditions, or in simulation models. However, such constancy rarely occurs in the real world, and there is large variation in reported levels of disease control for mixtures of crops with large plants (Mundt, 2002). Nevertheless, it appears possible to have substantial disease reductions in mixtures even for tree crops, despite the earlier cautions to the contrary by van der Plank (1968). Evidence for disease reductions of greater than 50% in mixtures has been reported for apple (*Malus* × *domestica*) scab (caused by *Venturia inequalis*) (Blaise and Gessler, 1994; MacHardy *et al.*, 2001) and *Melampsora* rust (caused by *Melampsora* spp.) of willow (*Salix* spp.) (McCracken and Dawson, 1998; Hunter *et al.*, 2002). More detailed descriptions of the use of mixtures for control of *Melampsora* rust are presented elsewhere in this volume.

Pathogen Evolution

Pathogen populations tend to be more diverse in mixtures than in pure stands (Browning and Frey, 1981; DiLeone and Mundt, 1994; Muller *et al.*, 1996; McCracken *et al.*, 2000; Koizumi, 2001), but will the extensive use of mixtures select over time for complex pathogen genotypes able to attack multiple host genotypes in a mixture? In small-scale experimental plots, the relative frequency of complex powdery mildew genotypes has been found to be greater in barley cultivar mixtures than in pure-stand controls (Munk, 1983; Chin and Wolfe, 1984a; Huang *et al.*, 1994). However, a different view may emerge when one considers both relative and absolute genotype frequencies. For example, Chin and Wolfe (1984a) found that the increased relative frequency of complex powdery mildew genotypes in barley cultivar mixtures was due to their effectiveness in controlling simple races, and that the absolute frequency of the most complex genotype was less in a three-way barley cultivar mixture than in the component pure stands. A very similar result was found when examining race frequency data after several years of extensive commercial use of barley cultivar mixtures in the German Democratic Republic (Wolfe *et al.*, 1992). Thus increased relative frequency of complex races in a mixture does not necessarily imply loss of disease control. It may be more difficult to determine the selective influence of mixtures for clonal pathogens, since virulence genes will not be randomly associated with other genes that influence pathogen fitness (Mundt, 1994a, 2002).

Pathogen complexity in mixtures has been the subject of many modelling efforts (Mundt, 2002). Most of these models assume that a fitness cost associated with virulence is the only mechanism to counter selection for increased virulence complexity in the pathogen. In general, the models suggest that it will be difficult to prevent complex races from eventually dominating the pathogen population, although the process may be sufficiently slow such that the pathogen population can be managed.

Although models of pathogen evolution usually require a cost of virulence for maintenance of pathogen diversity in mixtures, evidence supporting such costs is very weak (Parlevliet, 1981). Parlevliet (1981) explained the apparent lack of association between virulence and fitness with a conceptual model consisting of mutation to virulence, which is often associated with a fitness reduction initially, followed by selection for fitness modifiers that eventually ameliorate the initial fitness cost. Such fitness modifiers have been shown to reduce or eliminate fitness reductions initially associated with antibiotic resistance in bacteria (Cohan *et al.*, 1994; Morrell, 1997; Levin *et al.*, 2000), but have been largely ignored in plant pathology. Selection for fitness modifiers is likely to be slower than selection for virulence genes themselves (Wolfe and Barrett, 1980), thus slowing pathogen evolution towards complexity.

In addition, there are mechanisms other than a cost of virulence that may counter selection for pathogen complexity in mixtures (Leonard and Czochor, 1980; Mundt and Browning, 1985b; Mundt, 1994a, 2002; Lannou and Mundt, 1997). For example, quantitative adaptation to host genetic background can induce disruptive selection on complex races able to attack multiple host components, thus reducing their fitness relative to simple pathotypes (Chin and Wolfe, 1984a; Lannou and Mundt, 1996; Villareal and Lannou, 2000; Lannou, 2001). Other mechanisms include density-dependent selection, frequency-dependent selection, selection in heterogeneous environments, founder effects and migration (Mundt, 2002).

Although more data are needed, there is no evidence to date for strong selection of complex and highly aggressive pathogen genotypes in host mixtures. Similarly, in natural ecosystems, complex pathogen races can be the most frequent, but do not dominate the pathogen population (Segal *et al.*, 1982; Dinoor and Eshed, 1984) nor cause substantial damage to the host (Segal *et al.*, 1980, 1982). None the less, pathogen evolution should be managed by ensuring that no single mixture is grown exclusively in either time or space (Wolfe and Barrett, 1980), which can be a natural result of normal crop management in many cases (Mundt, 2002).

Conclusions

Large plant size, and its effect on autoinfection, is the factor most often cited as a deterrent to the use of mixtures to control tree diseases effectively (e.g. van der Plank, 1968; Miot *et al.*, 1999). As noted above, there are reports of substantial control of tree diseases with mixtures, including *Melampsora* rust of willow (McCracken and Dawson, 1998; Hunter *et al.*, 2002). On the other hand, much smaller disease reductions were noted for *Melampsora* rust in poplar mixtures (Miot *et al.*, 1999). There are several differences in biology, environment or experimental design that could explain the discrepant results between these similar pathosystems. Regardless, the contrasting results emphasize the need for designing mixtures to attain functional diversity, and for field experiments that are designed appropriately to provide an accurate evaluation of mixture efficacy for the epidemiological conditions in question.

Lack of agronomic and end-product uniformity have often been raised as limitations to the use of field crop mixtures, though these issues have caused little or no difficulty in many cases (Wolfe, 1985; Mundt, 1994b; Finckh *et al.*, 2000). Uniformity is likely to be even less crucial for crops produced primarily for biomass. In a hybrid poplar mixture, competition reduced the diameter of some clones. However, other clones compensated with increased growth, and the mean diameter and mean biomass was no different in the mixture as compared with its component monocultures (Berthelot, 2001). Under disease conditions, there is potential for more resistant willow clones to compensate for yield loss of diseased ones, as was found for a soybean (*Glycine max*) mixture in the presence of *Phytophthora sojae* (Wilcox and St Martin, 1998).

With field crops, it is generally assumed that a successful mixture will need to be composed of genotypes that yield well in monoculture (Pfahler and Linskens, 1979; Wolfe and Barrett, 1980; Mundt, 1994b), and the same is likely to be true for willow mixtures. Though small yield increases sometimes occur in the absence of disease (Wolfe and Barrett, 1980; Finckh and Mundt, 1992a; Smithson and Lenne, 1996), increased yield stability is more consistently associated with mixtures (Smithson and Lenne, 1996). Stability of production may be even more important for a perennial crop, for which the catastrophic loss of a single genotype to disease or other stress could occur after several years of economic investment in the plantation.

As with field crops, *Melampsora* populations were more diverse in mixtures than in monocultures, and there was no evidence for rapid selection of super-races. However, the life expectancy of a willow plantation is expected to be in the order of 25 years, which makes the potential for selection of a super-race more critical, and calls for incorporating a wide array of resistance within mixtures (McCracken *et al.*, 2000). Mixtures of differing composition could be deployed spatially, perhaps even within the same plantation. Further, as plantations will be established over a period of years within a region, regional diversity can be attained by altering mixture composition over time. As with field crops (Mundt, 2002) this will likely occur naturally as growers incorporate new, higher performing clones into mixtures.

References

Allan, R.E., Line, R.F., Peterson, C.J. Jr, Rubenthaler, G.L., Morrison, K.J. and Rohde, C.R. (1983) Crew, a multiline wheat cultivar. *Crop Science* 23, 1015–1016.

Barrett, J.A. (1980) Pathogen evolution in multilines and variety mixtures. *Zeitschrift für Pflanzenkrankheiten und Pflanzenschutz* 87, 383–396.

Berthelot, A. (2001) Mélange de clones entaillis à courtes rotations de peuplier: influence sur la productivité et l'homogénéité des produits recoltes. *Canadian Journal of Forest Research* 31, 1116–1126.

Blaise, P. and Gessler, C. (1994) Cultivar mixtures in apple orchards as a means to control apple scab? In: Butt, D.J. (ed.) Integrated control of pome fruit diseases. *Norwegian Journal Agricricultural Science Supplement* 17, 105–112.

Browning, J.A. and Frey, K.J. (1969) Multiline cultivars as a means of disease control. *Annuual Review. Phytopathology* 14, 355–382.

Browning, J.A. and Frey, K.J. (1981) The multiline concept in theory and practice. In: Jenkyn, J.F. and Plumb, R.T. (eds) *Strategies for the Control of Cereal Disease.* Blackwell, Oxford, pp. 37–46.

Burdon, J.J. and Chilvers, G.A. (1977) Controlled environment experiments on epidemic rates of barley mildew in different mixtures of barley and wheat. *Oecologia* 28, 141–146.

Calonnec, A., Goyeau, H. and de Vallavieille-Pope, C. (1996) Effects of induced resistance on infection efficiency and sporulation of *Puccinia striiformis* on seedlings in varietal mixtures and on field epidemics in pure stands. *European Journal Plant Pathology* 5, 733–741.

Chin, K.M. and Wolfe, M.S. (1984a) Selection on *Erysiphe graminis* in pure and mixed stands of barley. *Plant Pathology* 33, 535–546.

Chin, K.M. and Wolfe, M.S. (1984b) The spread of *Erysiphe graminis* f. sp. *hordei* in mixtures of barley varieties. *Plant Pathology* 33, 89–100.

Cohan, F.M., King, E.C. and Zawadzki, P. (1994) Amelioration of the deleterious pleiotropic effects of an adaptive mutation in *Bacillus subtilis*. *Evolution* 48, 81–95.

DiLeone, J.A. and Mundt, C.C. (1994) Effect of wheat cultivar mixtures on populations of *Puccinia striiformis* races. *Plant Pathology* 43, 917–930.

Dinoor, A. and Eshed, N. (1984) The role and importance of pathogens in natural plant communities. *Annual Review Phytopathology* 22, 443–466.

Elliot, V.J., MacKenzie, D.R. and Nelson, R.R. (1980) Effect of number of component lines on powdery mildew epidemics in a wheat multiline. *Phytopathology* 70, 461–462 [Abstract].

Finckh, M.R. and Mundt, C.C. (1992a) Plant competition and disease in genetically diverse wheat populations. *Oecologia* 91, 81–92.

Finckh, M.R. and Mundt, C.C. (1992b) Stripe rust, yield, and plant competition in wheat cultivar mixtures. *Phytopathology* 82, 905–913.

Finckh, M.R., Gacek, E.S., Goyeau, H., Lannou, C., Merz, U., Mundt, C.C., Munk, L., Nadziak, J., Newton, A.C., de Vallavieille-Pope, C. and Wolfe, M.S. (2000) Cereal variety and species mixtures in practice. *Agronomie* 20, 813–837.

Garrett, K.A. and Mundt, C.C. (1999) Epidemiology in mixed host populations. *Phytopathology* 89, 984–990.

Garrett, K.A. and Mundt, C.C. (2000) Host diversity can reduce potato late blight severity for focal and general patterns of primary inoculum. *Phytopathology* 90, 1307–1312.

Garrett, K.A., Nelson, R.J., Mundt, C.C., Chacon, G., Jaramillo, R.E. and Forbes, G.A. (2001) The effects of host diversity and other management components on epidemics of potato late blight in the humid highland tropics. *Phytopathology* 91, 993–1000.

Goleniewski, G. and Newton, A.C. (1994) Modelling the spread of fungal diseases using a nearest neighbour approach. *Plant Pathology* 43, 631–643.

Huang, R., Kranz, J. and Welz, H.G. (1994) Selection of pathotypes of *Erysiphe graminis* f. sp. *hordei* in pure and mixed stands of spring barley. *Plant Pathology* 43, 458–470.

Hunter, T., Peacock, L., Turner, H. and Brain, P. (2002) Effect of planting design on stem-infecting form of rust in willow biomass coppice. *Forest Pathology* 32, 87–97.

Koizumi, S. (2001) Rice blast control with multilines in Japan. In: Mew, T.W., Borromeo, E. and Hardy, B. (eds) *Exploiting Biodiversity for Sustainable Pest Management.* Proceedings of the impact symposium on exploiting biodiversity for sustainable pest management, Kunming, China. International Rice Research Institute, Los Banos, Philippines, pp. 143–157.

Koizumi, S. and Kato, H. (1987) Effect of mixed plantings of susceptible and resistant rice cultivars on leaf blast development. *Phytopathology Society of Japan* 53, 28–38.

Lannou, C. (2001) Intra-pathotype diversity for aggressiveness and pathogen evolution in cultivar mixtures. *Phytopathology* 91, 500–510.

Lannou, C. and Mundt, C.C. (1996) Evolution of a pathogen population in host mixtures. I Study of the simple race–complex race equilibrium. *Plant Pathology* 45, 440–453.

Lannou, C. and Mundt, C.C. (1997) Evolution of a pathogen population in host mixtures: Rate of emergence of complex races. *Theoretical and Applied Genetics* 94, 991–999.

Lannou, C., de Vallavieille-Pope, C. and Goyeau, H. (1995) Induced resistance in host mixtures and its effect on disease control in computer-simulated epidemics. *Plant Pathology* 44, 478–489.

Leonard, K.J. (1969) Factors affecting rates of stem rust increase in mixed plantings of susceptible and resistant oat varieties. *Phytopathology* 59, 1845–1850.

Leonard, K.J. and Czochor, R.J. (1980) Theory of genetic interactions among populations of plants and their pathogens. *Annual Review Phytopathology* 18, 237–258.

Levin, B.R., Veronique, P. and Walker, N. (2000) Compensatory mutations, antibiotic resistance and the population genetics of adaptive evolution in bacteria. *Genetics* 154, 985–997.

Luthra, J.K. and Rao, M.V. (1979) Multiline cultivars – how their resistance influences leaf rust epidemics in wheat. *Euphytica* 28, 137–144.

MacHardy, W.E., Gadoury, D.M. and Gessler, C. (2001) Parasitic and biological fitness of *Venturia inaequalis*: Relationship to disease management strategies. *Plant Disease* 85, 1036–1051.

Marshall, D.R. (1989) Modeling the effects of multiline varieties on the population genetics of plant pathogens. In: Leonard, K.J. and Fry, W.E. (eds) *Plant Disease Epidemiology,* Vol. 2. McGraw-Hill, New York, pp. 284–317.

McCracken, A.R. and Dawson, W.M. (1998) Short rotation coppice willow in Northern Ireland since 1973: development of the use of mixtures in the control of foliar rust (*Melampsora* spp.). *European Journal of Forest Pathology* 28, 241–250.

McCracken, A.R., Dawson, W.M., Watson, S. and Allen, C.Y. (2000) Pathotype composition in *Melampsora epitea* populations occurring on willow (*Salix*) growing in mixed and monoculture plantations. *European Journal of Plant Patholology* 106, 879–886.

Miot, S., Frey, P. and Pinon, J. (1999) Varietal mixture of poplar clones: Effect on infection by *Melampsora larici-populina* and on plant growth. *European Journal of Forest Pathology* 29, 411–423.

Morrell, V. (1997) Antibiotic resistance: road of no return. *Science* 278, 575–576.

Muller, K., McDermott, J.M., Wolfe, M.S. and Limpert, E. (1996) Analysis of diversity in populations of plant pathogens: the barley powdery mildew pathogen across Europe. *European Journal of Plant Pathology* 102, 385–395.

Mundt, C.C. (1990) Disease dynamics in agroecosystems. In: Carroll, C.R., Vandermeer, J.H. and Rossett, P.M. (eds) *Agroecology*. McGraw-Hill, New York, pp. 263–269.

Mundt, C.C. (1994a) Techniques to manage pathogen co-evolution with host plants to prolong resistance. In: Teng, P.S., Heong, K.L. and Moody, K. (eds) *Rice Pest Science and Management.* International Rice Research Institute, Los Banos, Philippines, pp. 193–205.

Mundt, C.C. (1994b) Use of host genetic diversity to control cereal diseases: Implications for rice blast. In: Leong, S., Zeigler, R.S. and Teng, P.S. (eds) *Rice Blast Disease*. CAB International, Wallingford, UK, pp. 293–307.

Mundt, C.C. (2001) Varietal diversification and disease control. In: Mew, T.W. Borromeo, E.

and Hardy, B. (eds) *Exploiting Biodiversity for Sustainable Pest Management.* Proceedings of the impact symposium on exploiting biodiversity for sustainable pest management Kunming, China. International Rice Research Institute, Los Banos, Philippines, pp. 135–142.

Mundt, C.C. (2002) Use of multiline cultivars and cultivar mixtures for disease management. *Annual Review Phytopathology* 40, 381–410.

Mundt, C.C. and Brophy, L.S. (1988) Influence of number of host genotype units on the effectiveness of host mixtures for disease control: a modeling approach. *Phytopathology* 78, 1087–1094.

Mundt, C.C. and Browning, J.A. (1985a) Development of crown rust epidemics in genetically diverse oat populations: effect of genotype unit area. *Phytopathology* 75, 607–610.

Mundt, C.C. and Browning, J.A. (1985b) Genetic diversity and cereal rust management. In: Roelfs, A.P. and Bushnell, W.R. (eds) *The Cereal Rusts,* Vol. 2. Academic Press, Orlando, Florida, pp. 520–560.

Mundt, C.C. and Leonard, K.J. (1985) Effect of host genotype unit area on epidemic development of crown rust following focal and general inoculations of mixtures of immune and susceptible oat plants. *Phytopathology* 75, 1141–1145.

Mundt, C.C. and Leonard, K.J. (1986) Effect of host genotype unit area on development of focal epidemics of bean rust and common maize rust in mixtures of resistant and susceptible plants. *Phytopathology* 76, 895–900.

Mundt, C.C., Leonard, K.J., Thal, W.M. and Fulton, J.H. (1986) Computerized simulation of crown rust epidemics in mixtures of immune and susceptible oat plants with different genotype unit areas and spatial distributions of initial disease. *Phytopathology* 86, 590–598.

Mundt, C.C., Brophy, L.S. and Kolar, S.C. (1996) Effect of genotype unit number and spatial arrangement on severity of yellow rust in wheat cultivar mixtures. *Plant Pathology* 45, 215–222.

Mundt, C.C., Cowger, C. and Garrett, K.A. (2002) Relevance of integrated management to resistance durability. *Euphytica* 124, 245–252.

Munk, L. (1983) Response of a powdery mildew population to a barley variety mixture. In: Lamberti, F., Waller, J.M. and Van der Graaf, N.A. (eds) *Durable Resistance in Crops.* Plenum, New York, pp. 105–107.

Newton, A.C., Ellis, R.P., Hackett, C.A. and Guy, D.C. (1997) The effect of component number on *Rhynchosporium secalis* infection and yield in mixtures of winter barley cultivars. *Plant Pathology* 45, 930–938.

Parlevliet, J.E. (1981) Stabilizing selection in crop patho-systems: an empty concept or a reality? *Euphytica* 30, 259–269.

Pfahler, P.L. and Linskens, H.F. (1979) Yield stability and population diversity in oats (*Avena* sp.). *Theoretical and Applied Genetics* 54, 1–5.

Robinson, R.A. 1976. *Plant Pathosystems.* Springer-Verlag, New York.

Schmidt, R.A. (1978) Diseases in forest ecosystems: the importance of functional diversity. In: Horsfall, J.G. and Cowling, E.B. (eds) *Plant Disease: an Advanced Treatise,* Vol. 2. Academic Press, New York, pp. 287–315.

Segal, A., Manisterski, J., Fishbeck, G., Wahl, I. (1980) In: Horsfall, J.G. and Cowling, E.B. (eds) *Plant Disease: an Advanced Treatise,* Vol. 5 *How Plant Populations Defend Themselves in Natural Ecosystems.* Academic Press, New York, pp. 75–102.

Segal, A., Browning, J.A., Fishbeck, G. and Wahl, I. (1982) Balance in indigenous plant populations. In: Heybroek, H.M. Stephan, B.R. and von Weissenberg, K. (eds) *Resistance to Diseases and Pests in Forest Trees.* PUDOC, Wageningen, Netherlands, pp. 361–370.

Smithson, J.B. and Lenne, J.M. (1996) Varietal mixtures: a viable strategy for sustainable productivity in subsistence agriculture. *Annals Applied Biology* 128, 127–158.

Van den Bosch, F., Frinking, F., Metz, J.A.J. and Zadoks, J.C. (1990) Focus expansion in plant disease. IV. Expansion rates in mixtures of resistant and susceptible hosts. *Phytopathology* 80, 598–602.

van der Plank, J.E. (1968) *Disease Resistance in Plants.* Academic Press, New York.

Villareal, L.M.M.A. and Lannou, C. (2000) Selection for increased spore efficacy by host genetic background in a wheat powdery mildew population. *Phytopathology* 90, 1300–1306.

Wilcox, J.R. and St. Martin, S.K. (1998) Soybean genotypes resistant to *Phytophthora sojae* and compensation for yield losses of susceptible isolines. *Plant Disease* 82, 303–306.

Wolfe, M.S. (1985) The current status and prospects of multiline cultivars and variety mixtures for disease resistance. *Annual Review Phytopathology* 23, 251–273.

Wolfe, M.S. (1992) Maintaining the value of our varieties. In: Munk, L. (ed.) *Barley Genetics VI,* Vol. 2. Munksgaard Int. Publishers, Copenhagen, Denmark, pp. 1055–1067.

Wolfe, M.S. and Barrett, J.A. (1980) Can we lead the pathogen astray? *Plant Disease* 64, 148–155.

Wolfe, M.S., Barrett, J.A. and Jenkins, J.E.E. (1981) The use of cultivar mixtures for disease control. In: Jenkyn, J.F. and Plumb, R.T. (eds) *Strategies*

for the Control of Cereal Disease. Blackwell, Oxford, pp. 73–80.
Wolfe, M.S., Brandle, U., Koller, B., Limpert, E,, McDermott, J.M., Muller, K. and Schaffner, D. (1992) Barley mildew in Europe: population biology and host resistance. *Euphytica* 6, 125–139.
Zhu, Y., Chen, H., Fan, J., Wang, Y., Li, Y., Chen, J., Fan J.-X., Yang, S., Hu, L., Leung, H., Mew, T.W., Teng, P.S., Wang, Z. and Mundt, C.C. (2000) Genetic diversity and disease control in rice. *Nature* 406, 718–722.

16 Short-rotation Coppice Willow Mixtures and Rust Disease Development

Alistair R. McCracken[1], W. Malcolm Dawson[2] and Diane Carlisle[3]
[1]*Applied Plant Science Division, Department of Agriculture and Rural Development, Newforge Lane, Belfast BT9 5PX, UK;* [2]*Applied Plant Science Division, Northern Ireland Horticultural and Plant Breeding Station, Department of Agriculture and Rural Development, Loughgall, Co. Armagh BT61 8JB, UK;* [3]*Department of Applied Plant Science, Queen's University of Belfast, Newforge Lane, Belfast BT9 5PX, UK*

Introduction

Rust, caused by *Melampsora epitea*, was first reported on short-rotation coppice (SRC) willow growing in N. Ireland, UK during the summer of 1985 (McCracken and Dawson, 1998). *Salix burjatica* 'Korso' had been grown in large mono-plots in the province for at least 10 years with no evidence of disease. In that same year rust was also observed in SRC willow plantations in other parts of the UK and the Republic of Ireland (McCracken and Dawson, 1998). In subsequent seasons rust infections occurred as early as mid-June, especially on regrowth from freshly coppiced stools. These early infections resulted in premature leaf fall which, in turn, was followed by invasion of the stems by secondary dieback organisms, including *Fusarium sambucinum* Fuchel and *Glomerella miyabeana* (Fukushi) von Arx., which caused high levels of stem dieback and ultimately stool death (McCracken, 1989).

There was evidence to suggest changing susceptibility of certain *Salix* spp. varieties to rust over a number of years. Hence varieties that were ostensibly free of rust for several years, became increasingly infected. This often occurred after 8–10 years' growth in mono-plots. For example, in N. Ireland *S. burjatica* 'Korso' was rust free for 10 years but rapidly succumbed to the disease during 1986. Similarly 38% of *Salix viminalis* 'Bowles Hybrid' stands examined in the UK in 1987 were free of rust and the remaining 62% showed only low to moderate levels. However, when the same stands were examined in 1988 only 22% had no symptoms, 67% had low to moderate levels and 13% were severely infected (Dawson and McCracken, 1994). Currently *S. viminalis* 'Bowles Hybrid' becomes heavily infected with rust wherever it is grown.

Early work (Dawson and McCracken, 1994) demonstrated that rust could be controlled with the intensive use of fungicides. In trials carried out at the Northern Ireland Horticulture and Plant Breeding Station, Loughgall (54°24′N, 7°35′W) benodanil, myclobutanil and captan/penconazol were applied to first year regrowth of *S. burjatica* 'Korso'. When benodanil was applied at 14-day intervals starting in mid-May, before the first appearance of rust pustules, there was a significant reduction in rust at all disease assessment dates. This was accompanied by increased dry matter yield at the end of the year and only 12% shoot death at the time of flushing the following spring, compared to over 60% in unsprayed plots (Dawson and McCracken, 1994). Similar but less pronounced trends were observed when the other fungicides were used.

Hence the use of large amounts of fungicide could result in good disease control with associated increase in dry matter yield and delayed leaf fall which, in turn, resulted in less shoot and stool death at flushing the following spring. However, it was clearly recognized that such use of fungicides was not a sustainable option, for a number of reasons:

- *Economic*: SRC willow is a low-value crop on which there are small financial margins. The cost of regular repeated fungicide applications throughout the growing season could not be sustained.
- *Environmental*: often willow is grown close to water courses or in areas where the intensive use of agrochemicals would be unacceptable.
- *Energy*: one of the end uses for SRC willow is as a source of renewable energy. It is therefore unacceptable to be applying agrochemicals which have high energy production demands.
- *Practical*: while it is relatively easy to apply fungicide to an SRC willow crop in its first few weeks regrowth, after coppicing it becomes impractical to get effective chemical cover of plants which are up to 3 m high (McCracken and Dawson, 1994).

It was therefore essential that a practical, cost-effective disease-control strategy be developed. In the early 1980s high yielding, rust-resistant *Salix* varieties were becoming commercially available from a breeding programme in Sweden, and laterally from a UK breeding programme. However, the experience with *S. burjatica* 'Korso' (Dawson and McCracken, 1994) and other *Salix* varieties suggested that, when grown in monoculture in conditions of high disease pressure, previously rust-resistant varieties could become increasingly susceptible. The projected life of a willow plantation is approximately 30 years.

In the early 1980s White (1982) reported that when barley was grown in cultivar mixtures there was a decrease in the incidence of powdery mildew caused by *Erysiphe graminis*, with an associated increase in yield. Wolfe (1985) reviewed the benefits of using multi-line cultivars and variety mixtures of cereals as a disease-control strategy, particularly for rusts and mildews. The use of varietal mixtures of cereals with as few as three components resulted in significantly reduced levels of diseases, which was in turn associated with increased yields. Certain barley mixtures gave increased yield in the presence of scald and net blotch (Mundt *et al.*, 1994), and where the mixture contained a susceptible cultivar along with relatively resistant cultivars the disease severity was reduced by up to 20–32%. However, in this trial some of the barley mixtures resulted in slight yield reduction, indicating the importance of careful evaluation of the mixture components, in order to choose those with a positive effect on disease control and yield (Mundt *et al.*, 1994). Similar results were obtained using wheat cultivar mixtures, with reduced leaf rust severity and increased yield (Mahmood *et al.*, 1991).

It was therefore proposed to test the concept of clonal/varietal/genotype mixtures of *Salix* spp. as a disease-control strategy for foliar rust caused by *M. epitea*. It should be noted that in early publications (e.g. McCracken and Dawson, 1994) the terms 'clonal' or 'mixed-clonal' were used. Subsequently the terms 'varieties' or 'species varieties' have been used (e.g. McCracken *et al.* 2001), and most recently individual mixture components have been called 'genotypes' (McCracken and Dawson, 2003). The term 'genotype' will therefore be used throughout this, and the following, chapter.

Mixtures with Few Components of Older Genotypes

In 1987 five *Salix* spp. genotypes, *S. viminalis* 'Bowles Hybrid', *S. viminalis* '683', *S* × *dasyclados*, *S. burjatica* 'Germany' and

S. burjatica 'Korso', were planted in a single replicate block, with a minimum plot size of 0.5 ha, as both mono-genotype and intimate mixtures (Site 1). In 1988 a 3.3 ha site was planted using *S. viminalis* 'Bowles Hybrid', *S. viminalis* '683', *S. × dasyclados*, *S. burjatica* 'Germany' and *S. mollissima-undulata* 'SQ83', in two replicate blocks both in mixture and mono-plots (Site 2). In 1989, a third area was planted using *S. viminalis* 'Bowles Hybrid', *S. viminalis* '683', *S. × dasyclados*, *S. burjatica* 'Germany', *S. mollissima-undulata* 'SQ83' with the addition of *S. × calodendron* in two replicate blocks, both in mixture and mono-plots (Site 3). In all cases the planting density was 20,000 cuttings/ha and stools were cut back at the end of their first year's growth and then allowed to grow in a 3-year harvest cycle (McCracken and Dawson, 1998). Random leaf samples were collected at fortnightly intervals from each *Salix* genotype growing in mono-plots or as part of the intimate mixtures in each site, starting in mid-May and continuing until mid-September. Rust levels were measured using a rust assessment key (McCracken and Dawson, 1992) and cumulative rust scores (Dawson and McCracken, 1994) calculated. This enabled disease progress curves to be drawn, in order to determine the effect of inclusion of the *Salix* genotypes in a mixture.

When relatively rust-susceptible *Salix* genotypes, e.g. *Salix × dasyclados*, *S. burjatica* 'Germany' and *S. viminalis* 'Bowles Hybrid', were grown as part of an intimate mixture, the level of disease was significantly reduced as the season progressed. Furthermore, the impact of the disease on the growth of stools was also greatly reduced. These disease reductions and increased growth in mixtures were similar to that achieved with the intensive use of fungicides (McCracken & Dawson, 1997). Three aspects of disease reduction were identified when susceptible genotypes were incorporated into mixtures:

1. Delay in the onset of disease: often the first significant numbers of rust pustules were observed on genotypes in mixtures 2–4 weeks after they were prevalent in mono-plots.
2. The build-up of disease was slower: even when genotypes growing in mixtures developed rust, the speed of progress of the disease was less.
3. Lower total rust scores at the end of the growing season: the outcome of the delay and slower build-up was that the level of disease at the end of the season was low enough not to induce premature leaf fall. This, in turn, meant that there was little or no infection with secondary pathogens that could potentially kill the stools (McCracken, 1989).

However, when *Salix × calodendron*, which only developed low levels of disease late in the growing season, was incorporated into a mixture, no benefits, in terms of disease reduction, were observed. In these trials *S. viminalis* '683' performed atypically. In most seasons the levels of rust on *S. viminalis* '683' were very similar irrespective of whether it was growing in mono-plots or incorporated within a mixture. It has been suggested (McCracken and Dawson, 1997) that one reason for this could be that *S. viminalis* '683' is made up of both male and female components and hence could achieve some of the benefits of a mixture.

In 1993 at least 11 pathotypes of *M. epitea* had been identified (Pei *et al.*, 1993). At this time it was thought probable that a genotype mixture would only be effective in reducing the impact of disease if the individual constituents of the mixture were not all susceptible to the same pathotype (Anon., 1994). Hence, genotypes, which were available for commercial SRC planting in the early 1990s, were attributed an 'incompatibility code' (Anon., 1994). These codes had been worked out using a differential set on willow genotypes. It was recommended that genotypes with the same incompatibility code should not be planted together in small (2–3 component) mixtures, whereas in larger mixtures only half of the genotypes included should have the same incompatibility code (McCracken and Dawson, 1997). As the European *Salix* breeding programmes released new planting material during the

1990s and into 2000, their susceptibility to similar pathotypes became an important issue. During this period almost all of the new genotypes were *S. viminalis* or *S. viminalis* hybrids, which were all theoretically susceptible to the same range of pathotypes.

A further concern relating to the use of genotype mixtures as a disease-control strategy was the potential for the selection of 'super-races' or 'super-pathotypes'. The term refers to the pathotypes which are capable of infecting all of the components of the mixture, thus making disease control even more difficult. Using mathematical models, Groth (1976) indicated that, in mixtures, complex races could develop, and that for polycylic pathogens such as rust, this could occur quickly. When random mixtures of spring barley were grown, complex pathotypes ('super-pathotypes') of the mildew pathogen, *Erysiphe graminis* f. sp. *hordei* were preferentially selected (Huang *et al.*, 1994). If similar cultivar mixtures were employed over long periods and large areas, then such super-races would eventually dominate the mildew population (Huang *et al.*, 1994). The pathogen adaptation in mixtures may be slow (Wolfe, 1985), although if the same mixture is grown over several seasons there will be selection for the modifier genes (Wolfe, 1985). This, in turn, would improve the performance of a pathogen genotype with complex virulence on that particular range of hosts. This would suggest that diversification of mixture components both in time and space will be a significant feature in the long-term sustainability of a willow plantation.

The pathotype composition of the *M. epitea* populations associated with the mixed and monoculture plots in this willow trial was analysed over 2 years, 1994 and 1995 (McCracken *et al.*, 2000). Six LET (*larici-epitea typica*) pathotypes and two LR (*larici-retusae*) pathotypes (Pei *et al.*, 1999) of *M. epitea* and *M. caprearum* were detected during the course of the study. In 1994, LET4, LET5 and LET6 were recorded in N. Ireland willow plantations for the first time. Furthermore, three new pathotypes, designated NI5, LET7 and LET8, were detected. On *S. burjatica* 'Germany' only pathotypes LR1 and LR2 were detected (McCracken *et al.*, 2000). There was no significant difference between the mean (Shannon–Weaver) diversity indices for all of the genotypes in each of the two years, 1994 (0.435) and 1995 (0.493). However, there was a significant difference between means in the mono-plots (0.372) and mixtures (0.556), indicating that there was a greater diversity of pathotypes on genotypes growing in mixtures (McCracken *et al.*, 2000). One pathotype clearly dominated on each of the *Salix* genotypes. Nevertheless, throughout the growing season, in each year, there was progression of pathotypes, with multiple pathotypes present on the same genotype or even the same leaf at the same time.

It is recognized that there is a real possibility that pathogens may be able to adapt to mixtures (Groth, 1976; Finckh and Mundt, 1992), resulting in populations dominated by 'super-races' or 'super-pathotypes'. This phenomenon has been reported with biotrophs such as rust and powdery mildews on cereal crop mixtures (Finckh and Mundt, 1992). The results from the willow mixture trial would seem to indicate that the evolution of such 'super-races' or 'super-pathotypes' is unlikely. Within *Salix* genotype mixtures there is a greater diversity of pathotypes, with the associated increase in competition between them, which in turn decreases the evolutionary pressure on each pathotype (McCracken *et al.*, 2000). Within any *Melampsora* rust population there is a wide diversity of individuals of carrying virulence genes that are capable of overcoming their corresponding host resistance genes. It was therefore recognized that if genotype mixtures were to continue as a commercially applicable disease-control strategy, then it was essential to have a range of components which contained as wide an array of multiple resistance genes as possible.

There were a number of key issues relating to the construction of *Salix* genotype mixtures that contributed to the management of rust disease, which gave increased yields and could be predicted to be sustainable over the average life of 7–10 harvest cycles of an SRC willow plantation. These were:

- optimum number of *Salix* genotypes included in a mixture
- how improved *Salix* genotypes, specifically bred for SRC willow production, would perform within mixtures
- planting configuration
- planting density
- effect of harvesting on disease development and yield.

Mixtures With Larger Numbers of Components, Including Improved Genotypes

A large two-block plantation was established at Castlearchdale, Co. Fermanagh, Northern Ireland (54°28′N, 7°43′W). The site was 60–70 m above sea level and comprised heavily gleyed soils with impeded drainage. Block 1 was planted in spring 1994 using standard planting protocols (Dawson and McCracken, 1995). It was cut back at the end of its establishment year in winter 1994/95. Block 2 was planted in spring 1995 and both blocks cut back in winter 1995/96. Hence in the 1996 growing season both blocks were at the same stage of growth – first year regrowth from freshly coppice stools. Twenty *Salix* genotypes (Table 16.1) were selected for inclusion within the trial.

Two criteria were used to select genotypes from clonal selection trials conducted in N. Ireland – high yield and reduced rust susceptibility. Three genotypes, *S. burjatica* 'Germany', *S. dasyclados* × *aquatica* 'V7511' and *S. dasyclados* × *caprea* 'V794' were known to be rust susceptible, but were included in the trial as a means of testing mixtures as a disease-control strategy. All varieties were planted in large mono-genotype plots, which were large enough to ensure that they functioned as mono-plots. Each plot was a split plot with cuttings being planted at three densities (10,000; 15,000 and 20,000 cuttings/ha) (McCracken *et al.*, 2001). Four mixture combinations (5-, 10-, 15- and 20-way) were employed, each of which was planted as a totally random mixture at the same three densities and which were also in proportionally larger plots.

Leaf samples were collected, at 2-weekly intervals, starting in mid-May through to mid-September in each of the 6 years, from each of the *Salix* genotypes growing in mono-plots and wherever they appeared within a mixture. Samples were assessed for rust (McCracken and Dawson, 1992) and disease progress curves were constructed for each genotype in each year at each density when growing in a mono-plot or when incorporated into a mixture.

There were considerable differences in the patterns of rust development on genotypes growing in mono-plots, in the 6 years

Table 16.1. Twenty *Salix* genotypes included in 5-, 10-, 15- and 20-way mixtures.

Included in 5; 10; 15; 20-way mixtures	Included in 15; 20-way mixtures
S. burjatica 'Germany'	*S. viminalis* × *caprea* 'V789'
S. mollissima-undulata 'SQ83'	*S. viminalis* '77683'
S. dasyclados × *aquatica* 'V7511'	*S. viminalis* '78101'
S. viminalis '77082'	*S. viminalis* '78195'
S. dasyclados × *caprea* 'V794'	*S. schwerinii* × *aquatica* 'V7534'
Included in 10; 15; 20-way mixtures	**Included in 20-way mixture**
S. viminalis × *aquatica* 'V7503'	*S. viminalis* '77699'
S. viminalis '78118'	*S. viminalis* 'Gigantea'
S. viminalis '78183	*S. viminalis* 'Gustav'
S. schwerinii × *viminalis* × *dasyclados* 'V7531'	*S. schwerinii* × *aquatica* 'V7533'
S. viminalis '870146 ULV'	*S. viminalis* '870082 ORM'

Table 16.2. Cumulative rust scores on 14 September on each of 20 *Salix* genotypes growing in mono-plots, 1996–2001. Least significant differences (LSDs: $P = 0.05$) between means within years are given (McCracken and Dawson, 2003. Reproduced with permission from the *Annals of Applied Biology*).

	1996	1997	1998	1999	2000	2001
Germany	8767	1733	3016	11319	676	2321
SQ83	5110	5607	5102	340	310	907
V7511	9110	1839	5287	11842	672	1835
77082	616	622	868	1993	800	358
V794	11458	1754	6565	14301	1306	2178
V7503	1868	283	59	2633	400	106
78118	2308	1082	762	740	693	317
78183	297	1035	1820	1219	1652	898
V7531	49	0	0	10	0	2
ULV	331	743	1360	266	1012	402
V789	926	2898	1451	2038	2300	105
77683	156	358	306	132	208	21
78101	1384	309	1243	305	1127	878
78195	8	375	370	82	239	12
V7534	365	0	5	493	1	47
77699	557	246	1516	1027	475	330
Gigantea	43	4	23	17	63	13
Gustav	5453	1442	2881	3447	3184	1117
V7533	80	0	0	240	2	23
ORM	741	184	172	856	85	209
LSD ($P = 0.05$)	5905	1299	2590	3347	569	1344

(Table 16.2). *Salix schwerinii* × *viminalis* × *dasyclados* 'V7531', *S. schwerinii* × *aquatica* 'V7533' and *S. viminalis* 'Gigantea' remained virtually rust free throughout the 6 years of the trial. However, there are still benefits in the inclusion of such genotypes within mixtures, not only to increase mixture diversity, but also their inclusion in mixtures should ensure that their rust resistance does not break down during the 30-year life of the willow plantation.

High levels of rust were recorded on *S. burjatica* 'Germany', *S. dasyclados* × *caprea* 'V794', *S. dasyclados* × *aquatica* 'V7511' and *S. viminalis* 'Gustav'. In general, the final level of rust was much higher in the year immediately following coppicing (1996 and 1999) than in subsequent years. In contrast, rust levels on *S. mollissima-undulata* 'SQ83' were highest during the first harvest cycle (1996–98) and low during the second cycle (1999–2001) to a point where the disease was having no impact on growth. This indicated that there had been a change in the balance of pathotype composition, with the previously dominant highly damaging pathotype becoming less prevalent. Rust levels on the remaining 12 genotypes growing in mono-plots were relatively low in all years and had little or no impact on growth or survival (Chapter 17). On the basis of rust development on genotypes growing in mono-plots, the 20 genotypes were ranked in terms of their relative susceptibility (Table 16.3).

Inclusion of rust-susceptible *Salix* genotypes in mixtures generally resulted in reduced rust, compared to the mono-plots. At the August assessment dates in 1996 and 1999, the rust scores on *S. burjatica* 'Germany' were significantly higher on plants growing on mono-plots than on plants in any of the mixtures. In September 1997 and 1998, when there were high levels of rust on *S. mollissima-undulata* 'SQ83' significantly lower rust scores were recorded on plants growing in the 5-, 10-, 15- and 20-way mixtures. On *S. dasyclados* × *aquatica* 'V7511' in 1997, 1998, 1999 and 2001 there was significantly more rust on plants in mono-plots compared to those in mixtures (Fig. 16.1). In 1996 the difference was only significant for the 20-way mixture. Rust levels on *S. dasyclados* × *caprea* 'V794' were generally highest in years immediately following coppicing, i.e. 1996 and 1999.

Table 16.3. Relative susceptibility to rust of the 20 *Salix* genotypes growing in mono-plots.

Rank	Genotype	
1	V7531	Most resistant
2	Gigantea	
3	V7533	
4	V7534	
5	78195	
6	77683	
7	ORM	
8	V7503	
9	77699	
10	ULV	
11	78101	
12	77082	
13	78118	
14	78183	
15	V789	
16	SQ83	
17	Germany	
18	Gustav	
19	V7511	
20	V794	Most susceptible

Mixtures with Limited Diversity

Concern had been expressed that almost all of the improved genotypes available for commercial planting were *S. viminalis* or *S. viminalis* hybrids (McCracken *et al.*, 2000; McCracken and Dawson, 2001). In 2000 a trial was planted at the Northern Ireland Horticulture and Plant Breeding Station, Loughgall, Co. Armagh. Seven *S. viminalis* genotypes were planted: *S. viminalis* '77082', *S. viminalis* 'Ulv', *S. viminalis* 'Orm', *S. viminalis* 'Jorr', *S. viminalis* 'Jorunn', *S. viminalis* 'Gigantea' and *S. viminalis* '78195' in large mono-genotype plots. *S. viminalis* '77082' and *S. viminalis* 'Ulv' were known to have a limited susceptibility to *M. epitea* (McCracken and Dawson, 2003) and so were both incorporated into a 2-way mixture, five 3-way mixtures with each of the other genotypes and a 7-way mixture including all of the genotypes. As in previous trials, leaf samples were collected at 2-week intervals, from in mid-May to mid-September, during each of the growing seasons 2000–2003.

In all 3 years, the highest levels of rust were recorded on *S. viminalis* '77082'. Rust

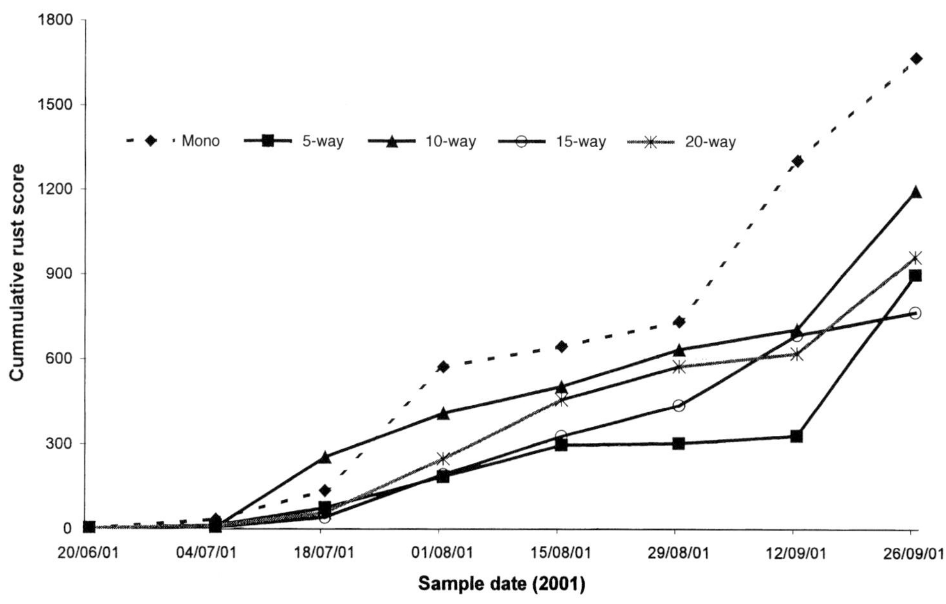

Fig. 16.1. Rust disease (June–September 2001) development on *Salix dasyclados* × *aquatica* 'V7511' grown in mono-plots and as part of 5-, 10-, 15- and 20-way mixtures.

was normally observed in mid-July and levels increased steadily during subsequent weeks. Highest levels of rust were observed in mono-plots in 2002 on *S. viminalis* '77082' and *S. viminalis* 'Orm'. In 2002 and 2003 there was a significant reduction in rust on *S. viminalis* '77082' in the 7-way mixture compared to levels in the mono-plot. A similar improvement was observed in 2002 on *S. viminalis* 'Orm'.

S. viminalis '77082' had been included in seven different combinations: in mono-plots, a 2-way mixture with *S. viminalis* 'Ulv', 3-way mixtures with *S. viminalis* 'Ulv' and one of *S. viminalis* 'Jorr', *S. viminalis* 'Jorunn', *S. viminalis* 'Gigantea' or *S. viminalis* '78195', or in a 7-way mixture. This meant that it was possible to identify whether the presence of certain willow genotypes influenced the degree of susceptibility of the rust susceptible genotype *S. viminalis* '77082'.

When growing with *S. viminalis* 'Gigantea' there was significantly less rust on *S. viminalis* '77082' compared to rust levels in the mono-plots. This occurred consistently in all 3 years, with differences being evident in late August and remaining throughout the rest of the growing season. Similar reductions were observed with *S. viminalis* '78195' and to a lesser extent *S. viminalis* 'Orm'. *Salix viminalis* 'Jorunn' was inconsistent with rust levels on *S. viminalis* '77082' being reduced in 2001 and 2003 but not in 2002. Inclusion with *S. viminalis* 'Jorr' resulted in no changes in amount of rust disease developing. The 2-way mixture with *S. viminalis* 'Ulv' had no effect on rust and the 7-way mixture containing all of the genotypes had a limited effect in reducing rust on *S. viminalis* '78195'.

It has been speculated (Dawson and McCracken, 1995; Mundt *et al.*, 2002; Chapter 15, this volume) that one of the primary mechanisms to explain the effectiveness of mixtures in reducing disease impact is a dilution of susceptible genotypes with less susceptible individuals, i.e. since the mixture contains a number of resistant components, the spread of the pathogen is slowed down. This does not seem to be the only mechanism functioning in this trial, otherwise the greatest reduction should have been observed in the 7-way mixture and there should have been little difference between each of the 3-way mixtures. It seems possible that another mechanism may be responsible. It is clear from the yield data (Chapter 17) that individual components can have a significant impact on the growth of neighbouring components, either beneficially or detrimentally. It is therefore postulated that the presence of a particular component, e.g. *S. viminalis* 'Gigantea' or *S. viminalis* 'Jorunn', induces a response in *S. viminalis* '78195'. This may be some form of induced systemic resistance to rust, but this hypothesis needs further investigation.

Conclusions

Growing SRC willow genotypes in monoculture in temperate wet climates carries a high risk, which these authors regard as unacceptable. While there are a number of high-yielding, rust-resistant *Salix* genotypes which may perform well over many harvest cycles, there does seem to be a genuine risk of previously resistant genotypes developing susceptibility over a relatively short period of time. Inclusion of such genotypes in a diverse mixture should guarantee their survival throughout the life of the plantation. In addition there are well-documented yield benefits.

The greater the diversity of the mixture, incorporating inter- and intra-species genotypes, it would seem the greater the benefits and long-term sustainability of the plantation. However, straight *S. viminalis* mixtures with a range of susceptibilities have reduced rust impact over the first 3-year harvest cycle. This trial needs to be run for at least two more harvest cycles to determine the effectiveness in the long-term.

The mechanism of the effectiveness of mixtures in reducing rust also merits further research. While the dilution effect is an obvious factor, it does seem that other, more complex mechanisms are operating.

Acknowledgements

The authors gratefully acknowledge funding for part of this work from the Framework5 R & D Programme of the European Union (QLRT-1999-01585). They are also grateful to the Department of Agriculture and Rural Development (Northern Ireland) for their support to the R & D programme over a number of years.

References

Anon (1994) *Plantation Design: Combating Disease and Pests*. Agriculture and Forestry Fact Sheet Short Rotation Coppice no. 14. Department of Trade and industry, Energy Technology Support Unit, London.

Dawson, W.M. and McCracken, A.R. (1994) Effect of *Melampsora* rust on the growth and development of *Salix burjatica* 'Korso' in Northern Ireland. *European Journal of Plant Pathology* 24, 32–39.

Dawson, W.M. and McCracken, A.R. (1995) The performance of polyclonal stands in short rotation coppice willow for energy production. *Biomass and Bioenergy* 8, 1–5.

Finckh, M.R. and Mundt, C.C. (1992) Stripe rust, yield and plant competition in wheat cultivar mixtures. *Phytopathology* 82, 905–913.

Groth, J.V. (1976) Multi-lines and super-races: a simple model. *Phytopathology* 66, 937–939.

Huang, R., Kranz, J. and Welz, W.G. (1994) Selection of pathotypes of *Erysiphe graminis* f. sp. *hordei* in pure and mixed stands of spring barley. *Plant Pathology* 43, 458–470.

Mahmood, T., Marshall, D. and McDaniel, M.E. (1991) Effect of winter wheat cultivar mixtures on leaf rust severity and grain yield. *Phytopathology* 81, 470–474.

McCracken, A.R. (1989) SEM study of fungi associated with cankers on willow. In: Rushton, B.S. and Flynn, A.M. (eds) *Proceedings of Irish Botanists Meeting 1988*. University of Ulster at Coleraine, p. 42.

McCracken, A.R. and Dawson, W.M. (1992) Clonal response in *Salix* to *Melampsora* rust in short rotation coppice plantations. *European Journal of Forest Pathology* 22, 19–28.

McCracken, A.R. and Dawson, W.M. (1994) Experiences in the use of mixed-clonal stands of *Salix* as a method of reducing the impact of rust disease. *Norwegian Journal of Agricultural Sciences* Suppl. 18, 101–109.

McCracken, A.R. and Dawson, W.M. (1997) Using mixtures of willow clones as a means of controlling rust disease. In: Bullard, M.J., Ellis, R.G., Health, M.C., Knight, J.D., Lainsbury, M.A. and Parker, S.R. (eds) *Aspects of Biology 49; Biomass and Energy Crops*. Association of Applied Biologists, Cirencester, UK, pp. 97–103.

McCracken, A.R. and Dawson, W.M. (1998) Short rotation coppice willow in Northern Ireland since 1973: development of the use of mixtures in the control of foliar rust (*Melampsora* spp.). *European Journal of Forest Pathology* 28, 241–250.

McCracken, A.R. and Dawson, W.M. (2001) Disease effects in mixed varietal plantations of willow. In: Bullard, M.J., Christian, D.G., Knight, J.D., Lainsbury, M.A. and Parker, S.R. (eds) *Aspects of Biology 65 – Biomass and Energy Crops II*. Association of Applied Biologists, Warwick, UK, pp. 255–262.

McCracken, A.R. and Dawson, W.M. (2003) Rust disease (*Melampsora epitea*) of willow (*Salix* spp.) grown as short rotation coppice (SRC) in inter- and intra-species mixtures. *Annals of Applied Biology* 143, 381–393.

McCracken, A.R., Dawson, W.M., Watson, S. and Allen, C.Y. (2000) Pathotype composition in *Melampsora epitea* populations occurring on willow (*Salix*) grown in mixed and monoculture plantations. *European Journal of Plant Pathology*106, 879–886.

McCracken, A.R., Dawson, W.M. and Bowden, G. (2001) Yield responses of willow (*Salix*) grown in mixtures in short rotation coppice (SRC). *Biomass and Bioenergy* 21, 311–319.

Mundt, C.C., Hayes, P.M. and Schon, C.C. (1994) Influence of barley mixtures on severity of scald and net blotch and on yield. *Plant Pathology* 43, 356–361.

Mundt, C.C., Cowger, C. and Garrett, K.A. (2002) Relevance of integrated management to resistance durability. *Euphytica* 124, 245–252.

Pei, M.H., Royle, D.J. and Hunter, T. (1993) Identity and host alternation of some willow rusts (*Melampsora* spp.) in England. *Mycological Research* 97, 845–851.

Pei, M.H., Hunter, T and Royle, D.J. (1999) Occurrence of *Melampsora* rusts in biomass willow plantaions for renewable energy in the United Kingdom. *Biomass and Bioenergy* 17, 153–163.

White, E.M. (1982) The effect of mixing barley cultivars on incidence of powdery mildew (*Erysiphe graminis*) and on yield in Northern Ireland. *Annals of Applied Biology* 101, 539–549.

Wolfe, M.S. (1985) The current status of multiline cultivars and variety mixtures for disease resistance. *Annual Review of Phytopathology* 23, 251–273.

17 Short-rotation Coppice Willow Mixtures and Yield

W. Malcolm Dawson[1], Alistair R. McCracken[2] and Diane Carlisle[3]
[1]*Applied Plant Science Division, Northern Ireland Horticultural and Plant Breeding Station, Department of Agriculture and Rural Development, Loughgall, Co. Armagh BT61 8JB, UK;* [2]*Applied Plant Science Division, Department of Agriculture and Rural Development, Newforge Lane, Belfast BT9 5PX, UK;* [3]*Department of Applied Plant Science, Queen's University of Belfast, Newforge Lane, Belfast BT9 5PX, UK*

Introduction

The principle on which crop yield is based is the conversion of light energy into the chemically bound energy contained in the economically important part of the crop, whether that is in root tubers, cereal grains or, in the case of short-rotation coppice willows, the above-ground woody stems. Not all of the total organic matter produced by the crop as a result of its photosynthetic activity is harvested. In short-rotation coppice, the stool, root system and leaves remain unharvested and, together with the woody stems (economic yield), make up the total biological yield. In willow coppice it has been estimated that the harvested stems represent 60% of the total biological yield, with the leaves (10%) and stool/root system (30%) making up the remainder. For short-rotation coppice the economic yield is normally quoted as an annual figure of tonnes of dry matter per hectare. To maximize the economic yield the crop canopy (leaf area index) must be maximized as early as possible and maintained for as long as possible through the control of negative biological influences, particularly pests and diseases. As a consequence, breeding, not just to improve productivity levels, but also to improve pest and disease resistance and plant form, is a major input into the area of yield from short-rotation coppice systems.

Genotype evaluation has been an ongoing commitment from the beginning of the programme on biomass. Initially this was with relatively unimproved genotypes, which had often been selected for purposes other than biomass production for energy. In 1989, a more formal link was forged with the breeding programme in Sweden and a trial protocol was agreed with the International Energy Agency. In 1996 a second dedicated breeding programme was initiated in the UK and potential operational genotypes were included in a revised testing protocol, drawn up with the Forestry Commission, to facilitate their inclusion in recommended listings for planting.

Yield Expectations

When the work on short rotation coppice as a renewable energy crop began, in the mid-1970s, yields of 12–15 tonnes DM/ha/year were recorded from experimental plots using the relatively non-specific and unimproved willow species and varieties available. Of the candidates assessed in these trials *Salix burjatica* 'Korso' emerged as the most productive over a range of sites, producing yields up to 17 tonnes DM/ha/year at high density (40,000/ha) planting. However, total dry matter (DM) yields from small experimental plots have to be treated with caution and early work indicated that these yields could be overestimated by as much as 30% (Stott *et al.*, 1987). This would objectively reduce these early yield estimates from 12–15 tonnes DM/ha/year to 8.4–10.5 tonnes DM/ha/year.

In the earliest collaborative trials, carried out at Long Ashton Research Station, Somerset, and in Northern Ireland (Anon., 1980), yield estimates for a range of *Salix* spp. genotypes indicated that *S. burjatica* 'Korso' and *S.* × *dasyclados* produced the highest yields across all sites. Fresh weight yields from *S. burjatica* 'Korso' ranged from 19.0 to 20.5 tonnes/ha/year equivalent to 8.7–9.4 tonnes DM/ha/year. Similarly *S.* × *dasyclados* gave 23–29 tonnes fresh weight = 10.6–13.3 tonnes DM/ha/year. However, both of these have shown an increasing susceptibility to *Melampsora* rust, to the extent that they can not now be recommended for planting.

In an effort, in particular, to improve dry matter yield and disease susceptibility, and more generally plant form, to facilitate mechanical harvesting, commercial willow breeding programmes were initiated in Sweden in 1987 and in the UK in 1996. The Swedish programme has focused on *S. viminalis, S. schwerinii* and *S. dasyclados* of European and central Russian and Siberian origin (Larsson, 1997, 2001). The Russian material has provided a good source of increased productivity and resistance to *Melampsora* rust. The UK programme has involved a much wider species range including *S. rehderiana, S. udensis, S. discolor* and *S. aegyptica* as well as *S. viminalis* and *S. schwerinii*. The first output from this breeding programme entered the final stages of comparative yield trials in 2002, with the earlier *S. viminalis* × *S. burjatica* crosses *S.* × *stottii* 'Ashton Stott' and 'Ashton Parfitt' coming into trials in 1992. Operational genotypes from these breeding programmes were tested at four sites, Yorkshire, N. Ireland, Devon and Avon, located across the UK (Lindegaard *et al.*, 2001), initially using a protocol designed by the International Energy Agency and later, in 1996, converting to a trial protocol agreed with Forestry Commission. The current protocol for carrying out these comparative trials involves establishing the cuttings and cutting them back at the end of the first growing season. Subsequently a 2-year yield and a 3-year yield are taken before consideration for inclusion in the recommended listing (Tabbush and Parfitt, 1996). These trials, using agreed standard varieties, gave a much more accurate picture of potential yield than the earlier trials. Initially the standard was *S. viminalis* '78183', the most widely planted species in Sweden, and later *S. viminalis* × *S. schwerinii* 'Tora' as representative of the newer, high-yielding hybrids.

When the mean yield of all varieties was compared, with the Avon site at 100%, Devon produced 79%, N. Ireland 75% and Yorkshire 68%. These differences were likely to be due to local topographical, soil and climatic factors. The yields from these four sites are collated and used by the Forestry Commission to produce a recommended listing for planting in the UK (Tabbush and Parfitt, 1996).

Most of the new operational varieties produced significantly higher yields than the *S. viminalis* '78183' standard. The earliest commercially available releases from the Swedish programme, *S. viminalis* 'Orm', 'Rapp' and 'Ulv', produced dry matter yields of 177%, 141% and 154%, respectively, of the standard. Later introduction of *S. viminalis* 'Jorr' and 'Joruun' and *S. viminalis* × *S. schwerini* 'Tora' and 'Sven' also gave improved yields, representing 156%, 142% and 202% and 167%, respectively. The most productive introductions from the UK programme, *S.* × *stottii* 'Ashton Stott' and 'Ashton Parfitt', have similarly given 243% and 220% of the standard. Even when compared with the new, higher-yielding standard, 'Tora', they show a significant increase in dry matter yield of 120% and 109%, respectively (Lindegaard *et al.*, 2001). Figure 17.1 shows the dry matter yield from two of these trials, established in 1996 and 1997, with 3-year yields in 2001 and 2002, respectively.

These yields show the highly significant improvements in dry matter yield not just on the early standard variety, *S. viminalis* '78183', but also on the new standard variety *S. viminalis* × *S. schwerinii* 'Tora'. For example, in 2002 *S. viminalis* × *S. schwerinii* × *S. viminalis* 'Sven' (930824) recorded 110% of the annual yield of *S. viminalis* × *S. schwerinii* 'Tora'. Additionally, these data also illustrate that annual dry matter yields from a first coppicing cycle are generally in the region of 20% lower than those from the second cycle. In this particular case, the improved yield in the second cycle is in cumulative response to site capture and to the fact that increasing harvesting interval gives increased annual dry matter increment (McElroy and Dawson, 1986).

This sustained improvement in dry matter (DM) and the use of a wide range of *Salix* species in the breeding programmes has given a situation currently where commercially sustainable yields of 10–12 tonnes DM/ha/year can be achieved, with the short-term objective of raising that to 14–15 tonnes DM/ha/year. In conditions where *Salix* species have been supplied with optimal nutrients and water, yields of over 30 tonnes DM/ha/year have been achieved (Stott, 1984; Christersson, 1987) from experimental plots, indicating that further improvements can be made before the biological potential of the species is approached.

It is generally accepted that, under the current commercial and energy supply conditions, that energy from short-rotation willow coppice is only marginally economic at best. However, a range of economic models haa shown (McBurney and Brigg, 1996; Rosenquist and Dawson, 2004) that yield has an important part to play in improving this situation and ensuring the economic sustainability of the enterprise. Rosenquist and Dawson (2004) demonstrated that, in Northern Ireland, assuming a £40 per tonne dry matter return, a yield increase of 20% to 14 tonnes DM/ha/year would produce a 40% increase in gross margin.

Yield Losses due to Rust

To assess the effect of *Melampsora* rust on the productivity of short-rotation coppice, willow fungicides have been used on the

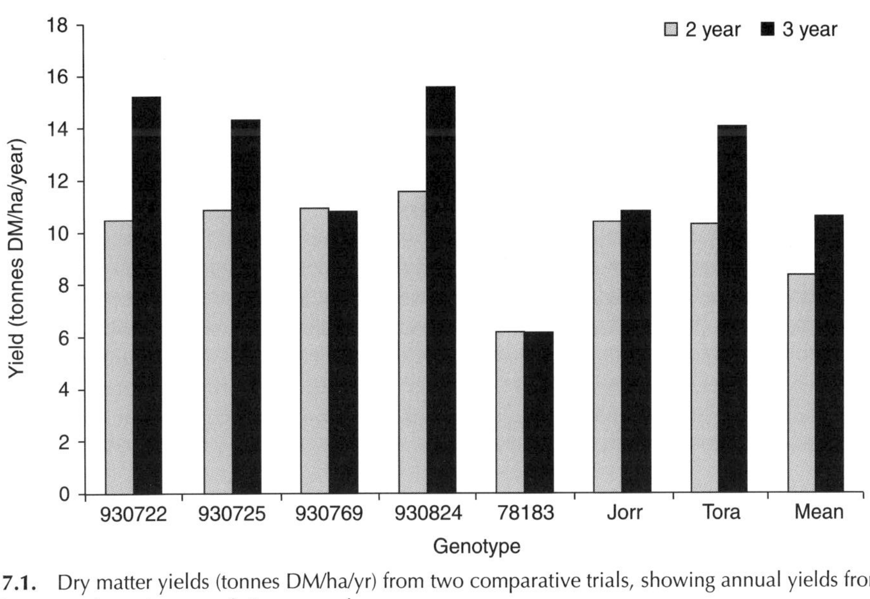

Fig. 17.1. Dry matter yields (tonnes DM/ha/yr) from two comparative trials, showing annual yields from a first, 2-year, cycle and a second, 3-year, cycle.

rust-sensitive variety *Salix burjatica* 'Korso'. Unlike other major agricultural commodities, the use of fungicides for the control of foliar rust in short-rotation coppice willow for energy is not an option principally on a cost basis. The economic burden of regular fungicide application could not be justified on such a high-volume, low-value crop. Additionally, the density of the growing crop, particularly in the second and third years of the rotation, produced a major practical problem in obtaining effective fungicide coverage. Also, the use of significant volumes of pesticide on an environmentally sensitive crop, and often in or near environmentally sensitive areas, would not be tolerated.

In these fungicide trials, benodanil (Calirus™) and myclobutanil (Systhane™) provided effective control of the rust, but only when applied in a protectant programme at 14-day intervals, beginning in mid-May before the rust had an opportunity to establish itself (McCracken and Dawson, 1998). In these circumstances, DM yield from the sprayed plots was significantly greater than from the unsprayed plots. In the Systhane™-treated plots, total DM yield was 78% higher than the unsprayed control, indicating the devastating effect on productivity of *Melampsora* rust on susceptible *Salix* varieties (Fig. 17.2).

Yield losses in rust-affected varieties result from premature leaf fall and a reduction in photosynthetic area, but also from shoot death in the years following infection. Leaf fall can reach significant levels early in the growing season, with the rust-sensitive variety *S. burjatica* 'Korso' showing in excess of 50% defoliation by mid-July (McCracken and Dawson, 1994). Where serious rust infection occurs in the year following coppicing, shoot death has been recorded at levels approaching 60% (Fig. 17.3). This was reduced to 5% in the fungicide-treated plots in this single-variety planting. However, where fungicides were not applied, the resulting weakening of the stool, and subsequent increased competition from weed growth, resulted in complete crop failure.

Mixtures: Early Trials

These serious reductions in yield and subsequent crop failures were not totally unexpected, given the dynamic nature of the pathogen and the clonally propagated nature of the crop. However, if the concept of producing a carbon dioxide neutral source of renewable energy from short-rotation coppice willow was to progress, an alternative disease-control strategy was

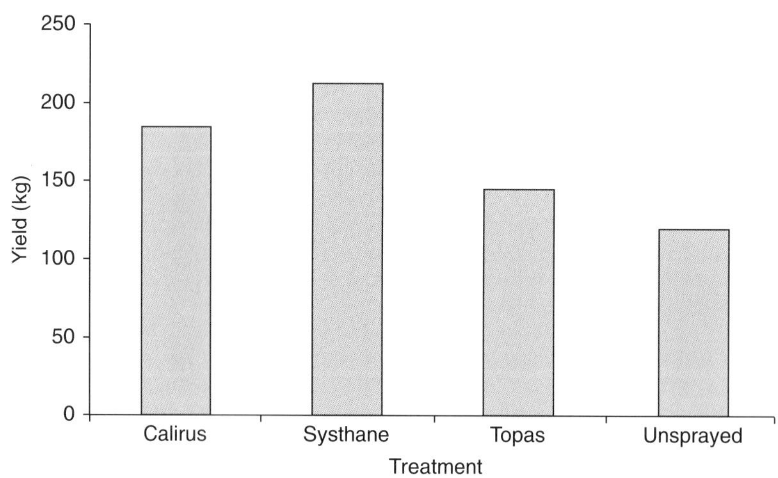

Fig. 17.2. Yield (kg DM) of *Salix burjatica* 'Korso' treated with a range of rust fungicides.

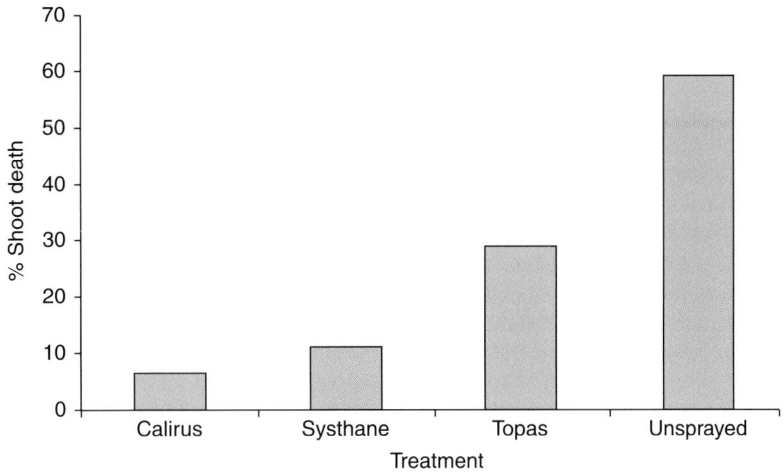

Fig. 17.3. Percentage shoot death in the year following coppicing in *Salix burjatica* 'Korso' treated with a range of rust fungicides.

required, and the potential of introducing diversity, and hence stability, into the system was investigated, and the use of mixtures began. Initially, these were limited mixtures made up of 4–6 components, and they were planted as structured intimate mixtures, i.e. the components were always planted in the same order rather than totally randomly (Anon., 1998). The first trials were established in 1987. Yield estimates taken in the dormant season throughout the life of the coppice system showed that final dry weight yields from the mixtures were always significantly higher (up to 40%) than the mean of its component varieties grown in monoculture. These differences were not apparent in the first year after coppicing, but became significant in year 2, maintaining and enhancing this difference until harvest in year 3 (Fig. 17.4) (McCracken and Dawson, 2001).

When the contribution of the individual component species genotypes was analysed, it was apparent that whereas the mixture, in total, outyielded its component parts, they individually varied in their response to inclusion in mixtures. In the early trials, *S. viminalis* 'Bowles Hybrid', *S. burjatica* 'Germany' and *S. viminalis* '683' all showed improved yields in mixtures, whereas *S.* × *dasyclados* did not show the same positive response (McCracken and Dawson, 2001). There is strong evidence to support the proposition that clonal yield response is highly site-specific; however, evidence from a range of sites indicates that the yield response of mixtures is maintained irrespective of site (Armstrong, 2000).

Having demonstrated, over a number of harvesting cycles, the yield advantages of mixtures containing relatively small numbers of components, work initiated in 1996 began to evaluate the effects of mixtures with up to 20 components, and planted at three densities: 20,000, 15,000 and 10,000/ha. A 5-way mixture was established with three components known to be highly rust susceptible. These were combined with a further five genotypes with reduced susceptibility, in a 10-way mixture. These ten were included with five more genotypes in a 15-way mixture, and a further five to form a 20-way mixture (McCracken and Dawson, 2001). Hence, the susceptible varieties formed 60% of the 5-way and 30%, 20% and 25% of the 10-, 15- and 20-way mixtures, respectively. The varieties included in the investigation were representative of breeding programmes in Finland and Sweden, and older varieties from various sources, which had been screened in selection trials at the Northern Ireland Horticulture and Plant Breeding Station. The 20 varieties comprised 11 straight *S. viminalis* varieties, including *S. viminalis* '78183' as the standard. The varieties and their yields are listed in Table 17.1.

Grown in monoculture, the highest yields were obtained from *S. schwerinii* × *aquatica* 'V7534', *S. schwerinii* × *aquatica* 'V7533' and *S. schwerinii* × *viminalis* × *dasyclados* 'V7531'. These yields were equivalent to 29.9, 25.9 and 24.9 tonnes DM/ha. At the other end of the scale,

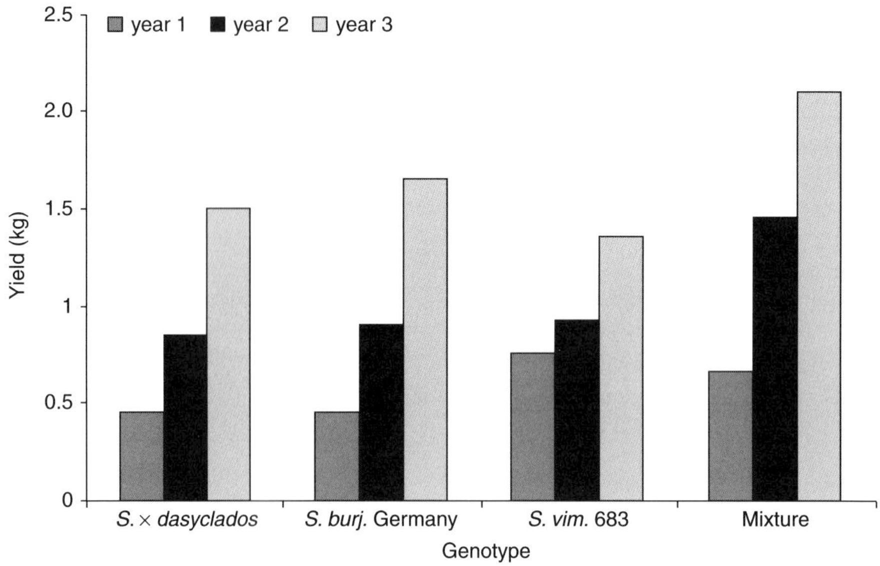

Fig. 17.4. Mean dry matter yields (kg per stool) from four *Salix* varieties grown in monoculture and as a 4-way mixture.

Table 17.1. Dry matter yield (first rotation) from 20 *Salix* species/varieties grown at three planting densities (kg 100 m^{-2}).

Genotype	10,000/ha	15,000/ha	20,000/ha	Mean	% of 78183
S. schwerinii × *aquatica* 'V7534'	306.8	81.2	308.6	298.9	173
S. schwerinii × *aquatica* 'V7533'	262.1	239.6	274.8	258.9	150
S. schwerinii × *viminalis* × *dasyclados*, 'V7531'	280.1	254.0	215.6	249.9	145
S. viminalis '78195'	249.0	227.0	236.6	237.5	137
S. viminalis '77683'	227.9	207.1	242.4	225.8	131
S. viminalis × *aquatica* 'V7503'	211.9	177.6	228.1	209.2	121
S. viminalis 'Orm'	241.6	191.0	188.8	207.1	120
S. viminalis 'Gigantea'	212.2	173.6	208.4	98.0	115
S. viminalis '77699'	97.5	169.8	192.8	186.7	108
S. viminalis 'Ulv'	180.3	165.7	196.8	180.9	105
S. viminalis '78101'	167.0	159.5	196.5	174.3	101
S. viminalis '78183'	174.0	151.7	192.3	172.7	100
S. viminalis '781181'	151.4	157.4	162.4	157.1	91
S. molissima undulata 'SQ83'	193.3	139.2	133.4	155.3	90
S. viminalis '77082'	160.2	125.4	166.9	150.8	87
S. viminalis × *caprea* 'V189'	105.4	136.8	139.4	124.9	72
S. viminalis 'Gustav'	105.1	96.1	123.5	108.8	62
S. dasyclados × *aquatica* 'V7511'	81.3	66.0	83.1	76.8	44
S. burjatica 'Germany'	71.1	64.9	78.4	71.5	41
S. dasyclados × *caprea* 'V794'	53.3	54.8	75.4	61.2	35

S. dasyclados × *aquatica* 'V7511', *S. burjatica* 'Germany' and *S. dasyclados* × *caprea* 'V794' produced yields equivalent to 7.7, 7.1 and 6.1 tonnes dry matter. These varieties were also the three rust-susceptible genotypes included in the mixtures.

The yields of these 20-component genotypes grown in monoculture were compared with equivalent areas grown in various mixture configurations. The yield from all of the mixtures was greater than the mean yield from an equivalent area of the constituent varieties grown in monoculture. Thus the results from these mixtures, containing up to 20 constituents, mirror the results from the earlier mixture trials. The increases in yield from the mixtures represented 163% in the case of the 5-way mixture, 111% in the 10-way mixture, 126% in the 15-way mixture and 119% in the 20-way mixture (Fig. 17.5). The yield increases recorded in the mixture plots were independent of planting density.

A number of other important issues also became apparent from this investigation. Where component genotypes were rust susceptible prior to their inclusion, in mixtures they did not improve their performance. In fact, due to competitive effects of the less-susceptible components and their proximity in randomly planted mixtures, their survival and productivity were often poorer in mixtures than in monoculture.

When the contribution of the various components to the mixture was analysed, it was clear that not all contributed positively to the increased yields obtained from mixtures. In all cases, the rust-susceptible genotypes produced significantly less in the mixture than their predicted value, calculated from the percentage of the mixture the component occupied. For example, it was postulated that an individual genotype within a 20-way mixture would contribute 5% of the total yield of that mixture. Where the productivity of a mixture was calculated as the yield from an equivalent area of the component genotypes grown in monoculture, these rust-susceptible genotypes performed marginally better. This was attributable to the competitive effects of the remaining, less-susceptible genotypes within the mixture (McCracken and Dawson, 2003).

Other varieties gave significantly higher yields than their predicted contribution, e.g.

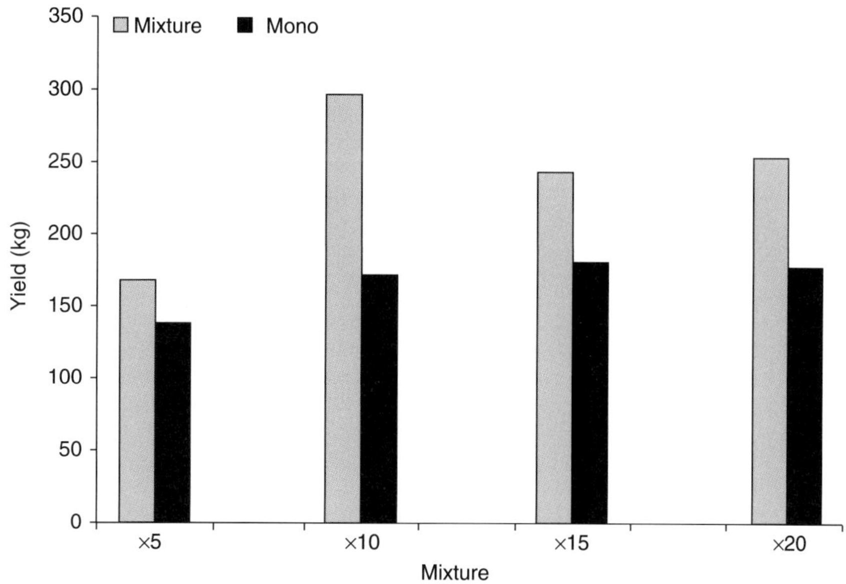

Fig. 17.5. Dry matter yield from 100 m^2 of 5-, 10-, 15- and 20-way mixtures, compared to an equivalent area of the components grown in monoculture.

S. viminalis 'Orm' contributed 11.5% of the yield in the 20-way mixture, compared with its expected 5%. Similarly *S. schwerinii* × *aquatica* 'V7533', *S. schwerinii* × *aquatica* 'V7534', *S. viminalis* '78195', *S. viminalis* '78683' all contributed a significantly greater proportion than expected of the yield in the 20-way mixture. These varieties are well suited to exploiting the diversity of growth habit within mixtures. They also have a greater vigour, being the highest yielding varieties in monoculture. The remainder of the components contributed at, or about, their expected levels, and although they did not add to the productivity of the mixture in terms of DM yield, they contributed to its sustainability by increasing the diversity of the mixture and thereby reducing the selection pressure in the disease

Yield Compensation

The dry matter yield figures for 5-, 10-, 15- and 20-way mixtures, compared with yield from an equivalent area of their components grown in monoculture, illustrates a further advantage of diverse mixtures. The 5-way mixture contained three genotypes, which were known to have significant susceptibility to rust. Subsequently, their susceptibility increased until the impact of the disease was sufficient to reduce their contribution to final dry matter figures to unacceptable levels. Although their inclusion in mixtures with more components reduced disease levels, DM yields still remained below acceptable levels and these varieties effectively made little or no contribution to overall mixture yields. In the case of the 5-way mixture, these susceptible varieties made up 60% of the components planted and this level was successively 'diluted' to 30%, 20% and 15%, with the addition of 5, 10 and 15 less-susceptible genotypes to the mixtures. The 5-way mixture was significantly less productive than the 10-, 15- or 20-way mixture, but the 10-, 15- and 20-way mixture yields did not differ significantly from each other. Thus, in the 5-way mixture the remaining two productive genotypes were unable to compensate fully for the losses of the three rust-susceptible components. Although they grew more vigorously, they were unable to make up the shortfall (Fig. 17.5). However in the 10-way mixture, where the rust-susceptible varieties constituted 30% of the components, the remaining varieties not only compensated for their loss but also produced a significantly higher dry matter yield, than the 5-way mixture. The same was true for the 15- and 20-way mixtures, where susceptible genotypes occupied progressively less of the space. The inclusion of more than 10 components in the mixture did not have any significant effect on final productivity. However, there is a strong argument that the increased diversity in the 15- and 20-way mixture would increase sustainability (McCracken and Dawson, 2001). Although more detailed work requires to be done, it is clear that where losses of components in a mixture, for whatever reason, remain below 30%, the remaining components can fully compensate in terms of dry matter yield. This is confirmed by earlier studies in planting density, where high planting densities of 20,000/ha tended, over a period of two or three rotations, to self-thin to 13,000–15,000/ha without materially affecting their productivity.

It is important to maximize this ability of mixtures to compensate for the loss of individual components through disease or other factors. In this respect, the planting pattern of the mixture has an important part to play. Ideally, the mixture should be planted totally randomly, with each component as widely separated from itself as the number of genotypes in the mixture allows. However, in commercial practice a compromise is necessary to deal with planting mechanization. Here, where full planting rods are used in automatic planters, randomly planted short runs (8–10 cuttings) of the same genotype offer an acceptable alternative. Although line mixtures or mosaics may be equally productive, they do not offer the same opportunities for yield compensation following the loss of component(s). Hence such mixtures will be less sustainable in the long term (Pei *et al.*, 1996).

Stool Survival

The ability of a genotype to survive within a mixture is obviously an important aspect of its contribution to overall yield. In early trials it became obvious that survival in monocultures was significantly higher than in mixtures in all of the genotypes investigated. Figure 17.6 shows the comparative survival rates for six component genotypes used in the early mixture investigations.

In monoculture, where there are uniform competitive effects from adjacent stools of the same genotype, survival is good – 90% for all varieties after the first harvesting cycle. However, unless the genotype is well adapted to exploiting the diversity of growth habit and conditions within a mixture, survival will be affected. All of these varieties showed a significant reduction in survival in mixtures. However, this reduction was at or below the critical 30% for yield compensation. These trials were planted at an initial density of 20,000/ha and survival of over 70% in all genotypes except *S.* × *calodendron* resulted in a reduced density of 14,000/ha. In the case of *S.* × *calodendron* the mixture survival of 40% indicated a genotype which was not well adapted to mixtures (Anon., 1998). This was due to its reduced competitive ability as a consequence of its later leafing up and shorter growth habit.

Similar survival data were recorded on later mixture plantations and, in these cases, the response varied with the genotypes' inherent susceptibility to rust. For resistant genotypes, or where rust has only been recorded late in the season and at very low levels, e.g. *S. schwerinii* × *viminalis* × *dasyclados* 'V7534', survival remained at over 80%, with no difference between the first and second harvests in 1999 and 2002. Where the genotype had a low to moderate susceptibility to rust, e.g. *S. viminalis* 77118, and its inclusion in mixtures had the effect of reducing rust to acceptable levels, survival was over 90%. Although reduced at the second harvest, it still maintained levels above 80% in mixtures. However, survival of highly susceptible varieties, e.g. *S. dasyclados* × *caprea* 'V794' was reduced after the first harvesting cycle, in some instances below the critical 70% where yield compensation can not make up the loss. This was further reduced after the second harvest (Fig. 17.7), and in all

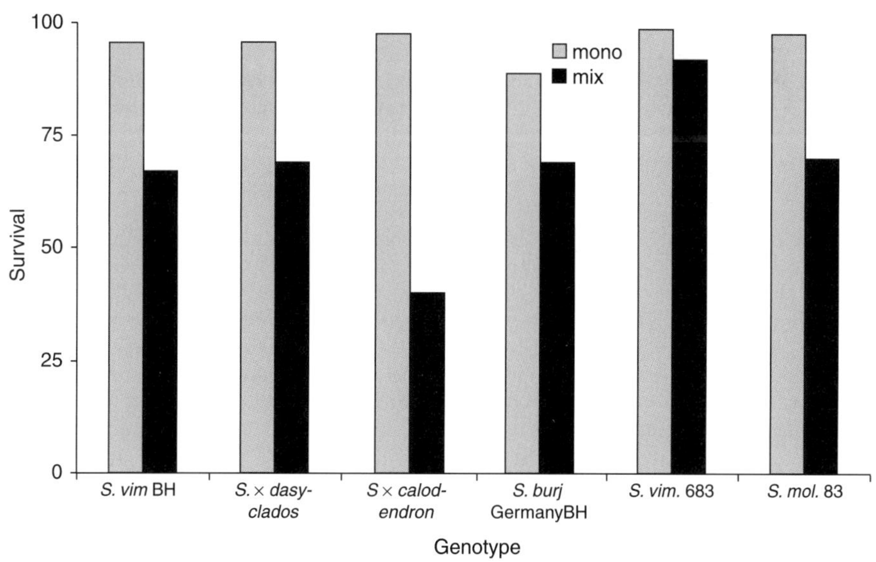

Fig. 17.6. Percentage survival after the first 3-year harvest of willow genotypes grown in monoculture and mixtures.

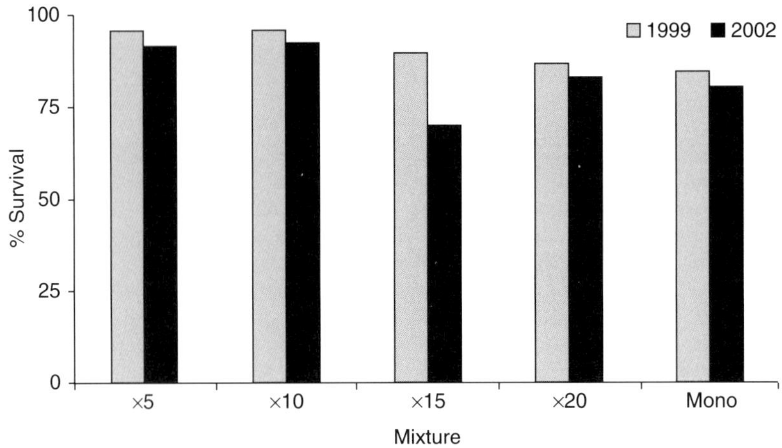

Fig. 17.7. Percentage survival of *Salix dasyclados* × *caprea* 'V794' in 1999 and 2002 after one and two harvesting cycles grown as a monoculture and in mixtures.

cases the survival rate was below 70% (McCracken *et al.*, 2001).

The competitive ability of highly rust-susceptible genotypes within mixtures is so impaired that survival is reduced below the level at which the remaining mixture components can compensate for the yield loss. Not only is survival adversely affected, but the impact of the disease reduces the capacity of those surviving to contribute effectively to overall yield.

Mixture Diversity

It has already been observed that diversity within mixtures is important in terms of the sustainability of the cropping system, particularly in respect of reducing the selection pressure on the rust disease organism. However, the Swedish Breeding programme, in particular, has tended to focus on *S. viminalis* and *S. viminalis* hybrids in producing new operational varieties for biomass production. In the early work carried out on genotype susceptibility to *M. epitea* pathotypes, all varieties, which had *S. viminalis* in their parentage, were susceptible to the same range of pathotypes (Pei *et al.*, 1996). It was apparent that the pathotype composition of the pathogen population varied from region to region. However, it was not at all clear whether mixtures composed of the commercially available, high-yielding, improved varieties, which were almost exclusively *S. viminalis* or *S. viminalis* hybrids, would function in the same way as more diverse mixtures. There are indications that, because of the wide range of susceptibility within these new varieties and the improved resistance introduced via *S. schwerinii*, these largely *S. viminalis* mixtures may function in the same way as more diverse mixtures. This is particularly so where the disease population is dominated by asexual reproduction and where, as a consequence, there is less opportunity for diversity through recombination as in sexually derived populations (Samils *et al.*, 2002). To evaluate mixtures containing genetically less diverse constituents, the focus of the work has turned to *S. viminalis*-only mixtures. These investigations included seven *S. viminalis* varieties from varying sources, including four of the commercially available improved varieties from the Swedish Breeding Programme (*S. viminalis* 'Orm', 'Ulv', 'Jorr' and 'Jorunn'). *S. viminalis* 'Ulv' was known to be rust susceptible, as was *S. viminalis* '77082'. The remaining two varieties *S. viminalis* 'Gustav' and *S. viminalis* '78195' had low levels of susceptibility. The trial was established in 2000 and in addition to the seven varieties grown as monocultures, they were also grown as a 7-way mixture. The rust-susceptible genotypes were combined in a range of 3-way mixtures with each of the remaining five genotypes (McCracken *et al.*, 2004). Total dry matter yields after the first harvesting cycle in 2003 are shown in Fig. 17.8.

Although the 7-way mixture was more productive than an equivalent area of the constituent varieties grown as monocultures, this difference was not statistically significant. This is in contrast to previous results where, without exception, mixtures were more productive statistically than the mean of their components grown in monoculture.

In the earliest mixture trials, established in Northern Ireland in 1987, the level of species diversity was high, with only two of the component varieties being *S. viminalis.* These trials indicated a highly significant increase in dry matter yield in mixtures, even when disease levels were generally low (Anon., 1998). This improvement in yield was also recorded in later trials, established in 1996, where the number of mixture components was higher, but the overall level of species diversity lower (McCracken and Dawson, 2003). However, in these current trials, where there is no species diversity, a significant yield advantage has not been recorded.

It was noted in these early mixture trials (Anon., 1998) that, where varieties were grown in monoculture, stool size and vigour were uniform but the same stools in mixed plantings were much more variable and were influenced by the size and vigour of adjacent stools. Additionally, the mixture provided a much more diverse canopy in terms of leaf characteristics and length and number of rods. This allowed the mixture to exploit the growing space to the full. However in the *S.viminalis* 'mixtures' the situation was more similar to the monoculture, with uniform stools and even canopy.

In another respect the *S viminalis* 'mixtures' behaved in a similar way to the more diverse mixtures: when the contribution of the various components to the mixture was analysed, it was clear that not all contributed positively to the yield (Fig. 17.9).

It was assumed as before that an individual genotype within a 3-way mixture could be expected to contribute 33.3% of the total yield of that mixture, because it occupied one third of the space. Similarly in a 7-way mixture individual components could be expected to contribute 14.3% of the DM yield. However, this was not the case

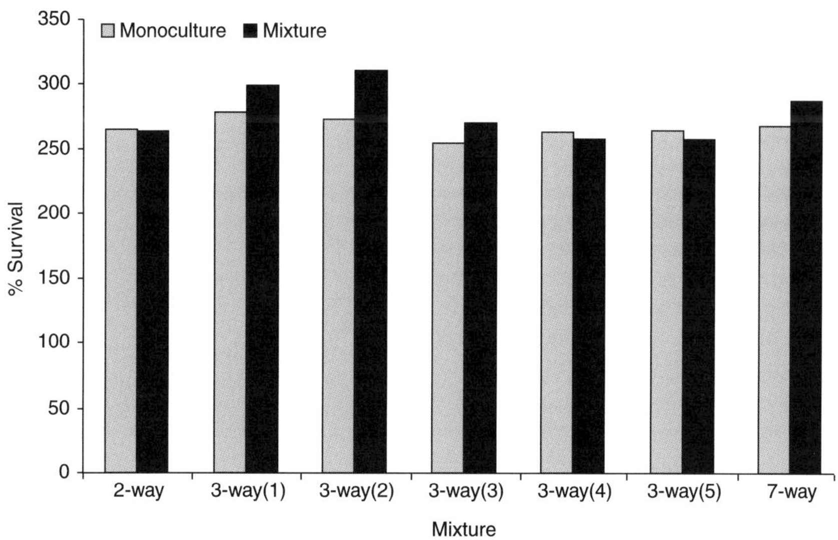

Fig. 17.8. Comparative dry matter yields (kg) from 100 m^2 of a range of *Salix viminalis* mixtures, compared with equivalent areas of the components grown in monoculture.

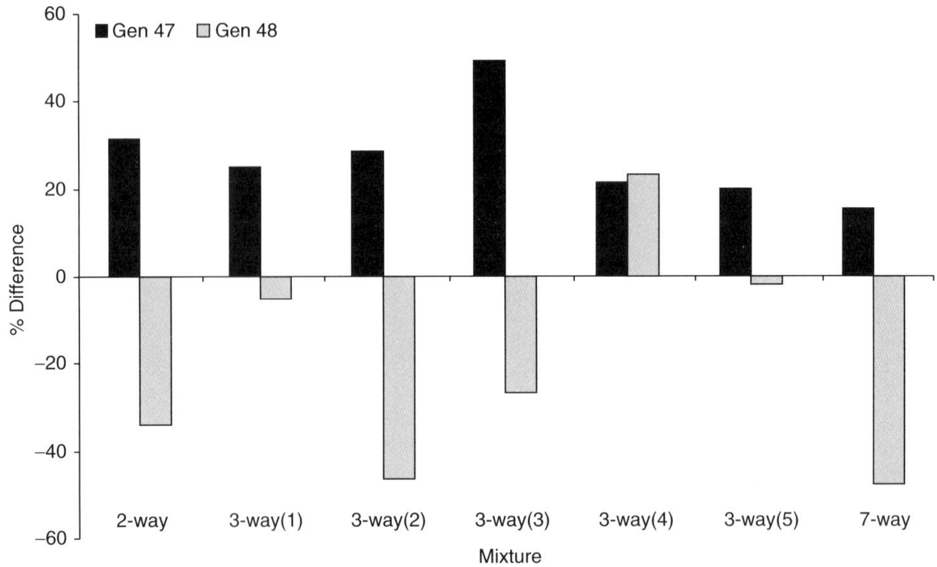

Fig. 17.9. Differences between expected and actual contribution to mixture yields from two *Salix viminalis* varieties included in 2-way (T8), 3-way (T9, 10, 11, 12 and 13) and 7-way (T14) mixtures.

with some varieties, e.g. *S. viminalis* '77082', which consistently contributed more to the mixture than expected in a range of mixtures, and *S. viminalis* 'Ulv', contributing less than expected. Generally only *S. viminalis* 'Orm' and *S. viminalis* 'Jorr' made positive contributions to the mixture above their expected value, with the others either contributing at or below their expected values. These types of genotype still make a positive contribution in terms of the diversity and their influence in reducing the impact of disease.

Further work will be necessary to unravel the interaction between mixture components and its influence on yield, disease impact and sustainability. It should be noted that non-uniform mixtures containing a number of heavy stools randomly within the plantation may be a practical difficulty for mechanical harvesting.

Conclusions

Where climatic conditions put additional disease pressure on short-rotation coppice willow plantations, such as the cool, wet oceanic conditions in western areas of the UK, the use of mixtures has been shown to offer a sustainable approach to disease control. With this in mind, the main conclusions are:

- Evidence is clear that where disease pressure is high, the planting of large areas of a single genotype, even where that genotype is less susceptible or resistant to rust, is not to be recommended.
- The yield of the newer genotypes from the dedicated breeding programmes in Sweden and the UK show highly significant yield increases over selected standards. Consequently, only these improved genotypes should be used in commercial developments.
- Generally, yield from diverse mixtures is greater than the equivalent yield of their component genotypes grown in monoculture.
- However, with less diverse mixtures these yield increases have not been recorded, but the disease-suppression aspect of the use of mixtures remains. This latter is important given that most of the improved genotypes

commercially available are *S. viminalis* or *S. viminalis* hybrids.

- Increased diversity in mixtures resulted in yield compensation when components of the mixture were lost due to disease or other factors. It has been shown that where a mixture loses up to 30% of its component genotypes it can fully compensate for the potential yield loss. Consequently, the inclusion of, at least, six genotypes in a mixture would appear to be prudent.
- Where genotypes have been shown to be significantly rust susceptible their inclusion in mixtures does not improve their performance or increase their sustainability.
- Completely random planting configurations in mixtures would seem to provide the greatest opportunity for yield compensation, rather than line mixtures or small block mosaic mixtures. In practical commercial terms, the short line random mixtures achievable with available commercial planters would seem to be the best practical compromise.
- There are differences in how individual component genotypes contribute to the overall yield of mixtures. However, provided this is not significantly negative, their inclusion is justified by the diversity they contribute to the plantation and the positive effect they therefore have on sustainability.

References

Anon. (1980) The willow as a genetic resource. *Annual Report of Long Ashton Research Station*, Bristol, pp. 45–49.

Anon. (1998) Establishment and monitoring of large-scale trials of short rotation coppice for energy. *Contractor report for Energy Technology Support Unit on behalf of the Department of Trade and Industry ETSU B/W2/00541/REP.*

Armstrong, A. (2000) National trials network – preliminary results and update. *Proceedings of the Short Rotation Coppice and Wood Fuel Symposium – From Research to Renewable Energy.* Alice Holt, Surrey, UK, pp. 12–18.

Christersson, L. (1987) Biomass production by irrigated and fertilised *Salix* clones. *Biomass* 12, 83–95.

Dawson, M. and McCracken, A. (1998) Clonal selection in willow (*Salix*) grown as short rotation coppice for energy production. Tests of Agrochemicals and Cultivars 19. *Annals of Applied Biology* 132 (suppl.), 56–57.

Larsson, S. (1997) Commercial breeding of willow for short rotation coppice. *Aspects of Applied Biology – Biomass and Energy Crops* 49, 215–218.

Larsson, S. (2001) Commercial varieties from the Swedish breeding programme. *Aspects of Applied Biology – Biomass and Energy Crops II.* 65, 173–182.

Lindegaard, K., Parfitt, R., Donaldson, G., Hunter, T., Dawson, M., Forbes, G., Carter, M., Whinney, C., Whinney, J. and Larsson, S. (2001) Comparative trials of elite Swedish and UK biomass willow varieties. *Aspects of Applied Biology – Biomass and Energy Crops II* 65, 183–198.

McBurney, S. and Brigg, J. (1996) Financial analysis of willow coppicing. In: *Proceedings of the Conference 'Willow a Commercial Opportunity'*, October, Enniskillen, Northern Ireland.

McCracken, A.R. and Dawson, W.M. (1994) Effects of *Melampsora* rust on the growth and development of *S. burjatica* 'Korso' in Northern Ireland. *European Journal of Forest Pathology* 24, 32–39.

McCracken, A.R. and Dawson, W.M. (1998) Short rotation coppice willow in Northern Ireland since 1973: development of the use of mixtures in the control of foliar rust (*Melampsora* spp.) *European Journal of Forest Pathology* 28, 307–311.

McCracken, A.R and Dawson, W.M. (2001) Disease effects in mixed varietal plantations of willow. *Aspects of Applied Biology – Biomass and Energy Crops* II 65, 255–262.

McCracken, A.R. and Dawson, W.M. (2003) Rust disease (*Melampsora epitea*) of willow (*Salix* spp.) grown as short rotation coppice (SRC) in inter- and intra-species mixtures. *Annals of Applied Biology* 143, 381–393.

McCracken, A.R., Dawson, W.M. and Bowden, G. (2001) Yield responses of willow (*Salix*) grown in mixtures in short rotation coppice (SRC). *Biomass and Bioenergy* 21, 311–319.

McCracken, A.R., Dawson, W.M. and Carlisle, D. (2004) Benefits of growing short rotation coppice (SRC) willow in genotype mixtures. *Proceeding of 2nd World Conference on Biomass for Industry and Energy*, Rome, May.

McElroy, G. and Dawson, M. (1986) Biomass from short rotation coppice willow on marginal land. *Biomass* 10, 225–240.

Pei, M.H., Royle, D.J. and Hunter, T. (1996) Pathogenic specialisation in *Melampsora epitea* var. *epitea* on *Salix*. *Plant Pathology* 45, 679–690.

Rosenquist, H. and Dawson, M. (2004) Economics of willow growing in Northern Ireland. *Biomass and Bioenergy* (in press).

Samils, B., McCracken, A., Dawson, M. and Gullberg, U. (2003) Host specific genetic composition of *Melampsora larici-epitea* populations on two *Salix viminalis* genotypes in a mixture trial. *European Journal of Plant Pathology* 109, 183–190.

Stott, K. (1984) Improving the biomass potential of willow by selection and breeding. In *Ecology and Management of Forest Biomass Systems.* Report 15. Swedish University of Agricultural Sciences, Uppsala, Sweden, pp. 233–260.

Stott, K., Parfit, R., McElroy, G. and Abernethy, W. (1987) Productivity of Coppice Willow in Biomass trials in UK. *Occasional publication of Long Ashton Research Station*, Bristol.

Tabbush, P. and Parfitt, R. (1996) *Poplar and Willow Clones for Short Rotation Coppice*. Research Information Note 278. Forestry Commission Research Division, Surrey, UK.

18 Effect of Preventative Fungicide Sprays on *Melampsora* Rust of Poplar in the Nursery

R.C. Sharma, S. Sharma and A.K. Gupta
Department of Mycology and Plant Pathology, Dr Y.S. Parmar University of Horticulture and Forestry, Nauni, Solan-173 230 Himachal Pradesh, India

Introduction

Poplars are prone to many insect pests and diseases which may account for losses of up to one-third total yield. *Melampsora ciliata* Barclay, a microcyclic autoecious rust confined to the Indian and Nepal Himalayas, is serious not only on *Populus ciliata*, an indigenous species, but also on exotic hybrids and species of *Populus* (Singh *et al.*, 1983; Vannini *et al.*, 1995; Sharma and Sharma, 2000, 2001). The pathogen is distributed between 700 and 3130 m above sea level and can cause moderate to severe losses in nurseries and plantations (Sharma and Sharma, 2002). The disease causes premature defoliation *c.* 2–3 months prior to normal leaf fall, thereby affecting plant growth (Sharma *et al.*, 2001). In light of the seriousness of the disease, the efficacy of five systemic and eight non-systemic fungicides, including three neem-oil formulations, was tested against leaf rust of *P. ciliata*.

Methodology

Plant material and fungicides

Cuttings (18–23 cm) of the highly rust-susceptible clone *P. ciliata* 'Theog' were planted in nursery beds at 33°N 77°E and 1500 m above sea level, in the first week of February 1999 and 2000. Twenty cuttings were planted at 45 × 45 cm spacing in each plot, replicated three times. Fungicides, carbendazim (0.05%), benomyl (0.05%), difenoconazole (0.02%), penconazole (0.06%), hexaconazole (0.1%), captan (0.2%), copper oxychloride (0.3%), foltaf (0.25%), mancozeb (0.25%), dinocap (0.1%), Nimbecidine (0.2%), Neem Jeevan Spray (0.3%) and Kranti spray oil (0.4%) were applied as sprays at fortnightly intervals. In all, six sprays were applied during the period 24 June to 2 September.

Disease incidence and severity

The percentage disease incidence was calculated by counting the infected and total number of leaves on each seedling. Disease severity was rated on a 0–7 scale on the basis of leaf area covered by rust pustules, as 0, 0.1–1.0, 1.1–5.0, 5.1–10.0, 10.1–25.0, 25.1–40.0, 40.1–65.0 and > 65%. Uredinial pustules per leaf were calculated by counting the uredinial pustules of 20 tagged leaves selected from top, middle and bottom positions of plants in each replication, and means calculated.

Apparent infection rate

The apparent infection rate (r) for different treatments was calculated according to the method of Van der Plank (1963), using the logistic equation:

$$r = [2.3/(t_2 - t_1)] \log_{10}[x_2/(1 - x_2)] \log_{10}[x_1/(1 - x_1)]$$

where r denotes the apparent infection rate per unit per day; $t_2 - t_1$ is the time interval between first and last observations; x_1 and x_2 are the proportions of diseased plant parts; and $1 - x_1$ and $1 - x_2$, the proportion of healthy plant parts at t_1 and t_2, respectively.

Area under the disease progress curve

From the severity data, the area under the disease progress curve (AUDPC) was calculated in accordance with Shaner and Finney (1977), as follows:

$$\text{AUDPC} = \sum_{i+1}^{n} (Y_{i+1} + Y_{i/2})(X_{i+1} - x_2)$$

where Y_i denotes disease severity (%) at the ith observation, X_i denotes time (days) at the ith observation, and n denotes total number of observations.

Results and Discussion

Minimum disease incidence was recorded following treatment with hexaconazole (30.39%) and was not significantly different ($P = 0.05$) on penconazole-treated plants (Table 18.1). The three ergosterol-biosynthesis inhibiting (EBI) fungicides, hexaconazole, penconazole and difenoconazole were superior to all other treatments. Ruaro and May (1996) reported the efficacy of difenoconazole, captan and mancozeb against urediniospore germination of *Melampsora medusae* under *in vitro* conditions. Among the three neem-oil formulations, Nimbecidine was most effective, with 50.64% disease incidence compared to 76.50% for the control.

Significant reductions in uredinial pustules per leaf were recorded when plants were sprayed with any of the fungicides (Table 18.2). The least number of uredinial pustules per leaf (3.8) was recorded on plants treated with hexaconazole, which was not significantly different from the numbers on plants sprayed with difenoconazole, penconazole, dinocap, Nimbecidine or Kranti spray oil. Foltaf was least effective, with 22.1 uredinial pustules

Table 18.1. Effect of preventive sprays of fungicides on leaf rust of nursery-grown poplar in nursery.

	Disease incidence (%)		
Fungicide	1999	2000	Mean
Carbendazim	27.62 (31.69)	69.20 (56.35)	48.41 (44.02)
Benomyl	22.10 (32.29)	75.92 (60.69)	55.40 (48.49)
Difenoconazole	29.49 (32.88)	51.75 (46.01)	40.42 (39.31)
Penconazole	29.49 (32.88)	41.47 (40.04)	35.48 (36.46)
Hexaconazole	28.90 (32.51)	31.87 (34.36)	30.39 (33.44)
Captan	56.60 (48.83)	70.61 (57.18)	63.61 (53.01)
Copper oxychloride	44.00 (41.55)	51.78 (46.02)	47.89 (43.79)
Foltaf	49.62 (44.79)	69.05 (56.24)	59.34 (50.51)
Mancozeb	38.75 (38.50)	54.55 (47.62)	46.65 (43.06)
Dinocap	47.00 (43.28)	37.90 (37.98)	42.45 (40.63)
Nimbecidine	40.27 (39.37)	61.01 (51.43)	50.64 (45.40)
Neem Jeevan spray	70.67 (58.13)	79.28 (63.34)	74.98 (60.74)
Kranti spray oil	58.00 (49.61)	76.94 (61.31)	67.47 (55.46)
Control	67.50 (55.26)	85.50 (67.77)	76.50 (61.51)
$CD_{0.05}$			3.60

Figures in parentheses are arcsine transformed values. $CD_{0.05}$ = critical difference at 5% level of significance.

per leaf compared to 35.1 on the control. Earlier efficacy of cyproconazole against *M. larici-populina* and carbendazim and cyproconazole against *M. ciliata* has been demonstrated (Khan *et al.*, 1988; Boudier, 1992; Khan, 1994).

Table 18.2. Effect of preventive sprays of fungicides on number of urediniopustules per leaf in nursery-grown poplar.

	Urediniopustules per leaf		
Fungicide	1999	2000	Mean
Carbendazim	5.9	12.5	9.2
Benomyl	15.6	13.2	14.4
Difenoconazole	4.2	4.8	4.5
Penconazole	6.5	5.3	5.9
Hexaconazole	5.3	2.2	3.8
Captan	16.8	16.5	16.7
Copper oxychloride	14.0	6.0	10.0
Foltaf	13.4	30.8	22.1
Mancozeb	7.5	16.5	12.0
Dinocap	15.1	2.5	8.8
Nimbecidine	10.6	7.2	8.9
Neem Jeevan spray	17.8	3.0	10.4
Kranti spray oil	12.5	5.5	9.0
Control	26.2	44.0	35.1
$CD_{0.05}$			5.3

The effect of fungicide sprays on the poplar leaf rust apparent infection rate and area under the disease progress curve is presented in Table 18.3. The least apparent infection rate (0.0044/unit/day) was recorded on plants treated with hexaconazole; this was not significantly different ($P = 0.05$) from that on plants treated with penconazole or difenoconazole. Maximum apparent infection rate was on foltaf-treated plants (0.0138/unit/day) compared to 0.0171/unit/day on control plants. The losses from diseased areas were also checked, by reducing the disease at the beginning of season or through decreasing the apparent infection rate during the growing period (Van der Plank, 1963). Measurement of the apparent infection rate was useful for rapid comparison of the effectiveness of various treatments (Sharma and Sharma, 1999).

The area under the disease progress curve (AUDPC) was least on hexaconazole-treated plants (4.01), and was not significantly different from penconazole- (4.67) or difenoconazole-treated plants. Of the neem-oil formulations, AUDPC was least on plants sprayed with Nimbecidine (6.87) as compared to 11.33 in unsprayed plants. AUDPC also provides a quantitative measure of disease development and its

Table 18.3. Effect of preventive sprays of fungicides on poplar leaf rust apparent infection rate (r) and area under the disease progress curve (AUDPC).

	Apparent infection rate (r)			AUDPC		
Fungicide	1999	2000	Mean	1999	2000	Mean
Carbendazim	0.0127	0.0077	0.0102	3.71	13.30	8.51
Benomyl	0.0054	0.0130	0.0092	4.84	10.22	7.53
Difenoconazole	0.0042	0.0063	0.0052	3.87	7.14	5.51
Penconazole	0.0049	0.0056	0.0052	4.16	5.19	4.67
Hexaconazole	0.0045	0.0043	0.0044	3.89	4.14	4.01
Captan	0.0091	0.0137	0.0134	7.43	10.25	8.84
Copper oxychloride	0.0034	0.0164	0.0099	5.97	7.09	6.53
Foltaf	0.0191	0.0086	0.0138	6.87	10.08	8.45
Mancozeb	0.0105	0.0046	0.0075	5.45	8.42	6.93
Dinocap	0.0116	0.0079	0.0098	6.68	4.65	5.67
Nimbecidine	0.0124	0.0025	0.0074	5.68	8.07	6.87
Neem Jeevan spray	0.0089	0.0042	0.0066	8.83	10.69	9.76
Kranti spray oil	0.0082	0.0056	0.0052	8.16	10.35	9.23
Control	0.0136	0.0207	0.0171	9.27	13.40	11.33
$CD_{0.05}$			0.0081			0.87

intensity (Lee Compbell, 1998; Sharma and Sharma, 1999). Lower AUDPC values in hexaconazole and penconazole are due to slow disease progress and poor terminal disease severity.

Conclusions

1. Hexaconazole was the most effective treatment in reducing leaf rust of *Populus ciliata*, followed by penconazole and difenoconazole.
2. Among neem-oil formulations tested, Nimbecidine was better in checking the disease.
3. The least number of uredinial pustules per leaf were recorded in hexaconazole-treated plants.
4. Apparent infection rates and area under the disease progress curve were least in hexaconazole-, penconazole- and difenoconazole-treated plants.

References

Boudier, B. (1992) Rusts of Salicaceae: remarkable efficacy of tebuconazole and cyproconazole. *Revue Horticulture* 323, 29–30.

Khan, Y. (1994) Studies on *Melampsora* leaf rust of *Populus* species. MSc thesis, Dr Y.S. Parmar University of Horticulture and Forestry, Solan, India.

Khan, S.N., Rehill, P.S., Tiwari, R.K., Rawat, D.S. and Misra, B.M. (1988) Control of poplar rust, *Melampsora ciliata* in nurseries. *Indian Journal of Forestry* 11, 253–255.

Lee Compbell, C. (1998) Disease progress in time: modelling and data analysis. In: Jones, D.G. (ed.) *The Epidemiology of Plant Diseases*. Kluwer Publishers, Dordrecht, pp. 181–206.

Ruaro, L. and May, L.L. (1996) Characterization of rust in poplar (*Populus* species) and the efficacy of fungicides *in vitro* over the germination of urediniospores. *Revista do Sector de Ciencias Agrarias* 15, 77–82.

Shaner, G. and Finney, R.E. (1977) Effect of nitrogen fertilization on the expression of slow mildewing resistance in Knox wheat. *Phytopathology* 67, 1051–1056.

Sharma, R.C. and Sharma, S. (1999) Effect of fungicides on the development of *Cladosporium* leaf spot of poplar in western Himalaya. *Acta Phytopathologica et Entomologica Hungarica* 34, 57–62.

Sharma, R.C. and Sharma, S. (2000) Status and distribution of foliar diseases of poplar in Himachal Pradesh. *Indian Phytopathology* 53, 57–60.

Sharma, R.C., Khan, Y., Sharma, S. and Malhotra, R. (2001) Development of *Melampsora ciliata* rust on nursery grown poplars in north western Himalayas. *Forest Pathology* 31, 313–319.

Sharma, S. and Sharma, R.C. (2001) Epidemiology of *Melampsora ciliata* leaf rust of poplars in India. *Zeitschrift für Pflanzenkrankheiten und Pflanzenschutz* 108, 337–344.

Sharma, S. and Sharma, R.C. (2002) Prevalence of poplar leaf rust in Himachal Pradesh. *Indian Phytopathology* 55, 81–83.

Singh, S., Khan, S.N. and Misra, B.M. (1983) Status of *Melampsora* rusts of poplars in India. *Indian Forester* 109, 743–747.

Van der Plank, J.E. (1963) *Plant Diseases: Epidemics and Control*. Academic Press, New York.

Vannini, A., Cecco, D., Monaci, L. and Anselmi, N. (1995) Main tree pathogens of western Himalayan forests in Nepal: description and risk of introduction. *Bulletin OEPP* 25, 455–461.

19 Biocontrol of Rust Fungi by *Cladosporium tenuissimum*

Salvatore Moricca[1], Alessandro Ragazzi[2] and Gemma Assante[3]
[1]*Dipartimento di Biologia Vegetale, Università di Firenze, Via La Pira 4, 50121 Firenze, Italy;* [2]*Dipartimento di Biotecnologie Agrarie, Sezione di Patologia Vegetale, Università di Firenze, Piazzale delle Cascine 28, 50144 Firenze, Italy;* [3]*Istituto di Patologia Vegetale, Università di Milano, Via Celoria 2, 20133 Milano, Italy*

Introduction

The use of chemical pesticides in agriculture is under increasing public scrutiny, with mounting concern over possible harmful effects on the environment and on human health. Agrochemicals pollute groundwater, enter the food chain, have a deleterious impact on many living organisms and may bring about pesticide resistance in plant parasites. Increasing awareness of such possible hazards is prompting a heightened interest in alternative strategies (Boland and Kuykendall, 1998).

One of the most promising approaches to control economically important crop pests and diseases is by exploiting naturally occurring antagonists. Over the past few years unprecedented advances have been made in the biological control of many damaging insect pests and phytopathogenic microorganisms (Butt and Copping, 2000).

Many fungal pathogens throughout the world have natural enemies that limit the harm they cause. Some of these competitors are non-fungal hyperparasites such as bacteria (Yuen *et al.*, 2001) or mycoviruses (Brasier, 1990), but most are other fungi. Antagonistic fungi are a major component of the microbial community on the plant rhizosphere and phyllosphere, and play a pivotal role in regulating many interactions between plants and parasitic microorganisms (Jeffries, 1997). Since they are an integral part of the ecosystem, no alien microbe species, toxic substances or chemicals are introduced into the environment by their use, and hence they appear to be more environmentally sustainable than other, more intrusive, control methods. As a result, fungal biocontrol agents (BCAs) are becoming a promising means to control fungal diseases and to reduce dependence on chemical pesticides (Butt *et al.*, 2001).

However, developing a reliable and effective method for the biocontrol of a fungal pathogen is not a straightforward process. Several biological control experiments that were successful *in vitro* produced inconsistent results in the field. Moving from the laboratory or the greenhouse to large-scale field-testing is difficult because, in the field, the antagonist is subject to environmental influences. Plant disease is the result of a dynamic interaction over time between a pathogen, a plant and the environment. The environmental component of the disease triad is crucial for the success of a BCA, the limited ecological amplitude of

(eds M.H. Pei and A.R. McCracken)

which might represent a major constraint. While an antagonist has to actively control a pathogen, over a period that may take from a week to several months, it must be able to withstand fluctuations in the physical environment and resist the competing action of other, established microorganisms. The insufficient ecological fitness of the antagonist inoculum may lower its effectiveness as a control agent, while a poor shelf-life may cause lack of persistence, which means that the control achieved does not last over a sufficiently long term (Whipps and Lumsden, 2001).

This chapter examines whether *Cladosporium tenuissimum* Cooke, a destructive hyperparasite of rust spores, can be exploited as a BCA of rust fungi. It reports on tests carried out to evaluate *C. tenuissimum* effectiveness against rusts in the genera *Melampsora*, *Cronartium*, *Peridermium*, *Uromyces* and *Puccinia*, focusing on its modes of action, the fine-level analysis of the fungal host–hyperparasite interface, the antifungal compounds it produces, and its ability to reduce disease, both *in vitro* and *in planta*. Since effective biological control is impossible without due consideration of the ecology of the BCA, as well as of the other partners involved, and an examination of their spatial relationships and interaction with the environment, attention is also paid to the biological nature of the fungus, in particular to those characteristics that enable it to survive in natural habitats and retain activity under varying environmental conditions.

The Rust Disease Problem

The rust fungi (*Uredinales*) are one of the largest groups in the *Basidiomycota*, with about 5000–6000 species described on a wide range of hosts, including ferns, gymnosperms, and mono- and dicotyledonous angiosperms (Alexopoulos *et al.*, 1996). Rust fungi occur worldwide, under varied climates, and include destructive species that are responsible for costly diseases in agriculture and forestry.

Rust epidemics became particularly severe in the 20th century, after extensive monocultures had been aggressively established in agriculture and forests as a result of advances in plant genetics and the modernization of agriculture. But such developments ignored a basic principle in epidemiology, which is that both the number of diseases and disease incidence increase in proportion to host abundance. The advantage of identifying, planting, harvesting and marketing particular genetically homogeneous crops was therefore, in many cases, nullified by sudden disease outbreaks, caused directly by low crop diversity. The limitless potential for pathogen spread in monocultures led to the rapid selection of rust pathotypes able to overcome host defences. As a consequence, destructive fungi such as the coffee leaf rust *Hemileia vastatrix*, the wheat stem rust *Puccinia graminis*, the *Melampsora* leaf rusts of *Salicaceae* (*Populus* and *Salix*), the *Cronartium* stem rusts of hard pines, to name just a few, spread epidemically over vast areas, causing enormous losses and often making it necessary to replace susceptible crops entirely with non-host species (Littlefield, 1981).

At present, the main strategies to limit populations of rust pathogens and maintain yield stability are the continual breeding for new disease resistances and the search for new fungicides (to deal with the selection of new pathogen genotypes that attack resistant crops and that are resistant to existing fungicides), as well as crop rotation and the adoption of spatial and/or temporal crop diversity (Wolfe, 1985; Zhu *et al.*, 2000).

However, the effectiveness of such measures is subject to a number of limiting factors. Constant improvements in crop resistance and the never-ending need for new fungicides are costly to the farmer, the consumer and to society at large. Economic, environmental and technical reasons often make it impractical to use fungicides to combat rusts, especially rusts of forest trees. Other measures, such as the eradication of alternate hosts, scheduled harvesting (particularly of hazard-rated crops and stands), pop-up spacing, crop rotation with non-hosts and standard sanitation procedures (removal of infected branches, burning of inoculum-bearing debris, etc.) are deficient because of the large geographic distribution of the pathogens, their wide host ranges, their high spore dispersal capabilities, and the autoecism of some rusts. Planting mixed genotypes with different disease-resistance profiles may be costly and may require particular expertise in cultivation practices, not to mention the fact that resistant plant varieties and fungicides are the major selective agents of pathogen virulence. Any single method of control thus has its limitations and it seems that a combination of measures in an integrated disease-management strategy is needed to counteract rust fungi.

Biological control is a non-chemical measure, exploiting microbiological agents to provide an additional method to combat rust disease. Among the common phylloplane mycobiota many fungi have been reported since the first half of the 20th century to be associated with rust fructifications (Arthur, 1929). Some of these fungi, eg. *Sphaerellopsis filum* (Biv.-Bern.F) Sutton (teleomorph: *Eudarluca caricis* (Fr.) O. Erikss.), *Scytalidium uredinicola* Kuhlman *et al.*, *Aphanocladium album* (Preuss) W. Gams, and several species of *Cladosporium*, *Tuberculina* and *Verticillium*, have already demonstrated a pronounced hyperparasitism against different spore states of a number of rust fungi (Kuhlman *et al.*, 1978; Tsuneda *et al.*, 1980; Sharma and Heather, 1981a; Wicker, 1981; Allen, 1982).

Cladosporium Species Hyperparasitic on Rusts

The genus *Cladosporium* is one of the most widespread and prevalent genera of fungi, containing over 500 species Some of its species have teleomorphs in the ascomycete *Mycosphaerella* (*Dothideales*), but a vast majority of taxa are known by their anamorphic state or by criteria such as host association or presumed host specificity. These additional criteria are used to support the traditional classification because the inadequacy or lack of description for several members, and the phenotypic plasticity within individual taxonomic entities, make typification of species difficult or impossible. Substrate differences, climate and geographic variations also influence morphological expression. As a consequence, morphological features such as conidium size, shape, septation, pigmentation, surface characteristics (smooth or with ornamentation), conidiophore size and morphology are often variable and inconsistent (Morgan-Jones and McKemy, 1990; Ho *et al.*, 1999).

Members of this genus display a variety of lifestyles: some commonly occur on their hosts as epiphytes, endophytes, or as pathogens; others thrive as saprophytes, or even as hyperparasites (Ellis, 1976; Petrini, 1991; Moricca *et al.*, 1999; Dugan and Lupien, 2002; Larran *et al.*, 2002; Abdel-Baky and Abdel-Salam, 2003).

Several taxa in this large genus are prevalently associated with rust sori, and some are assumed to be invariably hyperparasites of *Uredinales* (Table 19.1). *Cladosporium uredinicola* Speg. is a common necrotrophic hyperparasite that destroys rust hyphae and causes coagulation and disintegration of the cell cytoplasm of a number of hosts, such as *Puccinia cestri* Dietel and P. Henn. (Spegazzini, 1912), *Puccinia recondita* Roberge ex Desm. (Ellis, 1976), *Cronartium quercuum* (Berk.) Miyabe ex Shirai f. sp. *fusiforme* (Morgan-Jones and McKemy, 1990), *Puccinia violae* (Schum.) DC (Traquair *et al.*, 1984) and *Puccinia horiana* (Srivastava *et al.*, 1985). Spegazzini (1922) reported that *Cladosporium uredinophilum* colonized and destroyed *Uredo cyclotrauma* Speg. propagules in Paraguay. Steyaert (1930) described *Cladosporium hemileiae* Steyaert as an effective hyperparasite of the coffee rust fungus *Hemileia vastatrix* in Zaire (Democratic Republic of the Congo). Powell (1971) encountered *Cladosporium gallicola* in galls of *Cronartium comandrae* Pk. on *Pinus contorta* var. *latifolia* and considered the fungus parasitic on aeciospores and responsible for lowering aeciospore production. Sutton (1973) reported a close association between *C. gallicola* and *Endocronartium harknessii* (J.P. Moore) Hiratsuka on *Pinus banksiana*, and observed hyphae of the hyperparasite penetrating the rust aeciospores. Tsuneda and Hiratsuka (1979) found that *C. gallicola* parasitized *E. harknessii* both by simple contact, disintegrating the cell walls of the spores, and by actual penetration of the spores walls, with or without the formation of appressoria, and causing the coagulation and disappearance of the host cytoplasm. These authors also found evidence that the rust spores, acting by contact, secreted enzymes in order to disintegrate the cell walls. A similar behaviour was noted for *Cladosporium aecidiicola*, a fairly common hyperparasite of rusts in Europe and in the Mediterranean area (Hulea, 1939; Rayss, 1943), parasitizing *E. harknessii* on *Pinus contorta*, *Pinus muricata* and *Pinus radiata* in California (Byler *et al.*, 1972). This hyperparasite also heavily parasitized aecia of *Puccinia conspicua* (Arth.) Mains. in Arizona (Keener, 1954) and urediniospores of *Melampsora medusae* Thüm under storage conditions (Sharma and Heather, 1980). Srivastava *et al.* (1985) found that *Puccinia horiana* was regularly parasitized by *Cladosporium sphaerospermum*. Sharma and Heather (1981a,b) reported that *C. tenuissimum* was a common phylloplane fungus that actively colonized urediniospores of the rust *Melampsora larici-populina* Kleb. on *Populus* × *euro-americana*. More recently *C. tenuissimum* was also detected in collections of aeciospores of the two-needle pine stem rust *Cronartium flaccidum* and its non-host alternating form *Peridermium pini* (Moricca and Ragazzi, 1998; Moricca *et al.*, 1999).

This list of *Cladosporium* species parasitic on rusts is not exhaustive and is likely to increase in the near future. The lack of genuine morphological structures and the great instability of such characters not only hamper the placement of *Cladosporium* species in a congruent evolutionary context, but also make it impossible to identify consistently many individual taxa and distinguish them from related ones. As new molecular techniques become increasingly employed to confirm and refine known taxonomic relationships, it is predictable that new taxa now undistinguishable or considered synonyms, will be added to the list. The number of hyperparasites is also expected to grow once it is shown that some *Cladosporium* species, now thought to be common saprophytes because they are found on aged host structures, are in fact true hyperparasites.

Table 19.1. *Cladosporium* species hyperparasitic on rust fungi.

Parasite	Host
C. aecidiicola Thüm	*Endocronartium harknessii* (J.P. Moore), (Byler *et al.*, 1972); *Melampsora medusae* Thüm (Sharma and Heather, 1980); *Puccinia conspicua* (Arth.) Mains., (Keener, 1954)
C. gallicola Sutton	*Endocronartium harknessii* (J.P. Moore) (Sutton, 1973; Tsuneda and Hiratsuka, 1979); *Cronartium comandrae* Pk. (Powell, 1971)
C. hemileiae Steyaert	*Hemileia vastatrix* Berk. et Br. (Steyaert, 1930)
C. sphaerospermum Penzig	*Puccinia horiana* Henn. (Srivastava *et al.*, 1985)
C. tenuissimum Cooke	*Melampsora larici-populina* (Sharma and Heather, 1978); *Cronartium flaccidum* (Alb. et Schwein.) G. Winter; *Peridermium pini* (Pers.) Lév. (Moricca *et al.*, 1999)
C. uredinicola Speg.	*Puccinia cestri* Dietel and P. Henn. (Spegazzini, 1912); *Puccinia recondita* Roberge ex Desm (Ellis, 1976); *Cronartium quercuum* (Berk.) Miyabe ex Shirai f. sp. *fusiforme* (Morgan-Jones and McKemy, 1990); *Puccinia violae* (Schum.) DC (Traquair *et al.*, 1984); *Puccinia horiana* (Srivastava *et al.*, 1985)
C. uredinophilum Speg.	*Uredo cyclotrauma* Speg. (Spegazzini, 1922)

Identification and Distribution of *Cladosporium tenuissimum*

The dematiaceous hyphomycete *Cladosporium tenuissimum* has long been known as a polyphagous saprophyte occurring in the air, on the soil and on plant surfaces (Ellis, 1976). It is a frequent colonizer of senescent or dead plant material and is common on the phyllosphere and rhizosphere of plant species, but it is also reported as an endophyte (Fisher and Petrini, 1992) and a facultative plant pathogen, causing blights, leaf spots, and seed, fruit and blossom-end rots (Pandey and Gupta, 1983; Xiang *et al.*, 1989; Dohroo and Sharma, 1992; Sharma and Majumdar, 1993; Fujii *et al.*, 1995; Dhal *et al.*, 1997). Together with other *Moniliales*, it is the aetiological agent of human mycoses known as chromoblastomycosis, phaeohyphomycosis and eumycotic mycetoma (Gugnani and Okeke, 1989). The hyperparasitic nature of *C. tenuissimum* was first reported, as already mentioned, by Sharma and Heather (1981a,b) on the poplar rust *M. larici-populina* in Australia. It has recently attracted the attention of researchers because it suppresses rust under laboratory conditions and in glasshouses, and because it secretes compounds that are biologically active against various phytopathogens (Sharma and Heather, 1981a; Moricca *et al.*, 2001; Assante *et al.*, 2004).

The culture characteristics of this mitosporic fungus (colony appearance, texture and morphology, growth/temperature relationships, etc.), its micro-morphology (type of conidia and conidiogenesis) and other distinguishing features, are summarized in Moricca *et al.* (1999). The effect of nutrient composition on the production of secondary metabolites has also been investigated *in vitro* on different agar media and liquid cultures (Moricca *et al.*, 2001). Irrespective of the culture medium, the fungus produces typical, geniculate and sympodially elongated conidiophores, by virtue of which it is unambiguously ascribed to the anamorph genus *Cladosporium* Link (Domsch and Gams, 1980).

Since the abundance of heterogeneous taxonomic elements makes traditional classification at species level difficult, representative European isolates of *C. tenuissimum* were identified by matching mycological characteristics with nucleotide sequences from coding and non-coding regions of the ribosomal RNA operon (Moricca *et al.*, 1999) and by chemotaxonomic profiling (Moricca *et al.*, 2001). Molecular differences in nucleic acid sequences were instrumental particularly in *vis-à-vis* species recognition, since congeneric, morphologically similar taxa (*C. herbarum*, *C. cladosporioides* and other unidentified *Cladosporia*) were syntopic to *C. tenuissimum*, often sharing the same ecological niche (rust sori).

Several *Cladosporium* members have not yet been comprehensively described, many of its taxa are still ill-defined and the appropriateness of their separate recognition remains a vexed question (Morgan-Jones and McKemy, 1990). Many species are not considered good species because they have not been identified with sufficient confidence. As a result, a number of ecological, phytopathological and biomedical studies identify *Cladosporia* only at the generic level (Bolland, 1973; McKenzie and Hudson, 1976; Heather and Sharma, 1977; Sharma and Heather, 1978; Hamada and Fujita, 2002; Chew *et al.*, 2003). This means that information on the geographic distribution of *C. tenuissimum* in particular is scanty. Direct and indirect elements suggest, however, that it is an ubiquitous fungus with a worldwide distribution. It is known to be cosmopolitan and has been recovered from various matrices. In Europe it has been found on rust aeciospore samples of two-needled pines from various countries (Moricca *et al.*, 2001). On the other hand, a high incidence of airborne *Cladosporium* inoculum was reported in the London region by Ainsworth (1952). More importantly, that author also found that *Cladosporium* inoculum reached a peak in summer, coinciding with maturation of most rust spore stages. The possibility for the hyperparasite to colonize spores and control their numbers at developmentally just the right time is particularly attractive in the light of its potential exploitation for rust biocontrol.

The Hyperparasitic Relationship

In vitro antagonism

Interaction on glass slides

The great potential of *C. tenuissimum* as a BCA has been evident since the initial experiments of Heather and Sharma (1977) and Sharma and Heather (1978, 1981a,b, 1983, 1988). These authors observed both direct parasitism of urediniospores of *M. larici-populina* by this hyperparasite, and inhibition of rust spores without any physical contact, suggesting that antibiosis might also occur. This evidence induced these researchers to postulate that *C. tenuissimum* had significantly reduced the incidence and severity of rust disease in poplar plantations in the Canberra district for several years (Sharma and Heather, 1981a).

The antagonistic capability of *C. tenuissimum* was further confirmed when it was found to inhibit, *in vitro*, the germination of propagules of other rust fungi. Selected isolates of *C. tenuissimum* significantly reduced average percentage germination of aeciospores of *C. flaccidum* and *P. pini* at 12, 18 and 24 h from inoculation with conidial suspensions of the hyperparasite (33, 39 and 46% versus controls, respectively) (Moricca *et al.*, 2001). The germination of urediniospores of the bean rust fungus *Uromyces appendiculatus* treated simultaneously with a conidial suspension of the antagonist isolate 'Itt21' was reduced significantly from 56.4% to 36.9% after just 3 h of contact, a reduction of 35%. After 6 h, urediniospore germination was still lower than in the control (69.5 versus 83.3%), and at the end of the observations the difference was 17% (Assante *et al.*, 2004). Beyond this time, *U. appendiculatus* urediniospores no longer germinated. Freshly collected aeciospores of the pine twist rust *Melampsora pinitorqua* and of the common rust *Puccinia sorghi* inoculated with a mixture of conidia from different antagonist isolates displayed, after 24 h, reductions in germination of 19% and 21%, respectively, compared with the controls (Torraca, Italy, personal communication).

Timing of infection

The reduction in spore germination might also depend on the time of initial infection. Experiments on *C. flaccidum* and *P. pini* indicated that the order in which the rust and hyperparasite were deposited was the main factor causing variability in aeciospore germination. In these experiments three deposition sequences were tested: (i) aeciospores deposited 1 h prior to the hyperparasite conidia; (ii) conidia deposited 1 h prior to the aeciospores; and (iii) aeciospores and conidia deposited simultaneously. Maximum inhibition of aeciospore germination was achieved when conidia were inoculated 1 h before the rust aeciospores (Moricca *et al.*, 2001). This outcome suggests that, in nature, control is most effective if the antagonist establishes itself early on the host surfaces, before the rust, in order to build up a mass of inoculum sufficient to parasitize rust propagules as they burst from the plant epidermis somewhat later (Kranz, 1969a; Moricca *et al.*, 2001). However, the research data do not present a coherent picture. Other studies on *Puccinia recondita*, *Cronartium fusiforme* and *Cronartium strobelinum* found that with these rusts, infections became most severe when the antagonist *Darluca filum* was inoculated before them (Swendsrud and Calpouzos, 1972; Kuhlman *et al.*, 1978).

Interactions in storage

The recovery of *C. tenuissimum* at various latitudes, from separated geographic areas with varying climates, suggests that the fungus can survive and remain active under disparate environmental conditions. This adaptability is also shown by its ability to survive and parasitize aeciospores of *C. flaccidum* and *P. pini* in storage in a range of temperatures. In tests it was effective at

–20, 4 and 20°C. Control was greatest at 20°C but spore viability was decreased at all test temperatures, including –20°C. Viability also gradually decreased with storage time (Moricca *et al.*, 2001). Antagonist-treated spore lots were visibly discoloured. Stereoscope observations revealed deterioration of the rust propagules, which were densely intertwined with hyphae of the antagonist. The hyperparasite had proliferated extensively, giving rise to an appreciable mycelial biomass. Spore deterioration was the first indication that an exogenous enzymatic effect was probably involved in the disintegration process. The long shelf-life of *C. tenuissimum* inoculum at low temperatures provides evidence of ecological tolerance, a characteristic that is common to other hyperparasites (Kranz, 1969b). As in other fungal host–hyperparasite interactions, the antagonist thrives on the host spores, which represent an ideal *pabulum* (Swendsrud and Calpouzos, 1970). However, the host propagules are not vital for the antagonist which, being also a facultative saprophyte, a plant and animal parasite and an unspecialized hyperparasite, can survive on various materials such as plant debris, foliar exudates, small insects and other organic substances occurring in the environment.

Rust spores collected during adverse weather (high humidity, rainfall) and not properly dehydrated before storage, soon clumped together, and were impaired and discoloured. Many of these clumped spores were soon heavily overgrown with hyperparasite mycelium, which proliferated and sporulated profusely on them, markedly decreasing spore viability (Moricca *et al.*, 2001). High humidity is therefore favourable to infection, pathogenesis and sporulation of *C. tenuissimum*.

Effect of C. tenuissimum *culture filtrates*

Treatment of all rust spores with culture filtrates showed that enzyme(s) and/or toxic agent(s) had a role in *C. tenuissimum* parasitism. Aliquots (25 μl) of culture filtrates from four antagonist isolates spread separately on sterilized water-agar slides, dusted with aeciospores of two-needled pine rust fungi and incubated in the dark at 22°C in moist Petri dishes, strongly reduced spore germination after 12, 18 and 24 h, as compared with the controls. The rust provenances, the antagonist isolates and the interaction term (rust provenance × antagonist isolate) were not significant variables, indicating an absence of physiological specialization in the hyperparasite (Moricca *et al.*, 2001). Several of the spores examined individually under the microscope for viability were barely recognizable, with their spinules displaced and scattered all over the mounting medium, indicating an action of proteolytic enzyme(s) secreted into the medium.

The pronounced sterility of rust spores treated with *C. tenuissimum* culture filtrates suggested that the hyperparasite was a source of extracellular antifungal antibiotics, as it is in other hyperparasite–parasite interactions (Jackson *et al.*, 1997; Rodriguez and Pfender, 1997; Trejo-Estrada *et al.*, 1998). The toxicity of the culture filtrates was immediate, from the first inspection (after 12 h), and remained fairly constant for the duration of the experiment, with a slight increase in germination after 18 and 24 h.

Hyperparasitic secondary metabolites: a band of killers

C. tenuissimum actively produces metabolites with antifungal properties. These include a major pure common metabolite with the molecular formula $C_{20}H_{16}O_6$, and corresponding to M_r 352, and a series of related compounds, all of which were isolated from the ethylacetate crude extracts (EtOAc CEs) of several antagonist isolates (Moricca *et al.*, 2001). This metabolite was already known as cladosporol, a dimeric decaketide which had been isolated from *C. cladosporioides* (Fukushima *et al.*, 1993; Sakagami *et al.*, 1995). Cladosporol induces hyphal malformations in *Phytophthora capsici* when tested at 10 μg/disc (Fukushima *et al.*, 1993). It is also an inhibitor of β-1,3-glucan synthetase, the enzyme that synthesizes the fungal cell wall component β-1,3-glucan. In an *in vitro* assay with labelled UDP-glucose and β-1,3-glucan synthetase prepared from *Saccharomyces cerevisiae*, cladosporol showed an IC_{50} activity on the enzyme at 50 μg/ml (Sakagami *et al.*, 1995).

The series of related compounds purified from EtOAc CEs of *C. tenuissimum* cultures, consists of cladosporols B, C, D and E (Assante *et al.*, 2002; Nasini *et al.*, 2004). The major metabolite, now named cladosporol A, was isolated as a white powder and represented more than 30% of the crude extract. A second metabolite had the same ^1H and ^{13}C nuclear magnetic resonance (NMR) spectra as cladosporol A, except that it had a 4-oxo function instead of the C(4)HOH grouping. This metabolite was named cladosporol B. A third compound had the same basic skeleton as cladosporol A, but with a $C(2)H_2$-$C(3)H_2$ unit instead of the 2,3-oxirane ring. This compound was compatible with the molecular formula $C_{20}H_{18}O_5$ and was named cladosporol C. A fourth metabolite, cladosporol D, was a cream-coloured solid with the formula $C_{20}H_{18}O_6$. Its ^1H and ^{13}C NMR data, when compared with those of cladosporol C, indicated that it contained a C(3)HOH fragment instead of CH_2, the remaining signals being quite similar. The last metabolite, cladosporol E, was isolated as a brown solid with the formula $C_{20}H_{18}O_7$. Its ^1H and ^{13}C NMR spectra were very similar to those of cladosporol D, the only important difference being that this cladosporol had an additional hydroxy group at C-2 (Nasini *et al.*, 2004).

Cladosporols A–C, produced in an amount sufficient to enable some assays on their biological activity, inhibited, *in vitro*, a number of rust fungi, non-rust fungi, *Oomycota* and yeasts. They suppressed germination of urediniospores of *U. appendiculatus* and of aeciospores of *M. pinitorqua*, *C. flaccidum*, *P. recondita* and *P. sorghi*, in a range between 75 and 100%, when tested at 100 μg/ml. Cladosporol B was the most active of the group, completely suppressing germination of *U. appendiculatus* at 50 ppm, reducing it by more than 90% at 25 ppm, and lowering it even at 12.5 ppm. Cladosporol A, through less inhibitory than cladosporol B, was more active than cladosporol C, reaching an inhibition value higher than 80% at the highest concentration.

The cladosporols reduced radial growth of colonies of the phytopathogenic fungi *Alternaria alternata*, *Botrytis cinerea*, *Cercospora bieticola*, *Cercosporella herpotrichoides*, *Colletotrichum lindemuthianum*, *Fusarium roseum*, *Helminthosporium oryzae*, *Mucor* sp., *Rhizoctonia solani* and *Septoria tritici*; of the *Oomycota Phytophthora capsici*, *P. cinnamomi*, *P. erytroseptica*, *P. nicotianae* and *Pythium ultimum*; and of human-pathogenic strains of *Candida* sp. (Moricca *et al.*, 2001; Assante *et al.*, 2002; Nasini *et al.*, 2004; T. Kasuga, California, 2001, personal communication; Aloi and Fossati, Italy, 2002, personal communication). The strongest antagonistic effect was against the *Oomycota*.

Sensitivity of tested fungi varied in relation to concentration and differences in the functional groups bound in this family of metabolites to the tetralone skeleton. The antifungal activity of described cladosporols is likely to reside, as reported for other similar compounds (Arnone *et al.*, 1986; Fukushima *et al.*, 1993), in the intrinsic toxicity of the 2-tetralone chromophore and the occurrence of highly reactive substituents, like the epoxy group β-1,3-glucan, as a constituent of the fungal cell wall skeleton. By inhibiting the enzyme that synthesizes this constituent, the cladosporols directly affect the biochemistry and structural organization of the fungal cell. They thus strongly condition the pathogenicity of *C. tenuissimum* and play a major role in its hyperparasitism.

In vivo antagonism

In planta *assays*

Inoculation tests on whole rust-infected plants in a controlled environment (greenhouse or laboratory) give a first indication of how the rust may be controlled in nature. They therefore represent an important step in studying the mode of action of the hyperparasite. If a natural inoculation procedure and an objective disease evaluation protocol are followed, *in planta* assays can, in just a few weeks, provide valuable data on the biocontrol effectiveness of a tested microorganism. Furthermore, in such an artificial system, the tri-trophic interaction between host plant, rust parasite and hyperparasite can be more accurately investigated, since the effect of the environment, which in the field can positively or negatively affect each interacting partner, is eliminated.

In two consecutive growing seasons, 1999 and 2000, spermogonia of *C. flaccidum* that had developed on 2-year-old, rust-infected pine seedlings, were inoculated with mixed conidial suspensions from different *C. tenuissimum* isolates. Rust-infected control seedlings were sprayed only with sterile water/Tween 20. Disease evaluation, based on a standardized procedure, was completed after 5 months, and the incidence and severity of the rust infection were defined. The percentage of infected seedlings and the number of infections per seedling stem were significantly lower in the antagonist-treated seedlings than in the untreated controls. Percentage mortality was significantly lower in seedlings with antagonist inoculation than in those without. A mycelial biomass attributable to the hyperparasite and detectable as a felty, dark greenish-brown mycelium was observed on spermatial and aecial fructifications on the bark of some treated seedlings. The fungus was positively identified by microscope examination of the sporulating structures (erect, straight, regularly septate conidiophores; holoblastic, conidiogenous cells; cylindrical to clavate ramo-conidia with 2–3 flattened, thickened scars; intercalary and terminal conidia of various shapes and sizes) (Moricca *et al.*, 2001).

In an experiment exploring the hyperparasitism of *C. tenuissimum* on *U. appendiculatus*, a classical disease escape mechanism may have prevented the antagonist from parasitizing the rust. Primary leaves of bean plants of *P. vulgaris* L. cv. 'Borlotto nano Lingua di fuoco' were inoculated on their lower surface with a suspension of *U. appendiculatus* urediniospore and simultaneously treated with a suspension of *C. tenuissimum* conidia, or with a MPGG (malt extract, peptone, glucose, glycerol) culture filtrate of a 2-week-old liquid stationary culture of the hyperparasite. Controls were healthy bean plants inoculated with: (i) a water suspension of *C. tenuissimum* conidia; (ii) rust urediniospores, then treated with sterile water/Tween 20; or (iii) only a sterile uninoculated MPGG broth. Disease severity was scored by the number of pustules per square centimetre in a total of eight 1-cm^2 leaf areas on digitalized primary leaves, after 13 d from inoculation. After 1 month, plants inoculated simultaneously with *U. appendiculatus* and the conidial suspension developed rust in the normal way. By contrast, treatment with the antagonist culture filtrate provided total protection: the urediniospores did not germinate and the bean plants did not develop any infection (Assante *et al.*, 2004).

A possible explanation of these findings on bean rust is that in the simultaneous infections the time available for inter-fungus interaction was too short to enable the antagonist to establish a parasitic relationship with the pathogen. The specialized, biotrophic agent found refuge by rapidly penetrating into the living host tissues, occupying the ecological niche it has evolved since primordial times to colonize in order to gain access to nutrients and protection from natural enemies. The antifungal compounds and enzymes in the culture filtrate, on the other hand, acted immediately and prevented propagule germination and rust development.

Detached leaves

A simple test in a strictly controlled environment, using Petri dishes and a water-saturated atmosphere, can give an insight into the type of hyperparasitic interaction. Primary leaves of bean rust-infected plants, inoculated as above with a conidial suspension or a culture filtrate of *C. tenuissimum*, were immediately detached and incubated in 15-cm diameter Petri dishes. The leaf stems were dipped in a medium containing 1% water-agar (WA) supplemented with

5 ppm gibberellic acid (GA). Hyperparasite colonization of rust pustules and disease severity were monitored daily under a stereoscope, starting 1 week after hyperparasite inoculation and continuing until the end of the experiment. Rust regularly formed appressoria at the precise location of the stomata (indirect-type, dikaryotic penetration) but these appressoria began to collapse a few hours after formation. In spite of the positive tropism of *C. tenuissimum* conidia towards the rust propagules, as indicated by the many conidia closely attached to rust spores and appressoria, penetration of the bean leaves by the rust could not be prevented completely, and this can explain why, compared with the controls, disease severity was curtailed by 13% only. As in the experiment with the whole plants, the culture filtrate had a toxic effect on the detached leaves that prevented any rust spores from germinating (Assante *et al.*, 2004).

LM, SEM and TEM Examination of the Host/Parasite Interface

Examination of the interface between *C. tenuissimum* and the rust agents *C. flaccidum*, *P. pini* and *U. appendiculatus* with light (LM), scanning (SEM) and transmission electron microscopy (TEM) showed the strong antagonistic action of the hyperparasite, and the multiple strategies it employed to parasitize the rust fungi. The reproductive capacity of *C. tenuissimum* appeared greatly enhanced by the proximity of rust spores, most of which were inactivated and overgrown by the antagonist mycelium. The hyperparasite sporulated profusely on the spores, producing on their surface tufts of fructifications bearing numbers of conidiophores. These conidiophores generated a multitude of asexual propagation units, represented by ellipsoidal to limoniform ramo-conidia, oblong or fusiform intercalary conidia, and mostly globose or subglobose terminal conidia. Prolific antagonist sporulation gave rise to an enormous mass of conidia which, together with the germ tubes and hyphae growing from them, were attracted to the rust propagules, to which they became firmly attached.

The contact stimulus between the host and the hyperparasite – a reflection of recognition events among them – mediated the secretion of different substances at the host–parasite interface. These substances are believed to play a fundamental role in pathogenesis, either in interactions between plants and parasitic fungi (Chaubal *et al.*, 1991; Braun and Howard, 1994; Jones, 1994; Carver *et al.*, 1995; Nicholson, 1996) or between those fungi and their hyperparasites (Carling *et al.*, 1976; Tsuneda and Skoropad, 1977; Moricca *et al.*, 2001). Some of these substances, visible as a dense network of amorphous, fibrous material, were the direct product of hyperparasite metabolism and served to ensure close adhesion of the hyperparasite to the host cell wall (Moricca *et al.*, 2001). Amorphous material from a different source was observed adjacent to shrunken and eroded parts of the spore wall. This second type of extracellular, amorphous material characteristically accompanied spore penetration and was associated with the hyperparasite structures involved in the process, i.e. the variously shaped appressoria formed on the host surface and the infection hyphae. Ultrastructural examination at contact points provided evidence that lytic enzymes caused degradation of the host cell walls and released the amorphous material at the rust–hyperparasite interface. A decrease in electron density from the outer to the inner layers of the spore wall sometimes made the fibrillar chitin structure of the wall visible (Assante *et al.*, 2004).

An adhesive matrix pad intimately connected the host and the hyperparasite to each other. This pad serves a double function, to attach and support the appressorium and the penetrating hypha, and to be a reservoir for enzymatic penetration. Such functions have already been reported both in inter-fungus parasitism and in host plant–parasite interactions (Gold and Mendgen, 1984, 1991; Benhamou and Chet, 1996; Askary *et al.*, 1997; Moricca *et al.*, 2001). The growth of the antagonist on a synthetic medium containing the polymer laminarin as the sole carbon source, on the other hand, shows that it produces extracellular β-1,3-glucanases. There is extensive enzymatic degradation of the matrix wall polysaccharides in which the polymeric chitin microfibrils in rust fungi are embedded (Locci *et al.*, 1971; Trocha and Daly, 1974; Trocha *et al.*, 1974; Humme *et al.*, 1981; Maxemiuc-Naccache and Dietrich, 1981; Freytag and Mendgen, 1991), especially in the early stages of rust–hyperparasite interaction, when nutrients are vital for the antagonist. The involvement of β-1,3-glucanases in another hyperparasite interaction, that between *Fusarium solani* and *Puccinia arachidis*, has recently also been reported and discussed (Mathivanan, 2000). Other elements suggesting enzymatic activity besides the dissolution of the host cell wall, are the lack of indentation of the host wall at the contact site, and the minimal swelling of the infecting hyphal tip (Moricca *et al.*, 2001; Assante *et al.*, 2004).

Like congeneric hyperparasites (Tsuneda and Hiratsuka, 1979; Traquair *et al.*, 1984; Srivastava *et al.*, 1985), *C. tenuissimum* is endowed with alternative modes of penetration, since it can also invade propagules by physical destruction of the spore wall. This type of direct penetration uses a simple mechanical process of physical pressure against the host cell wall. Hyphae often coil around the rust spores, displacing the spinules as they advance, in some cases producing a swollen structure, and gaining access to the cell by breaching the spore wall, with or without the production of appressoria (Moricca *et al.*, 2001). A histological examination of the infection process reveals that when appressoria are produced, they generate penetration pegs that pierce the spore wall. Penetration pegs are narrower at the point where they pass through the spore wall, but once they have entered the cell lumen they swell out again. The hyperparasite destroys the protoplast of the host cells it invades and its mycelium proliferates inside the cells. Degradation of the spore contents is also evident from the many shrunk, collapsed and empty spores. While the inner wall layer of the rust spores therefore remains almost intact, the cell content is probably completely digested by the combined action of the β-1,3-glucanases and other lytic enzymes. It is supposed these enzymes work in cooperation with β-1,3-glucanases, there being no chitinase production by the antagonist when grown on medium with chitin as the sole carbon source. Moreover, the great number of ungerminated spores suggests that, especially in the early stages of penetration, toxic metabolites are secreted whose role is important since they kill the host cells and thus facilitate the colonization process (Moricca *et al.*, 2001; Assante *et al.*, 2004).

In brief, the parasitization of rust spores by the antagonist, as shown under the microscope, is divided into the following sequential events:

- pre-penetration (signal interplay with recognition, contact, adhesion, antibiosis, formation of infection structures);
- penetration (production of degrading enzymes, spore entry by mechanical pressure);
- post-penetration (evasion from the host cell, sporulation).

Ecological Fitness of *C. tenuissimum*

A precondition for the biological control of plant parasites is a full understanding of how the control agent operates. *C. tenuissimum* showed itself to be a destructive, unspecialized hyperparasite of rust fungi. Research has elucidated some of the basic principles underlying inter-fungus parasitism, clarified the fine structure of the rust–hyperparasite interface, shown that the hyperparasite inhibits rust propagules *in vitro* and rust diseases *in planta* under glasshouse conditions, and explored how the hyperparasite affects the target host. This last part of the research effort has led to the discovery of some lytic enzymes and toxic metabolites that are important pathogenicity determinants. Among the antagonist's 'weapons', these toxic metabolites are probably of prime importance. They are the cladosporols, a family of related compounds with strong antifungal activity, of which cladosporols B, C, D and E are described for the first time and reported as being produced by *C. tenuissimum* (Nasini *et al.*, 2004). The basic role of these substances in nature is to preserve the ecological niche of the hyperparasite: they protect the fungus against competing microorganisms; they prevent the growth of saprophytic microbes; and they displace plant pathogens from the plant surface (Vey *et al.*, 2001). The enormous potential of such molecules to control fungal parasites is underlined by recent findings in pharmacotherapy, where two compounds (caspofungin and micafungin), having the same biological activity as the cladosporols (inhibition of the synthesis of β-1,3-glucan), were reported as a new generation of antifungal drugs (Letscher-Bru and Herbrecht, 2003; Pawlitz *et al.*, 2003). These antifungal compounds have been patented and launched on the market by pharmaceutical companies Merck and Fujizawa as the first commercial inhibitors of β-1,3-glucan synthesis, and are claimed to be effective against several fungal infections in humans.

However, the fact that *C. tenuissimum* is a destructive hyperparasite that attacks and disintegrates rust spores, remains viable over a wide range of temperatures, possesses several aggression mechanisms, produces fungicidal metabolites and strongly suppresses rust development *in planta* does not guarantee it will be effective under natural conditions. A thorough understanding of hyperparasite biology, ecology and fitness is needed to obtain effective disease control and avoid inconsistency in efficacy. The antagonist has first to survive application, then to establish itself in the environment (forest or agroecosystem), and finally it must remain active until it is required for control. This means that it has to spend a significant period of time in a permanent habitat where it has to cope with environmental constraints and live side by side with the indigenous, competing microbial community. Inability to overcome such limitations may represent a crucial bottleneck for the hyperparasite.

The occurrence of *C. tenuissimum* at various latitudes and altitudes indicates, however, that the physical environment does not particularly affect its survival. Similarly, nutrient availability is not a problem for this microorganism, since it has a sufficiently broad range to exploit alternate hosts. It thrives on several rust species, on plant or animal hosts, it survives saprophytically on moribund or dead plant material, and it can overwinter on dead leaves, fallen flowers or fruits, or in necrotic spots. Such versatility indicates that its life in natural habitats is quite stable, as all these substrates are permanent or semi-permanent trophic reservoirs from which the microbe can disperse into the target host at its first appearance.

Among the attributes a good hyperparasite must have to be a successful BCA are, according to Wicker and Shaw (1968): a wide range (overlapping with that of its hosts); ecological amplitude (ensuring persistence within the host range); the production of abundant inoculum (necessary for epiphytotics to break out); an effective mode of action (to restrict the target disease); high infectivity; and virulence. An important attribute that should be added for the particular control of rust fungi is that the period of maximum sporulation of the hyperparasite should coincide with the maturation time of the rust spores. The aerobiological study of Ainsworth (1952) on the amount of *Cladosporium* inoculum in the air at different times of year, showed that it reached a peak in the spring and summer. These findings were confirmed by Cammack (1955), who reported that the release of *Cladosporium* air-spores from senescent or dead leaves was favoured by alternating wet and dry weather, a condition that frequently occurs in temperate regions of the world as a result of the diurnal variation in late spring and early summer. The peak of *C. tenuissimum* sporulation therefore coincides with multiplication of rust spores (aecidio-, uredinio-, teleuto- and basidiospores) and this is further evidence that rust biocontrol with *C. tenuissimum* is feasible.

Authors sceptical about biocontrol usually assert that antagonists already occur in natural habitats, and yet epidemics still continue to break out (Kranz, 1981). If we accept this argument, it would mean that the whole concept of disease control was vitiated, not only that of plants but that of all living organisms. Fortunately, however, the real successes achieved in controlling a number of important diseases shows that those suspicions are unfounded. The upsurge of a disease over time and space depends on a repeated cycle of infection, production of inoculum, and dispersal of inoculum to new sites. Pathogen inoculum spread is central to the development of any disease epidemic, and in the same way antagonist inoculum spread is fundamental in achieving control. Fluctuating climatic factors (temperature, relative humidity, precipitation, solar radiation, wind) and soil characteristics (texture, organic matter content, cation exchange capacity, moisture, pH) strongly affect propagule persistence in the epigeal and hypogeal milieu, as well as the dynamics of fungal populations. It is because of climatic and edaphic variations that the distribution of diseases in stands is patchy and the occurrence of hyperparasites erratic. For this reason the hyperparasite may need a long time to build up and maintain its biomass at levels that will control a target pathogen. If conditions conducive to high levels of infection are not forthcoming, the hyperparasite will be slow-acting and the pathogen will have time to cause disease symptoms and to reproduce on the crop. Purely epidemiological factors, therefore, often explain BCA ineffectiveness. A correct assessment of these, as well as of biological, ecological and technical (application methods) factors, are essential for the successful exploitation of BCAs in plant diseases.

Concluding Remarks

C. tenuissimum strongly reduces both the number and the longevity of rust spores, and also the amount of spores in the environment. Such selection pressure plays a prominent role in shaping pathogen population structure, the strong alteration of which may have dramatic effects on the epidemiology of the disease. Reduction of sporulation, infection period and dissemination causes, in fact, a restriction of the pathogen (Moricca *et al.*, 2001). If it is also considered that hyperparasites have a beneficial effect on the fitness of the host plants, which regain vigour when freed from fungal diseases (Kiss, 2001), it is clear that *C. tenuissimum* has an important role in the evolution of both the tree host and the rust pathogen.

Detailed examination of inter-fungus parasitism indicated that *C. tenuissimum* has potential to significantly curtail rust diseases and, as a consequence, the use of chemicals. This prospect is held out by knowledge of the biology and behaviour of *C. tenuissimum* in perennial habitats and agricultural settings. Nevertheless, recent research demonstrates that BCAs are not a cure-all to control all phytosanitary disorders. A single control measure can rarely provide effective and economically feasible levels of disease control. *C. tenuissimum* is a valuable resource to be used in an integrated pest management (IPM) framework where, together with other practices and control strategies, it can become a long-term, stable control measure.

Acknowledgements

The authors acknowledge the valuable information provided by Gianluca Nasini, CNR-ICRM, Milan, Italy, on the characterization of secondary metabolites; Dario Maffi and Marco Saracchi, University of Milan, and E. Bruno, University of Florence, for providing technical assistance with SEM and TEM observations; Novella Fossati and Claudio Aloj, Isagro, Italy, for data on the inhibition of *Puccinia recondita* and *Septoria tritici*; and G. Torraca, CNR-IPP, Florence, Italy, for data on the inhibition of *Melampsora pinitorqua* and *Puccinia sorghi*.

References

Abdel-Baky, N.F. and Abdel-Salam, A.H. (2003) Natural incidence of *Cladosporium* spp. as a bio-control agent against whiteflies and aphids in Egypt. *Journal of Applied Entomology* 127, 228–235.

Ainsworth, G.C. (1952) The incidence of air–borne *Cladosporium* spores in London region. *Journal of General Microbiology* 7, 358–361.

Alexopoulos, C.J., Mims, C.W. and Blackwell, M. (1996) *Introductory Mycology*, 4th edn. John Wiley & Sons, New York.

Allen, D.J. (1982) *Verticillium lecanii* on the bean rust fungus, *Uromyces appendiculatus*. *Transactions of the British Mycological Society* 79, 362.

Arnone, A., Nasini, G., Merlini, L. and Assante, G. (1986) Secondary mould metabolites. XVI. Stemphyltoxins, new reduced perylenequinone metabolites from *Stemphylium botryosum* f.s. *lactucum*. *Journal of the Chemical Society, Perkin Transactions* 1, 525–530.

Arthur, J.C. (1929) *The Plant Rusts (Uredinales)*. John Wiley & Sons, New York.

Askary, H., Benhamou, N. and Brodeur, J. (1997) Ultrastructural and cytochemical investigations of the antagonistic effect of *Verticillium lecanii* on cucumber powdery mildew. *Phytopathology* 87, 359–368.

Assante, G., Nasini, G., Moricca, S. and Ragazzi, A. (2002) Biological activity of the pine stem rust antagonist *Cladosporium tenuissimum* and its secondary metabolites. In: Elad, Y., Kohl, J. and Shtienberg, D. (eds) *Influence of A-Biotic and Biotic Factors on Biocontrol Agents*. IOBC/wprs Bulletin 25, 245–248.

Assante, G., Maffi, D., Saracchi, M., Farina, G., Moricca, S. and Ragazzi, A. (2004) Histological studies on the mycoparasitism of *Cladosporium tenuissimum* and urediniospores of *Uromyces appendiculatus*. *Mycological Research* 108 (2), 170–182.

Benhamou, N. and Chet, I. (1996) Parasitism of sclerotia of *Sclerotium rolfsii* by *Trichoderma harzianum*: ultrastructural and cytochemical aspects of the interaction. *Phytopathology* 86, 405–416.

Boland, G.J. and Kuykendall, L.D. (1998) *Plant–microbe Interactions and Biological Control*. Marcel Dekker, New York.

Bolland, L. (1973) Poplar rust in Queensland. *Australian Plant Pathological Society Newsletter* 2, 28.

Brasier, C.M. (1990) The unexpected element: mycovirus involvement in the outcome of two recent pandemics, Dutch elm and chestnut blight. In: Burdon, J.J. and Leather, S.R. (eds) *Pests, Pathogens and Plant Communities*. Blackwell Scientific Publications, Oxford, pp. 289–308.

Braun, E.J. and Howard, R.J. (1994) Adhesion of *Cochliobolus heterostrophus* conidia and germlings to leaves and artificial surfaces. *Experimental Mycology* 18, 211–220.

Butt, T.M. and Copping, L. (2000) Fungal biological control agents. *Pesticide Outlook* 11, 186–191.

Butt, T.M., Jackson, C. and Magan, N. (2001) Introduction – Fungal biological control agents: progress, problems and potential. In: Butt, T.M., Jackson, C. and Magan, N. (eds) *Fungi as Biocontrol Agents*. CAB International, Wallingford, UK, pp. 1–8.

Byler, J.W., Cobb, F.W. Jr and Parmenter, J.R. Jr (1972) Occurrence and significance of fungi inhabiting galls caused by *Peridermium harknessii*. *Canadian Journal of Botany* 50, 1275–1282.

Cammack, R.H. (1955) Seasonal changes in three common constitutes of the air-spora of Southern Nigeria. *Nature* 176, 1270–1272.

Carling, D.E., Brown, M.F. and Millikan, D.F. (1976) Ultrastructural examination of the *Puccinia graminis–Darluca filum* host–parasite relationship. *Phytopathology* 66, 419–422.

Carver, T.L.W., Ingerson–Morris, S.M., Thomas, B.J. and Zeyen, R.J. (1995) Early interactions during powdery mildew infection. *Canadian Journal of Botany* 73, 632–639.

Chaubal, R., Wilmot, V.A. and Wynn, W.K. (1991) Visualisation, adhesiveness and cytochemistry of the extracellular matrix produced by germ tubes of *Puccinia sorghi*. *Canadian Journal of Botany* 69, 2044–2054.

Chew, G.L., Rogers, C., Burge, H.A., Muilenberg, M.L. and Gold, D.R. (2003) Dustborne and air-borne fungal propagules represent a different spectrum of fungi with differing relations to home characteristics. *Allergy* 58, 13–20.

Dhal, N.K., Swain, N.C., Varshney, J.L. and Biswal, G. (1997) Etiology of mycoflora causing blossom-end rot of tomato. *Indian Phytopathology* 50, 587–592.

Dohroo, N.P. and Sharma, M. (1992) New host records of fungi from India. *Indian Phytopathology* 45, 280.

Domsch, K.H. and Gams, W. (1980) *Compendium of Soil Fungi*. Academic Press, London.

Dugan, F.M. and Lupien, S.L. (2002) Filamentous fungi quiescent in seeds and culm nodes of weedy and forage grass species endemic to the Palouse Region of Washington and Idaho. *Mycopathologia* 156, 31–40.

Ellis, M.B. (1976) *More Dematiaceous Hyphomycetes*. Commonwealth Mycological Institute, Kew, UK.

Fisher, P.J. and Petrini, O. (1992) Fungal saprobes and pathogens as endophytes of rice (*Oryza sativa* L.). *New Phytologist* 120, 137.

Freytag, S. and Mendgen, K. (1991) Surface carbohydrates and cell wall structure of *in vitro*-induced urediniospore infection structures of *Uromyces viciae-fabae* before and after treatment with enzymes and alkali. *Protoplasma* 161, 94–103.

Fujii, Y., Fukuda, A., Hamasaki, T., Ichimoto, I. and Nakajima, H. (1995) Twelve-membered lactones produced by *Cladosporium tenuissimum* and the plant growth retardant activity of cladospolide B. *Phytochemistry* 40, 1443–1446.

Fukushima, Y., Sakagami, Y. and Marumo, S. (1993) β-glucan biosynthesis inhibitors isolated from fungi as hyphal malformation inducer. *Bioorganic and Medicinal Chemistry Letters* 3, 1219–1222.

Gold, R.E. and Mendgen, K. (1984) Cytology of basidiospore germination, penetration and early colonization of *Phaseolus vulgaris* by *Uromyces appendiculatus var. appendiculatus*. *Canadian Journal of Botany* 62, 1989–2002.

Gold, R.E. and Mendgen, K. (1991) Rust basidiospore germling and disease initiation. In: Cole, G.T. and Hoch, H.C. (eds) *The Fungal Spore and Disease Initiation in Plants and Animals*. Plenum Press, New York, pp. 67–99.

Gugnani, H.H. and Okeke, C.N. (1989) Physiological characteristics of environmental isolates of pathogenic dematiaceous fungi. *Mycoses* 32, 78–83.

Hamada, N. and Fujita, T. (2002) Effect of air-conditioner on fungal contamination. *Atmospheric Environment*, 36, 5443–5448.

Heather, W.A. and Sharma, J.K. (1977) Some aspects of poplar rust research in Australia. *Australian Forestry* 40, 28–43.

Ho, M.H.M., Castaneda, R.F., Dugan, F.M. and Jong, S.C. (1999) *Cladosporium* and *Cladophialophora* in culture: descriptions and an expanded key. *Mycotaxon* 72, 115–157.

Hulea, A. (1939) Contributions à la connaissance des champignons commensaux des Uredinées. *Bulletin de la Section Scientifique de l'Academie de Roumanie* 22, 1–19.

Humme, B., Hoppe, H.H. and Heitefuss, R. (1981) Glucans from cell walls ofurediospore germ tubes of *Uromyces phaseoli* as elicitors of phytoalexin accumulation in *Phaseolus vulgaris*. *Phytopathologische Zeitschrift* 101, 51–64.

Jackson, A.J., Walters, D.R. and Marshall, G. (1997) Antagonistic interactions between the foliar pathogen *Botrytis fabae* and isolates of *Penicillium brevicompactum* and *Cladosporium cladosporioides* on faba beans. *Biological Control* 8, 97–106.

Jeffries, P. (1997) Mycoparasitism. In: Wicklow, D.T. and Söderström, B.R. (eds) *The Mycota. A Comprehensive Treatise on Fungi as Experimental Systems for Basic and Applied Research.* Springer-Verlag, Berlin, pp. 149–164.

Jones, E.B.G. (1994) Fungal adhesion. *Mycological Research* 98, 961–981.

Keener, P.D. (1954) *Cladosporium aecidiicola* Thuem. and *Tubercolina persicina* (Ditm). Sacc. associated with *Puccinia conspicua* (Arth.) Mains on *Helenium hoopesii* A. Gray in Arizona. *Plant Disease Reporter* 38, 691–694.

Kiss, L. (2001) The role of hyperparasites in host plant–parasitic fungi relationships. In: Jeger, M.J. and Spence, N.J. (eds) *Biotic Interactions in Plant–pathogen Associations*. CAB International, Wallingford, UK, pp. 227–236.

Kranz, J. (1969a) Das Verhaleten von *Darluca filum* (Biv-Bern.) Cast. *In vitro* unter verschiedenen Versuchsbeding-ungen. *Phytopathologische Zeitschrift* 65, 325–331.

Kranz, J. (1969b) Zur natürlichen Verbreitung des Rost-parasiten *Eudarluca caricis* (Fr.) *O. Erikss. Phytopathologische Zeitschrift* 65, 43–59.

Kranz, J. (1981) Hyperparasitism of biotrophic fungi. In: Blakeman, J.P. (ed.) *Microbial Ecology of the Phylloplane*. Academic Press, London, pp. 327–352.

Kuhlman, E.G., Mathews, F.R. and Tillerson, H.P. (1978) Efficacy of *Darluca* for biological control of *Cronartium fusiforme* and *C. strobelinum*. *Phytopathology* 68, 507–511.

Larran, S., Perelló, A., Simón, M.R. and Moreno, V. (2002) Isolation and analysis of endophytic microorganisms in wheat (*Triticum aestivum* L.) leaves. *World Journal of Microbiology and Biotechnology* 18 (7), 683–686.

Letscher-Bru, V. and Herbrecht, R. (2003) Caspofungin: the first representative of a new antifungal class. *Journal of Antimicrobial Chemotherapy* 51, 513–521.

Littlefield, L.J. (1981) *Biology of the Plant Rusts*. Iowa State University Press, Ames.

Locci, R., Ferrante, G.M. and Rodrigues, C.J. (1971) Studies by transmission and scanning electron microscopy on the *Hemileia vastatrix–Verticillium hemileiae* association. *Rivista di Patologia Vegetale* 7, 127–140.

Mathivanan, N. (2000) Chitinase and β-1,3 glucanase of *Fusarium solani*: effect of crude enzymes on *Puccinia arachidis,* groundnut rust. *Journal of Mycology and Plant Pathology* 30, 327–330.

Maxemiuc-Naccache, V. and Dietrich, S.M.C. (1981) Cell wall composition of spores of *Hemileia vastatrix* (coffee rust). *Revista de Microbiologia* 12, 61–64.

McKenzie, E.H.C. and Hudson, H.J. (1976) Mycoflora of rust-infected and non–infected plant material during decay. *Transactions of the British Mycological Society* 66, 223–238.

Morgan–Jones, G. and McKemy, J.M. (1990) Studies in the genus *Cladosporium sensu lato*: I. Concerning *Cladosporium uredinicola*, occurring on telial columns of *Cronartium quercuum* and other hosts. *Mycotaxon* 39, 185–202.

Moricca, S. and Ragazzi, A. (1998) Occurrence and significance of *Cladosporium tenuissimum* on two-needle pine rust aeciospores. In: *Proceedings of the First IUFRO 'Rusts of Forest Trees' WP Conference*, 2–7 August 1998, Saariselkä, Finland, pp. 171–182.

Moricca, S., Ragazzi, A. and Mitchelson, K.R. (1999) Molecular and conventional detection and identification of *Cladosporium tenuissimum* on two–needle pine rust aeciospores. *Canadian Journal of Botany* 77, 339–347.

Moricca, S., Ragazzi, A., Mitchelson, K.R. and Assante, G. (2001) Antagonism of the two-needle pine stem rust fungi *Cronartium flaccidum* and *Peridermium pini* by *Cladosporium tenuissimum in vitro* and *in planta*. *Phytopathology* 91, 457–468.

Nasini, G., Arnone, A., Assante, G., Bava, A., Moricca, S., Ragazzi, A. (2004) Secondary mould metabolites of *Cladosporium tenuissimum*, a hyperparasite of rust fungi. *Phytochemistry* 65, 2107–2111.

Nicholson, R.L. (1996) Adhesion of fungal propagules. In: Nicole, M. and Gianinazzi–Pearson, V. (eds) *Histology, Ultrastructure and Molecular Cytology of Plant–micro-organism Interactions*. Kluwer, Dordrecht, the Netherlands, pp. 117–134.

Pandey, K.N. and Gupta, R.C. (1983) A new leaf spot disease of banana caused by *Cladosporium tenuissimum* in India. *Madras Agricultural Journal* 70, 559.

Pawlitz, D., Young, M. and Klepser, M. (2003) Micafungin: a new echinocandin antifungal. *Formulary* 38 (6), 354–367.

Petrini, O. (1991) Fungal endophytes of tree leaves. In: Andrews, J.H. and Hirano, S.S. (eds) *Microbial Ecology of Leaves*. Springer Verlag, New York, pp. 179–197.

Powell, J.M. (1971) Fungi and bacteria associated with *Cronartium comandrae* on lodgepole pine in Alberta. *Phytoprotection* 52, 45–51.

Rayss, R. (1943) Contribution a l'étude de Deuteromycétes de Palestine. *Palestine Journal of Botany* Ser. III (1), 22–51.

Rodriguez, F. and Pfender, W.F. (1997) Antibiosis and antagonism of *Sclerotinia homoeocarpa* and *Drechslera poae* by *Pseudomonas fluorescens* Pf-5 *in vitro* and *in planta*. *Phytopathology* 87, 614–621.

Sakagami, Y., Sano, A., Hara, O., Mikawa, T. and Marumo, S. (1995) Cladosporol, β-1,3-glucan biosynthesis inhibitor, isolated by the fungus *Cladosporium cladosporioides*. *Tetrahedron Letters* 36, 1469–1472.

Sharma, J.K. and Heather, W.A. (1978) Parasitism of urediniospores of *Melampsora larici-populina* Kleb. by *Cladosporium* sp. *European Journal of Forest Pathology* 8, 48–54.

Sharma, J.K. and Heather, W.A. (1980) Effect of *Cladosporium aecidiicola* Thum. on the viability of urediniospores of *Melampsora medusae* Thum. in storage. *European Journal of Forest Pathology* 10, 360–364.

Sharma, J.K. and Heather, W.A. (1981a) Antagonism by three species of *Cladosporium* to three races of *Melampsora larici-populina* Kleb. *Australian Forest Research* 11, 283–293.

Sharma, J.K. and Heather, W.A. (1981b) Hyperparasitism of *Melampsora larici-populina* by *Cladosporium herbarum* and *Cladosporium tenuissimum*. *Indian Phytopathology* 34, 395–397.

Sharma, J.K. and Heather, W.A. (1983) Post-penetration antagonism of *Cladosporium tenuissimum* to uredinial induction by *Melampsora larici-populina*. *Transactions of the British Mycological Society* 80, 373–374.

Sharma, J.K. and Heather, W.A. (1988) Light and electron microscope studies on *Cladosporium tenuissimum*, mycoparasitic on poplar leaf rust, *Melampsora larici-populina*. *Transactions of the British Mycological Society* 90, 125–131.

Sharma, M. and Majumdar, V.L. (1993) Some new post-harvest diseases of ber fruits in India. *Indian Phytopathology* 46, 415.

Spegazzini, C. (1912) Mycetes Argentinenses. Ser. IV. *Anales del Museo Nacional de Buenos Aires* 23, 1–146.

Spegazzini, C. (1922) Fungi Paraguayenses. *Anales del Museo Nacional de Historia Natural de Buenos Aires* 31, 355–450.

Srivastava, A.K., Defago, G. and Kern, H. (1985) Hyperparasitism of *Puccinia horiana* and other microcyclic rusts. *Phytopathologische Zeitschrift* 114, 73–78.

Steyaert, R. L. (1930) *Cladosporium hemileiae* n. spec. Un parasite de l'*Hemileia vastatrix* Berk. & Br. *Bulletin de la Société Royal de Botanique du Belgique* 63, 46–48.

Sutton, B.C. (1973) Hyphomycetes from Manitoba and Saskatchewan, Canada. *Mycological Papers* 132, 1–143.

Swendsrud, D.P. and Calpouzos, L. (1970) Rust urediospores increase the germination of pycnidiospores of *Darluca filum*. *Phytopathology* 60, 1445–1447.

Swendsrud, D.P. and Calpouzos, L. (1972) Effect of inoculation sequence and humidity on infection of *Puccinia recondita* by the mycoparasite *Darluca filum*. *Phytopathology* 62, 931–932.

Traquair, J.A., Meloche, R.B., Jarvis, W.R. and Baker, K.W. (1984) Hyperparasitism of *Puccinia violae* by *Cladosporium uredinicola*. *Canadian Journal of Botany* 62, 181–184.

Trejo-Estrada, S.L., Paszczynski, A. and Crawford, D.L. (1998) Antibiotics and enzymes produced by the biocontrol agent *Streptomyces violaceusniger* YCED-9. *Journal of Industrial Microbiology and Biotechnology* 21, 81–90.

Trocha, P. and Daly, J.M. (1974) Cell walls of germinating uredospores. II. Carbohydrate polymers. *Plant Physiology* 53, 527–532.

Trocha, P., Daly, J.M. and Lagenbach, R.J. (1974) Cell walls of germinating uredospores. I. Amino acid and carbohydrate constituents. *Plant Physiology* 53, 519–526.

Tsuneda, A. and Hiratsuka, Y. (1979) Mode of parasitism of a mycoparasite, *Cladosporium gallicola*, on western gall rust, *Endocronartium harknessii*. *Canadian Journal of Plant Pathology* 1, 31–36.

Tsuneda, A. and Skoropad, W.P. (1977) The *Alternaria brassicae–Nectria inventa* host–parasite interface. *Canadian Journal of Botany* 55, 448–454.

Tsuneda, A., Hiratsuka, Y. and Maruyama, P.J. (1980) Hyperparasitism of *Scytalidium uredinicola* on western gall rust, *Endocronartium harknessii*. *Canadian Journal of Botany* 58, 1154.

Vey, A., Hoagland, R.E. and Butt, T.M. (2001) Toxic metabolites of fungal biocontrol agents. In: Butt, T.M., Jackson, C. and Magan, N. (eds) *Fungi as Biocontrol Agents*. CAB International, Wallingford, UK, pp. 311–346.

Whipps, J.W. and Lumsden, R.D. (2001) Commercial use of fungi as plant disease biological control agents: status and prospects. In: Butt, T.M., Jackson, C. and Magan, N. (eds) *Fungi as Biocontrol Agents*. CAB International, Wallingford, UK, pp. 9–22.

Wicker, E.F. (1981) Natural control of white pine blister rust by *Tuberculina maxima*. *Phytopathology* 71, 997.

Wicker, E.F. and Shaw, C.G. (1968) Fungal parasites of dwarf mistletoes. *Mycologia* 60, 372–383.

Wolfe, M.S. (1985) The current status and prospects of multiline cultivars and variety mixtures for disease resistance. *Annual Review of Phytopathology* 23, 251–273.

Xiang, C.T., Chang, G.B., Zhao, H.F., Li, Y.L. and Li, B.L. (1989) Study on biological characters of *Cladosporium tenuissimum* Cook. *Journal of Northeast Forestry University* 17, 38–41.

Yuen, G.Y., Steadman, J.R., Lindgren, D.T., Schaff, D. and Jochum, C. (2001) Bean rust biological control using bacterial agents. *Crop Protection* 20, 395–402.

Zhu, Y., Chen, H., Fan, J., Wang, Y. and Li, Y. (2000) Genetic diversity and disease control in rice. *Nature* 406, 718–722.

20 Biology and Genetic Diversity of the Rust Hyperparasite *Sphaerellopsis filum* in Central Europe

Mirko Liesebach[1] and Irmtraut Zaspel[2]

[1]*Federal Office and Research Centre for Forests, Department of Forest Genetics, Hauptstrasse 7, A-1140 Vienna, Austria;* [2]*Federal Research Centre for Forestry and Forest Products, Institute for Forest Genetics and Forest Tree Breeding, Eberswalder Chaussee 3A, D-15377 Waldsieversdorf, Germany*

Introduction

Sphaerellopsis filum (Biv.-Bern. ex Fr.) Sutton is a widely distributed hyperparasite of rust fungi, occurring naturally on about 30 genera. It is specific to rust fungi (*Uredinales*), but has an extremely broad host range within this diverse group of important plant pathogens. A host list of rusts, compiled from literature and herbarium material, comprises 369 species (Kranz and Brandenburger, 1981). Most records, by far, were made on species of the genus *Puccinia*, but it was also common on *Uromyces* spp., *Melampsora* spp. and *Phragmidium* spp.

Most commonly, the hyperparasite has been recorded in its imperfect stage on uredial sori and seldom in its perfect stage, *Eudarluca caricis* (FR.) Eriksson (*Ascomycotina, Dothideomycetidae, Pleosporales, Venturiaceae*). *Sphaerellopsis filum* has been recorded on several or single species of the genera *Puccinia, Uromyces, Schroeteriaster, Sphaerophragmium, Phragmidium* and the form-genus *Uredo* (Kranz, 1973; Yuan *et al.*, 1998). The anamorphic stage of the hyperparasite has been observed only on *Melampsora* spp. rusts of *Salicaceae*, whereas no reports exist on the appearance of the teleomorph, *E. caricis*, on this plant family.

The efficacy of *S. filum* is based on its ability to degrade uredial sori, which stops the micro-cycle of propagation of uredospores and so prevents new rust infections. The potential of the fungus as a method of controlling rust diseases, which cause great economic damage, was soon recognized. The first experiments using artificial inoculations of the hyperparasite to control *Puccinia triticina* were carried out by Schroeder and Hassebrauk (1957).

More recently, the use of *S. filum* as a biological protectant to control rust diseases has again raised interest. Its use may be of particular value in the cultivation of fast-growing tree species, e.g. willows and poplars used for biomass production in short-rotation coppice plantations, where chemical control methods are not applicable for technical and environmental reasons, and hence there is a need for alternative rust disease control strategies (Whelan *et al.*, 1997). Quantitative investigations on the suppressive effect of *S. filum* on willow rust, carried out by Yuan *et al.* (1999) and Pei *et al.* (2003) demonstrated its potential for disease control. These studies on the efficacy of *S. filum* against willow rust showed that rust spore production decreased as the *S. filum* conidia inoculum increased. Under experimental conditions certain strains of the hyperparasite were able to reduce rust spore production by up to 98%.

Despite the evidence for the widespread distribution of *S. filum*, there were only limited records for its occurrence in central Europe on willow rusts. Furthermore there were no references found describing its complete biological cycle, causes of pathogenicity and efficiency of rust control under natural conditions. Therefore a screening of rust samples was started to determine the frequency of the mycoparasite on rusts of the genus *Salix*. The mode of action of *S. filum* on willow rusts and the genetic diversity of the hyperparasite were also investigated.

Occurrence and Frequency of *S. filum* on Willow Rusts in Central Europe

In summer and autumn 2000–2002 leaf material from 36 willow species and hybrids (478 samples) infected with *Melampsora* spp. rusts was assessed for hyperparasitism by *S. filum*. The leaf samples were collected from willows growing in clone collections, short-rotation coppice (SRC) plantations and several natural sites in Germany and related regions (Table 20.1). They were viewed under a light microscope (Olympus SZH10 Research Stereo) at 25- to 30-fold magnification and the conidiomata of *S. filum* were detected as black spots inside the yellow uredia (Fig. 20.1).

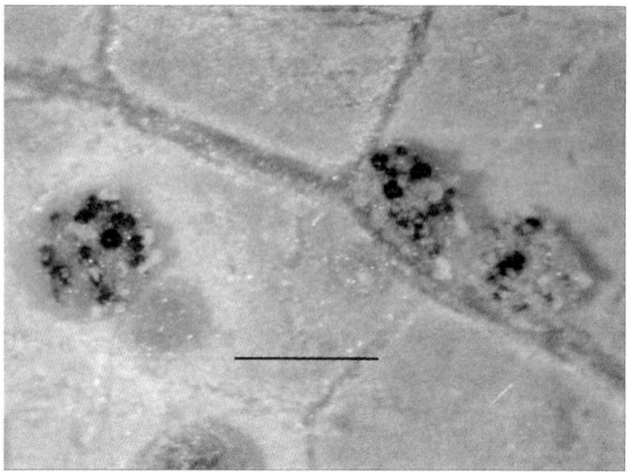

Fig. 20.1. Pycnidia of *Sphaerellopsis filum* formed on willow rust uredias on *Salix daphnoides* (bar = 10 mm).

Table 20.1. Sites where leaf samples were collected from willow plants and the number of *Sphaerellopsis filum* isolates obtained.

Site	Latitude	Longitude	Elevation (m asl)	Isolates (number)
I. Clone collections:				
Waldsieversdorf	52°32′N	14°05′E	46	20
Großhansdorf	53°39′N	10°15′E	50	8
Eberswalde	52°49′N	13°47′E	35	26
Hann. Münden	51°28′N	09°38′E	140	3
Arnsberg	51°24′N	08°03′E	210	7
Long Ashton (UK)	51°25′N	02°35′W	40	1
II. SRC plantations:				
Berlin-Buch	52°39′N	13°28′E	50	2
Berchem (Belgium)	51°10′N	04°50′E	5	2
III. Natural sites:				
district MOL	52°38′N	14°07′E	50	1
river site Elbe	53°04′N	11°21′E	20	2
Wächtersbach	50°14′N	09°15′E	170	1
Germering	48°08′N	11°20′E	540	2
Merbachfeld	51°29′N	09°37′E	120	1
Arnsberg	51°24′N	08°03′E	210	2

asl, above sea level; SRC, short-rotation coppice.

The levels of infection for each of the 3 years, 2000–2002, were 15%, 25% and 26%, respectively. There was no correlation between the level of infection in any given year and the number of *Sphaerellopsis* pycnidia detected on uredia, which differed significantly from sample to sample. While the overall infection levels were low in the population assessed, 75% of willow species and hybrids investigated were infected with *S. filum*. On rust on only nine species/hybrids was there no infection with the hyperparasite visible (Table 20.2). Additionally, *S. filum* was observed in uredia of the stem-infecting form of rust on hybrids, *S. caprea* × *viminalis* and *S.* × *sericans*.

Cultural and Microscopic Studies

Some authors have classified *S. filum* as an obligate, biotrophic hyperparasite, dependent on its host for its development and unable to be be cultured on artificial media (Leinhos and Buchenauer, 1992). However, older studies (Keener, 1933; Schroeder and Hassebrauk, 1957), as well as more recent ones (Pei *et al.*, 2003) and those of our own (Zaspel and Liesebach, 2004), have demonstrated that the fungus can be cultivated on the most common mycological media (Malt Extract Agar, Potato Dextrose Agar), using standard microbiological practices. Often the isolation of *S. filum* was complicated by the occurrence of secondary saprophytic bacteria. To reduce the growth of such bacteria the agar was supplemented with the antibiotics chloramphenicol (10 μg/ml) and oxytetracyclin (15 μg/ml), which did not inhibit the germination of *S. filum* conidia and enabled the establishment of single-spore isolates of the fungus.

A culture collection of *S. filum* comprised 77 isolates from *Melampsora* spp. causing rust on willow, supplemented with four isolates from *Melampsora* spp. causing poplar rust and six isolates from *Puccinia* spp. (Liesebach and Zaspel, 2004). The hyperparasite was cultured successfully on Malt Extract Agar Complete Medium (SERVA) containing 3% malt extract and 0.3% peptone from soy, with the pH adjusted to 7.0. Cultures were maintained at 16°C under continuous ultraviolet (UV) light at 370 nm (Blacklight Blue Philips TLD 36W/08). The growth of *S. filum* isolates *in vitro* was slow, with a mean radial growth of 40–50 mm in 25 days. After 5–7 months in culture many isolates had restricted growth with the fungus building up a black plectenchyma, without any production of conidia. At this stage mycelial production of a range of isolates could be stimulated by

Table 20.2. Frequency of occurrence of *S. filum* on rusts on willow species and hybrids.

Frequently (50% or more of assessed samples)	Rarely (< 50% of assessed samples)	No findings
S. purpurea	*S. alba* + hybrids	*S. adenophylla*
S. daphnoides	*S. aurita*	*S. aurita* hybrids
S. × *aquatica*	*S. caprea* + hybrids	*S. cinerea* + hybrids
S. acutifolia	*S. cordata*	*S. kazbekensis*
S. megeriana	*S. fragilis* + hybrids	*S. purpurea* hybrids
S. magnifica	*S. nigra*	*S. schwerinii*
	S. nigricans hybrids	*S. triandra* hybrids
	S. pentandra	*S.* × *smithiana*
	S. repens + hybrids	
	S. triandra	
	S. viminalis + hybrids	
	S. × *dasyclados*	
	S. × *hirtei*	
	S. × *sericans*	
	S. × *stipularis*	

cultivation in liquid medium, Malt Extract Broth Complete Medium (SERVA) on a shaker with slow rotation (80 rpm) under similar conditions.

The colour and growth behaviour *in vitro* were extremely variable between isolates. Schroeder and Hassebrauk (1957) reported that the appearance of *S. filum* isolates obtained from *Puccinia* spp. rusts of cereals and isolates from *Melampsora* rusts could be categorized into three groups. Group I had white aerial mycelium and pale pink- or brown-coloured submerged mycelium, with sparse sporulation. Group II had whitish, pale pink, or pale beige superficial mycelium without or with restricted submerged mycelium, and formation of conidia was sparse. Group III showed grey or black aerial mycelium and a black submerged mycelium. This group showed regular conidia formation in whitish drops on the surface. The appearance of some isolates changed following longer cultivation periods and therefore an assignment of isolates to a determined growth type was not possible.

In vivo the *S. filum* isolates developed black pycnidia emerging from well-developed rust uredia. Under conditions of high relative humidity, whitish cirrhi and drops of conidia appeared from the pycnidial conidiomata. Conidia were ellipsoid, 1-septate, rarely 2-septate, hyaline and later pale brown with a size of 16–18 m × 5–6 μm (Fig. 20.2). These data confirm the descriptions given by Schroeder and Hassebrauk (1957). They differ marginally from the description given by Sutton (1980) in that the conidia were marked by an apical, fan-shaped, gelatinous appendage that disappeared with proliferation, making this trait unsuitable as a specific character.

Scanning electron micrographs (SEMs) illustrated the infection process of rust uredia by *S. filum*. A fine net of non-specialized hyphae passed through the rust uredium, and conidia became directly fastened on rust spores. In an advanced stage of parasitism, rust spore walls and spore content showed enzymatic degradation and the invaded uredia subsequently collapsed (Fig. 20.3A–D). This extreme disorganization of cytoplasma in penetrated rust spores suggested rapid diffusion of toxic substances, as well as degrading enzymes, and confirmed the observations on the *Puccinia graminis*/*S. filum* host–parasite relationship (Carling *et al.*, 1976).

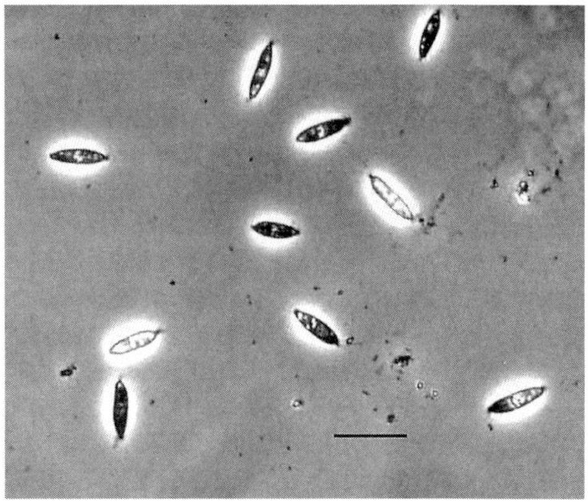

Fig. 20.2. Conidia formed by *Sphaerellopsis filum* (bar = 20 μm).

Mode of Action of Parasitism of *Sphaerellopsis filum*

The mode of action of the parasitization by *S. filum* is unclear for most parts. Older studies described appressoria-like structures of the fungus penetrating uredospores of *Puccinia* sp. rust (Schroeder and Hassebrauk, 1957). Later works indicated biochemical interactions of the rust–mycoparasite combinations and suggested lytic exoenzymes and toxic metabolites (Koc and Defago, 1983; Leinhos and Buchenauer, 1992). Two antibiotics of the xanthocillin type, Darlucin A and B, which exhibited antibacterial, antifungal and weak cytotoxic activity, have been isolated (Zapf *et al.*, 1995). Furthermore, it can presumed that activity by degrading exoenzymes, such as chitinase, cellulase, xylanase, and/or 1,3-β-glucanase complexes, can contribute to the efficacy of the mycoparasitic fungus. In order to determine the relative importance of enzymes, Wirth and Wolf (1992) used a method based on the colorimetric assay for the determination

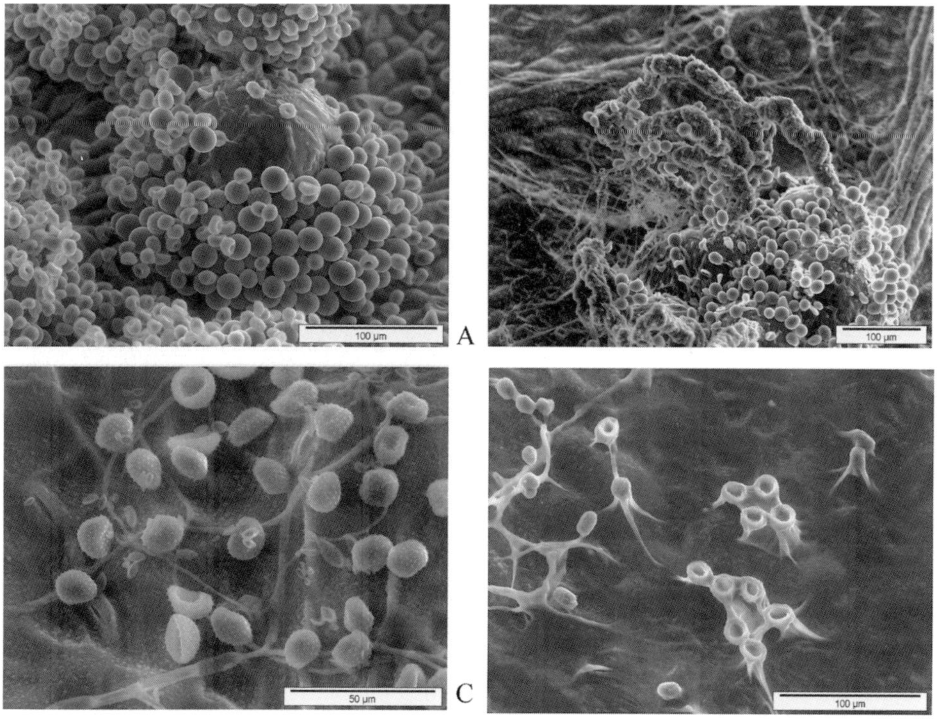

Fig. 20.3. Scanning electron micrographs of the infection process of a willow rust uredinium by *Sphaerellopsis filum*. (A) Section of a uredial sorus of *Melampsora* spp. willow rust bearing a *S. filum* pycnidium. (B) Long cirrhi of conidia are developing from a pycnidium. (C) Conidia of *S. filum*, with their appendages fixed on uredospores. (D) Late stage of parasitism, a uredium with collapsed rust spores is embedded in mucous substances.

of the polysaccharide *endo*-hydrolases, cellulase (E.C.3.2.1.4), xylanase (E.C.3.2.1.8), 1,3-β-glucanase (E.C.3.2.1.39) and chitinase (E.C.3.2.1.14). The substrates used in these assays were carboxymethyl-substituted, and water-soluble polysaccharide derivatives labelled with Remazol Brilliant Blue R (CM-Cellulose-RBB, CM-Curdlan-RBB, CM-Xylan-RBB) or Remazol Brilliant Violet 5R (CM-Chitin-RBV) (Loewe Biochemical, Sauerlach, Germany). The determination of the enzyme activity was carried out on culture filtrates from isolates grown in liquid culture at 21°C for 20 days. Assays were performed by adding an aqueous solution of dye-labelled substrate to culture filtrates and buffer, followed by incubation at 35°C. The reaction was terminated with HCl and cooling, and the supernatants containing soluble dye-labelled degradation products measured spectrophotometrically at 600 nm for substrates labelled with RBB, and at 550 nm with RBV. The fresh weight of *S. filum* isolates was determined in order to calculate enzyme production for each 1 g fungal fresh weight.

Enzyme activity was measured for 17 *S. filum* isolates obtained from *Melampsora* spp. willow rust and one from *Melampsora* sp. poplar rust. Chitinase showed the highest level of activity per gram mycelia fresh weight. Furthermore, differences in activity of enzyme complexes, xylanase, cellulase and 1,3-β-glucanase could be demonstrated (Table 20.3).

There was a high level of variation between the enzyme activities of individual isolates. These large differences could have been the consequence of the artificial cultivation techniques that resulted in the development of mutants lacking the ability to produce exoenzyme. On the other hand,

Table 20.3. Sample mean of enzyme activity g^{-1} fungal fresh weight of the *S. filum* collection isolated from *Melampsora* willow and poplar rusts and comparisons between enzyme complexes (U test).

		Wilcoxon rank sum test[a]		
Substrate	Mean (SD)	Cellulase	1,3-β-Glucanase	Xylanase
Cellulase	1.45 (2.30)	–		
1,3-β-Glucanase	1.36 (2.56)	0.201	–	
Xylanase	1.45 (1.84)	0.045	0.151	–
Chitinase	2.55 (1.56)	< 0.0001	< 0.0001	< 0.0001

[a]Data analysis was carried out by SAS program (SAS Inc, Cary, NY). Calculated data of enzyme activities of several enzyme complexes were computed for comparison with the Wilcoxon rank sum test (U test) and small *P* values for |Z| to support the alternative hypothesis of H_0 were described.

differences have been demonstrated to be due to *S. filum* populations being composed of pathogenically specialized isolates differing widely in their virulence (Yuan *et al.*, 1999). The correlation between mycelial fresh weight in liquid culture and absorbance for chitinase ($r = 0.429^{***}$, $P < 0.0001$) and for xylanase ($r = 0.532^{***}$, $P < 0.0001$) showed only a weak relation calculated for all isolates tested. A significant correlation was demonstrated between fresh weight and activity of enzyme complexes of cellulase and xylanase ($r = 0.885^{***}$, $P < 0.0001$).

Genetics

The use of polymerase chain reaction (PCR)-based genetic techniques, such as internal transcribed spacer-restriction fragment length polymorphism (ITS-RFLP) and sequencing of determined DNA fragments, offered the possibility of studying the genetic diversity at the molecular level and the phylogenetic relationships of isolates of the mycoparasite *S. filum*. For this investigation, DNA was extracted from the collection of 87 isolates (Liesebach and Zaspel, 2004), which included 77 isolates from willow rust. For comparative studies four isolates from *Melampsora* spp. poplar rust and six from *Puccinia* spp. rusts were included. Of the latter, one was obtained from *P. obscura* on *Bellis perennis* and five from *P. abrupta* on *Parthenium hysterophorus* from Ethiopia.

Protocols for PCR amplification of ITS regions and RFLP analysis are described by Liesebach and Zaspel (2004). The primer pair ITS1 (5′ TCC GTA GGT GAA CCT GCG G 3′) and ITS4 (5′ TCC TCC GCT TAT TGA TAT GC 3′) universal primers (White *et al.*, 1990) were used for the amplification of the ITS regions based on the repetitive units of the nuclear rDNA consisting of conserved coding and variable non-coding sequences. This method permitted the amplification of a single DNA fragment containing ITS1, ITS2, 5.8S and partial sequences of 18S and 25S rDNA genes, of a total size between 494 and 537 bp. The RFLP analyses were carried out by restriction digestion of the ITS region with the enzymes *Taq*I, *Eco*RI, *Hha*I, *Rsa*I or *Mbo*I. Using the clear patterns of the five restriction enzymes in combination, in total five RFLP profiles could be distinguished clearly. Table 20.4 shows the frequency of the multilocus patterns. The length of the uncut amplicon using the primer pair ITS1 and ITS4 and the number of the fragments generated by restriction digestion with *Taq*I, *Eco*RI, *Mbo*I, *Rsa*I and *Hha*I are listed in Table 20.5.

A set of 19 DNA samples, representing the patterns detected by RFLP and the resulting subgroups, were sequenced. The identified ITS1, ITS2 and 5.8S regions of *Sphaerellopsis filum* isolates were used to infer phylogenetic relationships among these isolates. The sequence data were aligned using the 'CLUSTALX' program (Thompson *et al.*, 1997). For calculating a tree the neighbour-joining (N-J) method was used (Saitou and Nei, 1987). Bootstrap

Table 20.4. RFLP profiles of ITS region, frequency, host plant species of isolates, number of sub-groups and sampling sites.

RFLP profile	Frequency	*S. filum* from rusts from species/hybrids (number of isolates)	Isolated from collections/ sites	Number of sub-groups
I	27	*S. alba* (1); *S. caprea* hybrid (1); *S. cordata* (1); *S. daphnoides* + hybrid (9); *S. fragilis* (1); *S. purpurea* (2); *S. triandra* (1); *S. viminalis* + hybrid (8); *Populus* sp. (3)	Waldsieversdorf; Eberswalde; Hann. Münden; Long Ashton/UK; district MOL; riversite Oder, riversite Elbe; Germering	3 (A, B,C)
II	6	*Bellis perennis* (1); *Parthenium hysterophorus* (5)	District MOL; Ethiopia	1 (L)
III	44	*S. alba* + hybrid (2); *S. aurita* (2); *S. acutifolia* (1); *S. caprea* + hybrids (6); *S. daphnoides* (5); *S. fragilis* + hybrid (3); *S. megeriana* (1); *S. nigra* (1); *S. nigricans* hybrid (1); *S. purpurea* (1); *S. viminalis* (8); *S.* × *aquatica* (4); *S.* × *dasyclados* (4); *S.* × *sericans* (2); *S.* × *stipularis* (2); *Populus* sp. (1)	Waldsieversdorf; Großhansdorf; Eberswalde; Hann. Münden; Arnsberg; district MOL; Wächtersbach; Berlin-Buch; Berchem/B	2 (D, E)
IV	9	*S. caprea* (1); *S. daphnoides* (2); *S. pentandra* (1); *S. viminalis* + hybrid (5)	Eberswalde; Arnsberg; Germering; Berchem/B	5 (G, H, I, J, K)
V	1	*S. repens* hybrid (1)	Eberswalde	1 (F)

RFLP, restriction fragment length polymorphism; ITS, internal transcribed spacer.

Table 20.5. Lengths of the uncut amplicons and number of the fragments generated by restriction digestion with *Taq*I, *Eco*RI, *Mbo*I, *Rsa*I and *Hha*I.

Profile	Sub-group	Number of isolates	Uncut amplicon	Number of fragments				
				*Taq*I	*Eco*RI	*Mbo*I	*Rsa*I	*Hha*I
I	A	2	520	4	2	4	2	6
I	B	2	519	4	2	3	2	6
I	C	1	520	3	2	3	2	6
III	D	3	536	4	2	4	2	6
III	E	1	537	4	2	3	3	6
V	F	1	497	3	3	3	1	2
IV	G	2	495	6	2	4	1	3
IV	H	1	495	5	2	4	1	4
IV	I	1	499	5	2	4	1	3
IV	J	1	498	5	2	4	1	3
IV	K	1	494	5	2	3	1	3
II	L	3	512	4	2	4	2	5

sampling was carried out on 1000 replicate data sets to assess the confidence limits of branch pattern (Felsenstein, 1985). The program TreeView (Page, 1996) was used to construct phylogenetic trees, which were rooted using the sequences of ITS1, ITS2 and 5.8S of *Colletotrichum acutatum* (GenBank accession EM_FUN:CAC536212) as the outgroup.

Based on the sequence analyses, two main lineages were distinguished. The profiles between these main groups differed at 32% or more of their nucleotide positions (Table 20.6). Within the first main group there were three profiles (I, II, III). The differences between those profiles were about 12%. The variation within each profile was very low (less than 2%). The second main

Table 20.6. Percentage divergence between two sequences of 19 *Sphaerellopsis* isolates (I–V: RFLP profiles).

	s27	s42	s23	s45	s15	s13	s68	s101	s18	s40	s76	s21	s9	s71	s99	s16	s14	s17	s8
s27					I														
s42	00																		
s23	> 00	> 00																	
s45	02	02	01																
s15	01	01	01	02															
s13	11	11	12	13	11			II											
s68	11	11	12	13	11	00													
s101	11	11	12	13	11	00	00												
s18	11	11	11	12	11	11	11	11				III							
s40	11	11	11	12	11	11	11	11	00										
s76	12	12	12	13	12	11	11	11	> 00	> 00									
s21	12	12	12	13	12	11	11	11	> 00	> 00	01								
s9	33	34	34	35	33	34	34	34	34	34	34	34						IV	
s71	34	34	34	36	34	34	34	34	34	34	34	34	01						
s99	32	32	32	33	32	33	33	33	33	33	32	33	05	05					
s16	32	32	33	34	32	33	33	33	33	33	32	33	05	05	03				
s14	32	32	33	34	32	33	33	33	33	33	33	33	05	04	02	01			
s17	33	33	33	34	33	34	34	34	33	33	33	33	05	04	03	01	> 00		
s8	39	39	40	40	39	39	39	39	40	40	40	40	16	16	14	13	12	12	V

group comprised two profiles (IV, V), which differed in between 12 and 16% of their nucleotide positions. The larger of the two profiles (IV) had a higher variation (up to 5%) than those of the first main group. The other profile (V) was only represented by a single isolate.

A phylogenetic tree considered these results (Fig. 20.4). The analysis showed, on the one hand, two clear lineages and, on the other hand, a separation of the five profiles detected by RFLP. There was also a close phylogenetic relationship of *S. filum* isolated from rust on *Salix* spp., *Populus* spp., *Bellis perennis* and *Parthenium hysterophorus*. The division into the two main groups, with the large distance, suggested the existence of two separate genetic units, indicating different species or sub-species (Liesebach and Zaspel, 2004). Furthermore, the results demonstrate, that there is no geographical concentration in a special profile of the mycoparasite, or any relation to the rust provenence (Liesebach and Zaspel, 2004). *S. filum* isolates from *Melampsora* spp. rusts on *Salix* and *Populus* are grouped into both main types. Isolates from the same willow clones at the same site in different years belonged partly to the same profile and partly to another profile compared to the previous year. Isolates from leaves and from the stem infection form of the same willow clone all belonged to profile III. Only profile II was the exception because this group contained unique strains from *Puccinia* rusts originating from the African continent and from Germany.

A high genetic diversity could be detected within a restricted population of *S. filum* isolates. This indicated specific rust–hyperparasite relationships and could be associated with the high number of *Melampsora* rust genotypes on willow (Chapter 6). Until now, we have not been able to detect any relations between them, and currently it has been impossible to establish any specific relationship between incidence of specific willow rust and *S. filum* genotypes. The sequence analyses of selected strains showed that the *S. filum* isolates belonged to two main types, which differed in more than a third of their

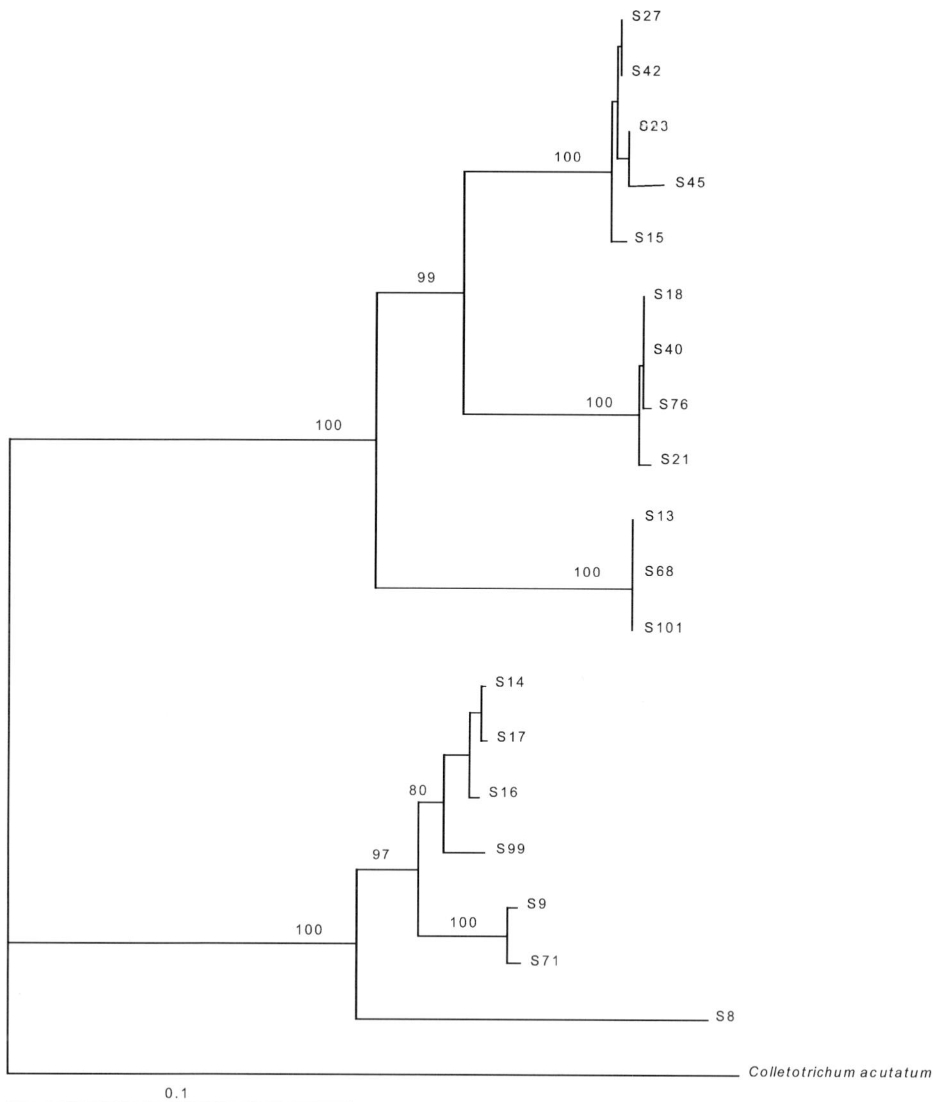

Fig. 20.4. Phylogenetic tree of 19 isolates of *Sphaerellopsis*. The tree was constructed by the neighbour-joining method on the basis of the ITS1, ITS2 and 5.8S region sequences. The sequence of *Colletotrichum acutatum* was used for rooting the tree. Bootstrap values for 1000 re-samplings are indicated on each node.

nucleotide positions. This finding indicated the existence of two separate taxonomic forms of the hyperparasite, which existed side by side. It is possible that these are two separate species or sub-species. There was no evidence indicating either a host preference of these types or a different geographic distribution connected with different ecological demands. The most conspicuous difference was the profile II, which contained all strains obtained from *Puccinia* rusts.

Since hyperparasitic mechanisms are diverse and complex, the development of the mycoparasite is closely dependent on the development of rust fungi. It was evident that strains of *S. filum* were able to produce polysaccharide-degrading exoenzymes, which could destroy the important polymers of cellulose and chitin of the cell walls of rust spores. SEM micrographs gave insight into the host–mycoparasite interface and showed clearly the degradation of rust uredia by the destructive hyperparasitic fungus. In contrast cellulase-degrading enzymes did not appear to have a role in the parasitizing process on rust uredia. Furthermore, the enzyme complex played a certain role for a possible saprophytic phase on plant surfaces during overwintering and would be needed for the phyllosphere competence of *S. filum* in the absence of a rust host.

In conclusion, the occurrence of *S. filum* on rusts of *Salicaceae* in Germany and neighbouring regions was evident, as well as the appearance of a range of genotypes of the hyperparasite. The existence of two different taxonomic, species or sub-species, units can be assumed, but to date it has been impossible to assign these to host plant species and/or rust genotypes. This is an area needing further investigation. To exploit this information as a means of of biocontrol, it is important to establish multiclonal short-rotation coppice (SRC) plantations as a way of increasing the biodiversity of fungal communities of the hyperparasite complex. This would seem to be more effective than those biomass plantations with a limited number of willow genotypes.

Acknowledgements

The research was funded by the EC and is part of the joint project 'Integrated, non-fungicidal control of *Melampsora* rusts in renewable energy willow plantations' (QLK5–1999–01585). Ms H. Mattauch is gratefully acknowledged for technical assistance with this work. The fungal material from *Parthenium hysterophorus* was kindly supplied by Dr T. Tessema (Humboldt University Berlin), the samples from Belgium were supplied by Drs R. Ceulemans and I. Laureysens (University of Antwerpen), those from Long Ashton by Dr M.H. Pei and T. Hunter (IACR Long Ashton Research Station). Dr F. Ehrig (BAZ Aschersleben) is thanked for SEM photographs.

References

Carling, D.E., Brown M.F. and Millikan D.F. (1976) Ultrastructural examination of the *Puccinia graminis–Darluca filum* host–parasite relationship. *Phytopathology* 66, 419–422.

Felsenstein, J. (1985) Confidence limits on phylogenies: an approach using the bootstrap. *Evolution* 39, 783–791.

Keener, P.D. (1933) Some characteristics of *Darluca* in culture. *Proceedings Academy of Natural Sciences* (Philadelphia, Pennsylvania) 7, 1–8.

Koc, N.K. and Defago, G. (1983) Studies on the host range of the hyperparasite *Aphanocladium album*. *Phytopathologische Zeitschrift* 107, 214–218.

Kranz, G. (1973) A host list of the rust parasite *Eudarluca caricis* (Fr.) O. Eriks. *Nova Hedwigia* 24, 169–180.

Kranz, G. and Brandenburger, W. (1981) An amended host list of the rust parasite *Eudarluca caricis*. *Journal of Plant Disease Protection* 88, 682–702.

Leinhos, G.M.E. and Buchenauer, H. (1992) Hyperparasitism of selected fungi on rust fungi of cereal. *Zeitschrift für Pflanzenkrankheiten und Pflanzenschutz* 99, 482–498.

Liesebach, M. and Zaspel, I. (2004) Genetic diversity of the mycoparasite *Sphaerellopsis filum* acting on *Melampsora* willow rusts. *Forest Pathology* 34, 292–305.

Page, R.D.M. (1996) TREEVIEW: an application to display phylogenetic trees on personal computers. *Computer Applications in the Biosciences* 12, 357–358.

Pei, M.H., Hunter, T., Ruiz, C., Bayon, C. and Harris, J. (2003) Quantitative inoculation of willow rust *Melampsora larici-epitea* with the mycoparasite *Sphaerellopsis filum* (teleomorph *Eudarluca caricis*). *Mycological Research* 107, 57–63.

Saitou, N. and Nei, M. (1987) The neighbour-joining method: a new method for reconstructing phylogenetic trees. *Molecular Biology and Evoution* 4, 406–425.

Schroeder, H.V. and Hassebrauk, K. (1957) Beiträge zur Biologie von *Darluca filum* (Biv.) Cast. und einigen anderen auf Uredineen beobachteten Pilzen. *Zentralblatt für Bakteriologie und Parasitenkunde* II, 110, 676–696.

Sutton, B.C. (1980) *The Coelomycetes*. Commonwealth Mycological Institute, Kew, UK, pp. 470–471.

Thompson, J.D., Gibson, T.J., Plewniak, F., Jeanmougin, F. and Higgins, D.G. (1997) The ClustalX windows interface: flexible strategies for multiple sequence alignment aided by quality analysis tools. *Nucleic Acids Research* 24, 4876–4882.

Whelan, M.J., Hunter, T., Parker, S.R. and Royle, D.J. (1997) How effective is *Sphaerellopsis filum* as a biocontrol agent of *Melampsora* willow rust? *Aspects of Applied Biology* 49, 143–148.

White, T.J., Bruns, T., Lee, S. and Taylor, J. (1990) Amplification and direct sequencing of fungal ribosomal RNA genes for phylogenetics. In: Innis, M.A., Gelfand, D.H., Sninsky, J.J. and White, T.J. (eds) *PCR Products: a Guide to Methods and Applications*. Academic Press, San Diego, California, pp. 315–322.

Wirth, S.J. and Wolf, G.A. (1992) Micro-plate colourimetric assay for *endo*-acting cellulase, xylanase, Chitinase. 1,3-β-glucanase and amylase extracted from forest soil horizons. *Soil Biology and Biochemistry* 24, 511–519.

Yuan, Z.W., Pei, M.H., Hunter, T. and Royle, D.J. (1998) *Eudarluca caricis*, the teleomorph of the mycoparasite *Sphaerellopsis filum* on blackberry rust *Phragmidium violaceum*. *Mycological Research* 102, 866–868.

Yuan, Z.W., Pei, M.H., Hunter, T., Ruiz, C. and Royle, D.J. (1999) Pathogenicity to willow rust, *Melampsora epitea*, of the mycoparasite *Sphaerellopsis filum* from different sources. *Mycological Research* 103, 509–512.

Zapf, S., Hossfeld, M., Anke, H., Velten, R. and Steglich, W. (1995) Darlucin-A and Darlucin-B, new isocyanide antibiotics from *Sphaerellopsis filum* (*Darluca filum*). *Journal of Antibiotics* 48, 36–41.

Zaspel, I. and Liesebach, M. (2004) Interaction of rust fungi and rust hyperparasite *Sphaerellopsis filum*. *Acta Physiologia Plantarum* 26, Suppl. 3, 129–130.

21 Mycoparasite *Sphaerellopsis filum* and its Potential for Biological Control of Willow Rust

Ming Hao Pei[1] and Zhiwen W. Yuan[2]
[1]*Rothamsted Research, Harpenden, Hertfordshire AL5 2JQ, UK;* [2]*Institute of Applied Ecology, Academia Sinica, PO Box 417, Shenyang, China*

Background

Sphaerellopsis filum (Biv.-Bern. ex Fr.) Sutton (*Darluca filum* (Biv.-Bern. ex Fr.) Cast.) is a mycoparasite of rust fungi (*Uredinales*). The fungus can be seen, often in large numbers, on rust pustules (sori). Since some of the most important plant diseases are caused by rust fungi, the possibility of using the mycoparasite for biological control has long been of interest. We are interested in *S. filum* as a possible means of biological control of *Melampsora* rusts of willows (*Salix* spp.). When grown in short-rotation coppice (SRC) plantations, willow is a potential renewable source of energy and fibre. In coppice plantations, *Melampsora* spp. cause the most widespread and damaging disease of willows, reducing biomass yields by up to 40% in susceptible clones (Parker *et al.*, 1993). Because of economic, technical and environmental considerations, control of rust using fungicides is not favoured and alternative approaches, including biological control, are needed.

S. filum was first reported on *Melampsora* spp. on *Salix* in 1873 in Germany by Fuckel (cited by Morelet and Pinon, 1973). Since then, it has been reported on 28 species of *Melampsora* on willows (Kranz and Brandenburger, 1981). Our notice was first drawn to the mycoparasite in 1990, when a rust epidemic in an established plantation of *S. viminalis* 'Bowles Hybrid' at Long Ashton became suppressed without external intervention. Subsequent studies showed that *S. filum* can reduce the spore production in willow *Melampsora* by more than 50% (Morris *et al.*, 1995; Whelan *et al.*, 1997).

In this chapter, we give a brief review of the existing information on *S. filum* and discuss the potential of using the mycoparasite for biological control of willow rusts in short-rotation coppice plantations.

Nomenclature and Taxonomy

The fungus was first described by Bivona-Bernadi in 1813 as *Sphaeria filum* (Bivona-Bernadi, 1813). Two of the hosts described were rusts. Later, the fungus was transferred to the genus *Phoma* by Fries (1823) and given the name *Phoma filum*. The genus *Darluca* was established by Castagne in 1851 (Castagne, 1851) using the previously named *Sphaeria filum* as type species and accordingly, the fungus was

 Rust Diseases of Willow and Poplar (eds M.H. Pei and A.R. McCracken)

named *Darluca filum* (Biv-Bern. ex Fr.) Cast. Before 1851, the same fungus had been sent out from the Castagne herbarium to several European botanists as *Darluca vagans*. Since the name *Darluca vagans* as used by Castagne prior to 1851 was not published, it cannot be recognized as valid. *Darluca filum* was the most widely used name for the fungus until Sutton (1977) proposed adopting the genus name *Sphaerellopsis* Cooke (Cooke, 1883, *Grevillea* 12:23) for the fungus.

The genus *Eudarluca* was established by Spegazzini (1908) for a pyrenomycete occurring on uredinia of a rust on *Canna* sp. in Brazil, citing *Eudarluca australis* as type species. It was assumed that *E. australis* was the perfect state of *Darluca filum*. The connection between imperfect and perfect states of the fungus had not been proven until Keener (1951) produced pycnidia and conidia of *S. filum* from cultures derived from ascospores of *Eudarluca* from *Puccinia extensicola oenotherae* (Mont.) Arth. on *Carex* sp. in Pennsylvania, USA.

Eriksson (1966) examined a collection described by Fries (1823) as *Sphaeria caricis* and found asci and ascospores which were identical to the type collection of *E. australis* (Eriksson, 1966). The same collection contained typical pycnidia of *S. filum* on the leaves of *Carex* sp. heavily infected by *Puccinia caricina* DC. After studies of some 300 herbarium collections and extensive literature search, Eriksson (1966) combined several synonyms of the perfect state of *Darluca* into *Eudarluca caricis* (Fr.) O. Erik. He listed the following as synonyms of *E. caricis*: *Dothidea genistalis* Pers. ex. Fr., *Dothidella appendiculata* deLacr. ex. Br. & Har., *Eudarluca australis* Speg., *Eudarluca indica* Ramakr., *Leptosphaeria nigrificans* Bub. & Wróbl., *Myrmecium cannae* Dearn. & Barth., and *Uleodothis pasipali* Stev.

Two varieties of *E. caricis* were recognized by Eriksson (1967): var. *caricis* with poorly developed stromata and 2–3 septate ascospores, and var. *indica* with well-developed stromata and 1-septate ascospores.

Morphology

Anamorph

The fungus forms pycnidia on rust sori. Occurrence of *S. filum* is most common on uredinia but it can frequently be found on telia and aecia. The pycnidia of *S. filum* are black when mature, sub-globose, 90–200 µm in diameter, within or upon the rust sori, and often had distinct ostioles through which the spores are exuded in thin, white mucilaginous cirrhi. Conidia are hyaline, 1-septate, fusiform, 13–18 × 3–5 µm and often with a small gelatinous cap at one or both ends. Small (2–3 × 1 µm) microconidia can sometimes be found mixed with conidia in herbarium collections of the fungus (Eriksson, 1966). Yuan *et al.* (1998) described microconidia in cultures derived from ascospores of *E caricis* from *Phragmidium violaceum* on blackberry *Rubus fruiticosus*. The fresh microconidia were cylindrical to oblong–ellipsoid, hyaline, non-septate and measured 4.0–5.5 × 1.5–2.0 µm.

Teleomorph

Blackish-brown stromata of *E. caricis* can be found on rust sori which often contain pycnidia of *S. filum*. Stromata are prominent in var. *indica* but poorly developed in var. *caricis*. Stromata are comprised of compact cells, brown to dark-coloured outside and hyaline inside. Pseudothecia occur in a few to numerous locules and are variable in size. Pseudothecia are completely or partly immersed, sub-globose or ampulliform, often with protruding, conical necks, and measure 100–260 µm across. Asci have two layers of wall, cylindric-clavate, short-stalked, 55–90 × 7–11 µm, 8-spored. Ascospores are irregularly biseriate, spindle-shaped, slightly inequilateral, 1-septae in var. *indica* and 2–3 septate in var. *caricis*, hyaline in var. *indica* and pale honey-yellow in var. *caricis*, and measure 15–24 × 3.5–5 µm. Pseudoparaphyses are branched and 2–3 µm thick.

Natural Occurrence of *S. filum*

The fungus is distributed globally, occurring in all parts of the world except Antarctica. It appears that the teleomorph of the fungus is more common in tropical regions (Eriksson, 1966).

Much information on the behaviour of natural populations of *S. filum* came from field surveys or observations. Early studies on *Peridermium peckii* on hemlock (*Tsuga canadensis*) indicated that natural infection of aecia by *S. filum* prevented their maturing and suppressed spore release (Adams, 1920). In a survey of ten maize fields in a location in the highlands of Kenya, *S. filum* occurred widely on uredinia of *Puccinia sorghi*, colonizing up to 40% of rust uredinia (Kranz, 1969a). In a pure forest stand in Florida, USA, Kuhlman *et al.* (1978) found that *S. filum* colonized 93% of uredinia of *Cronartium strobilinum* on *Quercus minima* and less than 1% of the uredinia contained telia. In contrast, in an adjacent mixed stand, more sparsely populated with the same oak, only 32% of rust uredinia were infected and 26% of them had telia.

In a larch forest stand in north-eastern China, field assessments showed that the frequency of rust sori colonized by *S. filum* increased over the 3 consecutive years of study, 38% for 1986, 57% for 1987 and 62% for 1988 (Yuan *et al.*, 1988). At the same time, there was a decrease in the frequency of needles bearing rust sori, 68% for 1986, 58% for 1987 and 37% for 1988. Yuan and Han (2000) reported that, over a period of 3 years, the frequency of plants infected by *Uromyces hedysari-mongolici* in a stand of *Hedysarum mongolicum* decreased from 65% to 27% while rust pustules colonized by *S. filum* increased from 36% to 65%.

In England, the stem-infecting form (SIF) of *Melampsora* rust (Pei *et al.*, 1995) caused severe infections in an established coppice planting of *Salix viminalis* 'Bowles Hybrid' in May–July 1990. By mid-August, however, the great majority of rust pustules became discoloured and inactive. Close inspections revealed that over 90% of the rust uredinia were colonized by *S. filum* (M.H. Pei, unpublished results). Assessment in the same planting in 1992 showed 60–80% of the uredinia on stems were colonized by *S. filum* by the end of July (Morris *et al.*, 1995).

Life Cycle and Overwintering

Occurrence of the sexual stage

From available information, teleomorph of the fungus appears to be rare in temperate regions. No perfect state of *S. filum* was known in Europe until 1962 when Eriksson found *E. caricis* at two locations in Sweden (Eriksson, 1966). In Finland, only four samples of *E. caricis*, three mature and one immature, were found (Eriksson, 1967). Relatively recently, *E. caricis* was found on *Phragmidium violaceum* on blackberry *Rubus fruticosus* in south-west England (Yuan *et al.*, 1998). The morphology of the *E. caricis* on *Ph. violaceum* conformed to var. *caricis* offered by Eriksson (1966). Cultures derived from ascospores produced pycnidia and conidia typical of *S. filum*, confirming the connection between the sexual and asexual stages of the fungus.

Microconidia were also found together with conidia in the cultures derived from ascospores of *E. caricis* from *Ph. violaceum* (Yuan *et al.*, 1998). Actual functions of the microconidia of *S. filum* are not certain but they appear to be equivalent to spermatia, which play a significant role in sexual recombination in fungi.

Life cycle study in willow plantations

In willow plantations, asexually cycling *S. filum* conidia are responsible for epidemics of *S. filum* on rusts in the growing season. Morris *et al.* (1995) also showed that *S. filum* is capable of overwintering as conidia on infected rust pustules of the stem-infecting form (SIF) of *Melampsora* on *Salix viminalis*. It is not yet clear whether conidia are the main form of spores that cause primary infections early in the season. A low frequency of suspected *E. caricis* stromata was observed in rust pustules on midribs and stems of a biomass willow, *Salix viminalis* L. 'Bowles Hybrid' (Morris *et al.*, 1995). To determine whether the sexual stage occurs during its life cycle, monthly samples were taken between August 1994 and March 1995 and examined after sectioning. The result was inconclusive because all the stromata examined were immature and no recognizable asci or ascospores were found.

In a separate study, abundant *S. filum* pycnidia containing microspores were found on the leaves of *S. burjatica* 'Germany' heavily infected by *M. larici-epitea* in late autumn in 1999 and in 2000 (M.H. Pei and T. Hunter, unpublished results). These leaves were collected and placed outside in muslin bags in late autumn. At the same time, suspected stromata on the remains of rust pustules on 'Bowles Hybrid' stems were also collected from the field and placed outside in 20-ml Sterilin tubes (bottom ends were cut off and both ends were wrapped with muslin). The fruiting bodies on the leaves and the stromata on 'Bowles Hybrid' stems were examined by squashing the fruiting bodies on slides or making hand sections at 1–2-monthly intervals until July. Each time, at least ten fruiting bodies from the leaves or the stems were examined. Again, no recognizable asci or ascospores were found over the 2-year period of study.

Amplification fragment length polymorphism (AFLP) analyses of the *S. filum* isolates collected from willow and poplar rusts in the UK showed that there is considerable variation (Nei and Li's similarity coefficients as low as 0.52) and that the *S. filum* on *Salicacea* does not have a clonal lineage (M.H. Pei and C. Ruiz, unpublished results). To date, it is not certain whether the field populations of *S. filum* occurring on willow *Melampsora* undergo a sexual life cycle.

Pathogenic Variation

The fungus is found associated with at least 369 species of 30 genera of rusts worldwide (Kranz and Brandenburger, 1981). Such an extensive host range raises the question whether *S. filum* is composed of pathogenically specialized groups that are equivalent to the *formae speciales* in biotrophic fungi such as rusts and powdery mildews. The issue of host specificity bears great significance in biological control as more virulent strains of *S. filum* would be more effective in suppressing rust disease development.

Keener (1934) carried out experiments in a greenhouse, in which 19 species of rust belonging to five genera were inoculated with 11 *S. filum* isolates obtained from *Puccinia* spp. and *Uromyces* spp. Of a total of 175 inoculations, 101 (58%) resulted in formation of *S. filum* pycnidia on the rusts. Although there were indications that *S. filum* from different sources may differ in their pathogenicity to different rusts, firm evidence of pathogenic specialization was lacking because of ill-defined experimental conditions. As acknowledged by Keener, many factors, such as the age and type of rust sori, inoculation methods, greenhouse temperature and the condition of plant hosts may have contributed to the failed infections by *S. filum*. In a review on the hyperparasitism of biotrophic fungi, Kranz (1981) summarized that 'all attempts have failed to reveal such a specificity'.

Conclusive evidence on the host specificity in *S. filum* came to light more recently. Yuan *et al.* (1999) examined pathogenicity of nine isolates of the mycoparasite to *M. larici-epitea* under strictly controlled experimental conditions. Five of the tested isolates were derived from *Melampsora* spp. on willows and poplar, one from *Puccinia coronata* on couch grass (*Agropyron repens*), two from *Phragmidium violacearum* on blackberry (*Rubus fruticosus*) and one from *Triphragmiopsis laricinum* on larch (*Larix kaempferi*). Two inoculation experiments were carried out. In the first, *S. filum* and rust were applied simultaneously on to leaf discs of *Salix burjatica* 'Korso'. In the second, leaf discs were inoculated with the rust initially, and then resulting rust uredinial pustules were inoculated with *S. filum*. In both experiments, all *S. filum* isolates from *Melampsora* spp. and that from

P. coronata, developed pycnidia on the willow rust. No pycnidia were produced from the other three isolates. Figure 21.1 shows the results from the first experiment, in which Isolate E from *M. larici-epitea* on *S. viminalis* reduced rust spore production by 98% compared with the control (without application of *S. filum*), while Isolate H from *Puccinia coronata* on couch grass reduced rust spore production by 64%. These experiments clearly demonstrated that *S. filum* is composed of pathogenically specialized populations differing widely in their virulence. When a further inoculations were carried out with 13 *S. filum* isolates derived from *Melampsora* on willow and poplars, all the isolates proved to be effective against *M. larici-epitea*, reducing rust sporulation by 88–99% 13 days after simultaneous inoculation with the rust and *S. filum* (M.H. Pei and C. Ruiz, unpublished results).

Mode of Action

The mode of infection of the mycoparasite is not yet well understood. Hulea (1939) suggested that the fungus colonizes rust sori simply through the opening of host epidermis resulting from the development of the sori. Although *S. filum* hyphae can attach themselves to rust germ tubes or even envelop them, no penetration of germ tubes was found (von Schroeder and Hassebrauk, 1957). Using scanning and transmission electron microscopy, Carling *et al.* (1976) observed that, after inoculation with *S. filum*, the majority of mycelial development occurred in the sori below the exposed layers of urediniospores. As mycelia of *S. filum* and rust develop inside plant tissue (see Kranz, 1981), it is certain that the mycoparasite absorbs nutrients from rust sori. However, the actual metabolism involved in the colonization by *S. filum* is not clear.

Conditions for Colonization of Rust

High humidity appears to be essential for infection of rusts by *S. filum*. Optimum incubation time under moist conditions for *S. filum* infection was 16–24 h for *Cronartium flaccidum* on oak leaves and 48 h for *Puccinia recondita* on wheat seedlings (see Kranz, 1972).

S. filum can establish infections at a range of temperatures, 8°C as minimum, 22–24°C optimum and 30°C maximum (Kranz, 1981). Inoculation experiments of *M. larici-epitea* with *S. filum* showed that the fungus can readily establish infection of rust pustules at 10–26°C (Morris *et al.*, 1995).

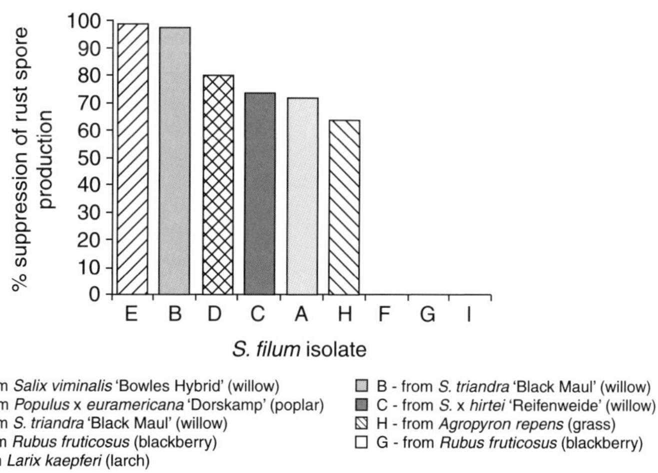

Fig. 21.1. Suppression of rust spore production of *Melampsora larici-epitea* in simultaneous inoculation with the rust and nine *S. filum* isolates (from Pei *et al.*, 1999).

Survival on Plant Surface and Timing of Infection

It was reported that with or without the presence of rust, *S. filum* can infect wheat leaves outside uredinia (von Schroeder and Hassebrauk, 1957; Kranz, 1972). When applied on to leaf discs of a willow (*S. burjatica*), *S. filum* remained viable on the willow leaves in the absence of rust for up to 14 days (Whelan *et al.*, 1997). Examination using scanning electron microscopy revealed that *S. filum* spores germinated and produced a hyphal network, which spread across the leaf surface.

Swendsrud and Calpuzos (1972) reported that application of *S. filum* to wheat seedlings 3 days prior to inoculation with *P. recondita* resulted in less infection by *S. filum* compared with simultaneous inoculation with both *S. filum* and rust spores. With *M. larici-epitea*, more rust pustules were colonized by *S. filum* and fewer rust spores were produced when *S. filum* and rust were inoculated simultaneously, compared to the inoculations of rust pustules with *S. filum* 3–4 days after rust pustules had appeared (Yuan *et al.*, 1999). It was reported that *S. filum* failed to infect *P. recondita* in inoculation experiments with *S. filum* later than 4 days after rust inoculation (von Schroeder and Hassebrauk, 1957). Keener (1934), Kranz (1972) and Kuhlman *et al.* (1978) found that infections by the mycoparasite increased with age of rust sori.

Storage of the Fungus

On dried rusted wheat leaves, *S. filum* remained viable for 36 weeks in a refrigerator, 10 weeks in the laboratory and from 6 to 12 weeks on water in the laboratory (Kranz, 1969b). With *S. filum* on willow and poplar rusts, we found that keeping dried leaf samples carrying both *S. filum* and rust sori in a freezer is an effective way of long-term storage of the fungus (M.H. Pei, unpublished results). In our lab, willow or poplar leaves carrying *S. filum* were dried at room temperature for 2–3 days and stored at –15°C. To recover the *S. filum*, dry cirrhi were suspended in water and spread over agar or inoculated on to rust sori. Some of the *S. filum* samples stored at –15°C for several years proved to be viable.

Growth and Sporulation on Media

S. filum can be grown readily on agar media such as Potato Dextrose Agar (PDA). It also grows well on chemically defined media and utilizes a wide range of carbon and nitrogen compounds, including inorganic nitrogen sources (Calpouzos *et al.*, 1957; Bean, 1968). Mycelial cultures of the fungus can also be grown in liquid media (T. Hunter, unpublished results).

Near-ultraviolet (UV) light was considered to be helpful for sporulation of *S. filum* (Kranz, 1969a). However, we often experienced difficulties in inducing sporulation on agar media with relatively old cultures of *S. filum*. In such cases, mycelial cultures of *S. filum* were homogenized and applied, together with rust urediniospores, on to detached leaves of hosts (see Pei *et al.*, 2003). After 2 weeks of incubation in a growth cabinet, the leaves were re-inoculated by brushing the newly produced conidial cirrhi over the same leaves. After a further 2 weeks, fresh cirrhi were removed using fine sterile forceps under a stereomicroscope and used for inoculation.

Incubation Period

Incubation periods (time lapse between inoculation and appearance of pycnidia) for *S. filum* have been reported in a range of 4–9 days, mostly 5–7 days at summer temperatures (see Kranz, 1981). When *S. filum* isolates obtained from *Melampsora* spp. on willow and poplars were inoculated simultaneously with rust, pycnidia developed 6–7 days after inoculation (Yuan *et al.*, 1999).

'Infection Efficiency' in *S. filum*

It appears that relatively small numbers of *S. filum* are required to establish infection of rust pustules. Kranz (1972) inoculated *P. recondita* on wheat seedlings with *S. filum* using seven spore concentrations between 30×10^3 and 3000×10^3 spores/ml and produced infections on 17–63% rust pustules. Kuhlman *et al.* (1978) inoculated uredinia and telia of *Cronartium fusiforme* on water oak (*Quercus nigra*) with six concentrations of *S. filum*, ranging from 31×10^3 to 1000×10^3 spores/ml, and obtained 6–73% infection of uredinia.

In a recent study, *S. filum* was applied using different spore concentrations on to willow leaf discs inoculated with *M. larici-epitea* (Pei *et al.*, 2003). In this experiment, willow leaf discs were inoculated using a standard inoculum of rust (inoculum density = 262 viable spores per 0.95 cm^2 leaf disc) and, as a result, an average of 46 uredinia formed on each leaf disc. Results from this study showed that the ratios of infected rust pustules/*S. filum* conidia applied were in a range of 0.25–0.31 when less than 20 *S. filum* spores were inoculated on to a leaf disc. The ratios were 0.8–0.12 when 60–200 *S. filum* spores were inoculated on to a leaf disc. There was a close correlation between *S. filum* inoculum and percentage infection of rust uredinia/the number of pycnidia (Fig. 21.2). Suppressive effects of *S. filum* on rust spore production were more obvious in the second assessment, carried out 23 days after inoculation.

In plant–rust pathogen systems, the proportion of inoculum that produces pustules/lesions is referred to as infection efficiency (see Zadoks and Schein, 1979). With rusts, it has been proven that one rust spore produces only one pustule on the hosts (Zadoks and Schein, 1979; Pei *et al.*, 2002) and, therefore, the ratio between the inoculum density and the number of rust pustules produced is equivalent to the infection efficiency. With *S. filum*, however, it is not certain whether a conidium can produce more than one pycnidium. Therefore, the ratio of the number of pycnidia/*S. filum* conidia applied is not entirely synonymous with the infection efficiency in plant–rust systems.

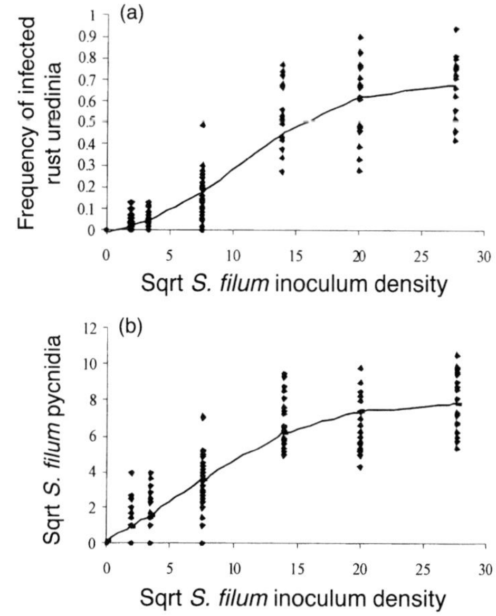

Fig. 21.2. Effects of inoculum densities (the number of conidia applied per leaf disc) of *Sphaerellopsis filum* on (a) the frequency of rust uredinia infected and (b) the number of *S. filum* pycnidia produced, per leaf disc 13 days after inoculation. Sqrt, square root. (From Pei *et al.*, 2003.)

Effects of *S. filum* on Rust Spore Production and Disease Severity

Both rust fungi and *S. filum* produce large numbers of spores during the season. Quantitative inoculations with *M. larici-epitea* suggested that each uredinium can produce up to 13×10^3 urediniospores per day (Pei *et al.*, 2002). When assessed 13 days after inoculation of *M. larici-epitea* with *S. filum*, the average number of *S. filum* conidia produced per pycnidium was 8920 ± 880 ($P = 0.05$) (Pei *et al.*, 2003). Given the assumption that the average incubation period of *S. filum* had been 8 days, more than 1000 conidia would have been produced per pycnidium per day.

Kranz (1981) observed that, compared with the uredinia free of *S. filum*, one *S. filum* pycnidium per uredinium of *P. recondita* could reduce urediniospore production by 10–70%, and nearly complete coverage of rust uredinia by pycnidia reduced urediniospores by 50–95%. Kuhlman *et al.* (1978) reported 25–50% reduction in basidiospore production in *Cronartium flaccidum* compared to the telial leaves free of *S. filum*. With larch rust *Triphragmiopsis laricinum*, Yuan (1991) found that germination of teliospores from the telia colonized by *S. filum* was reduced by some 50% compared with those from the telia free of *S. filum* pycnidia.

Hau and Kranz (1978) described a computer model using *P. recondita* f. sp. *tritici* to assess the effectiveness of *S. filum* in suppressing rust epidemics. The model comprised functions for green leaf biomass of wheat corrected for phase-dependent susceptibility, interactions between host growth and disease development, etc. Using the model, various assumptions were tested, i.e. the extent of colonization of rust sori by *S. filum*, time to reach certain coverage (e.g. 40% of pustule area), the time between the first appearance of rust and the first appearance of *S. filum*, and the effect of rust density. The model estimated that rust severity would be reduced by 87% when 40% of pustule area is covered by *S. filum* 2 days after rust pustules appeared. If 40% of pustule area was covered by *S. filum* 4, 6, 8 or 10 days after rust pustules appeared, rust severity would be reduced by 78%, 64%, 42% or 24%, respectively. Hence the model showed that early establishment of infection by *S. filum* may be more effective in suppressing disease levels.

Yuan *et al.* (1998) conducted two inoculation experiments with *M. larici-epitea* and *S. filum*. In the first experiment, *S. filum* and rust were applied simultaneously on to leaf discs of *Salix burjatica* 'Korso'. In the second experiment, leaf discs were inoculated with rust initially and then resulting rust uredinia were inoculated with *S. filum*. In the first experiment, 55–99% pustules were colonized and rust spore production was suppressed by 64–98%. In the second, the same figures were 33–97% and 53–73%, respectively.

In quantitative inoculations of *M. larici-epitea* with *S. filum* (Pei *et al.*, 2003), there was a reasonable linear correlation ($y = -2.9353x + 423.75$, % variance accounted for = 51.1) between the square root number of rust spores produced (y) and the frequency of rust uredinia infected by *S. filum* (x). According to the relationship, the % reduction in spore production (compared with the control) can be calculated as in Table 21.1.

Spread in Plantations

How *S. filum* spreads in a plantation is an important question to be answered in deployment of the mycoparasite for biocontrol. The fungus colonizes rust pustules and produces masses of conidia (cirrhi), which are easily suspended in water. The gelatinous caps of conidia may indicate a splash-borne mechanism of spore dispersal. Such a structure may also help conidia to be attached to rust spores or to insects to facilitate airborne dispersal. In nature, dry weather conditions are certainly detrimental to the spread of *S. filum*. In southern India, *S. filum* is absent during the dry season but may occur after heavy showers (Ramakrishnan and Narasimhalu, 1941; Sundaram, 1962).

Table 21.1. Estimated percentage reduction in rust spore production at different levels of infection by *Sphaerellopsis filum* 13 days after simultaneous inoculations with rust and *S. filum*.

Frequency of uredinia colonized by *S. filum*	% reduction in rust spore production
0	0
20	26
40	48
60	66
80	80
100	91

To investigate the spread of *S. filum* in plantations, Morris *et al.* (1995) conducted a field experiment using three sub-plots in a *S. viminalis* 'Bowles Hybrid' plantation at Long Ashton, south-west England. Early in the season (May), *S. filum* was introduced to two of the sub-plots by placing willow plants previously inoculated with both *M. larici-epitea* and *S. filum*. A lattice of 12 × 12 stools from each subplot was examined for the presence of *S. filum* at 6- to 10-day intervals until late July, and the data were analysed using two-dimensional distance class analysis (2D analysis) (Grey *et al.*, 1986). It was shown that *S. filum* disperses in a non-random, concentric pattern from the central inoculum source (introduced) as opposed to the random pattern found in the control sub-plot where *S. filum* occurred naturally without introduction. The average rate of *S. filum* dispersal from the inoculum source was slow initially (in the first 4 weeks of observation) and became more rapid thereafter.

The dispersal of *S. filum* in willow plantations was further examined using molecular markers (M.H. Pei and C. Ruiz, unpublished results). As *S. filum* reproduces asexually in the growing season, DNA fingerprinting provides an effective way of examining the genotype distribution in plantations. Sampling was done from *S. burjatica* 'Germany' in a willow planting that was surrounded by cereal crops. Natural infections were detected in the early stages of rust epidemics (July 1998). *S. filum* samples were made consecutively from 46 stools (0.5 m apart) of 'Germany' in the autumn (September). A single-spore isolate was made from each sample and fingerprinted using AFLP. A preliminary analysis showed that two genotypes were most frequent, one of which was found on 43% of the stools. The same genotypes often occurred in adjacent stools, indicating that rain splash or leaf contact may be an important means of spread in *S. filum*. However, somewhat scattered distribution of the same genotypes also suggested that other mechanisms, dispersal by insects or rust spores, for example, might play a significant role in the spread of *S. filum*.

Potential for Biological Control of Rust in Coppice Plantations

Laboratory inoculation experiments have consistently proved that, given optimum conditions, *S. filum* can be remarkably effective in suppressing rust disease on willow. It was also revealed that some genotypes of *S. filum* were more virulent than others. As knowledge on the pathogenic variation in *S. filum* became available only very recently, it has not been possible to verify whether the numerous reports that rust disease was suppressed markedly by *S. filum* were associated with the occurrence of highly virulent populations of *S. filum* in the field.

In a low-input crop system such as SRC willow, sustaining the mycoparasite populations in the field is important in the deployment of *S. filum* for biocontrol. Previous work by Morris *et al.* (1995) and our unpublished results showed that *S. filum* is capable of overwintering as conidia on infected rust pustules of the stem-infecting form of *Melampsora* (SIF) on *S. viminalis*. *Melampsora* rusts on SRC willows are highly specialized in their pathogenicity (Pei *et al.*, 1996). For example, the pathogenicity of SIF is confined to certain clones within *S. viminalis*. By carefully designing host genotype mixtures, it is possible to minimize the impact of certain forms of rust on which *S. filum* can overwinter. Rapid spread of the mycoparasite to keep pace with rust development can only be achieved in mixed clonal plantings, in which the clones carrying *S. filum* are planted in intimate mixtures with others. The 'carrier' of *S. filum* may not be confined to the clones that are susceptible to SIF, since some other *Melampsora* species, such as *M. capraearum*, can cause stem infections and may harbour overwintering *S. filum*.

Morris *et al.* (1994, 1995) successfully introduced *S. filum* into willow planting using potted willow plants. These studies showed that effective biocontrol can be achieved only if the mycoparasite is introduced or sustained at many sources throughout the plantation. Practical ways of achieving this require further investigation.

Concluding Remarks

In this chapter, we have given a brief review of some of the existing information on taxonomy, life-cycle, pathogenicity, and suppressive effects of *S. filum* on rusts. We have also discussed the potential of using the mycoparasite for biological control of willow rusts in short-rotation coppice plantations. The development of a practicable rust control strategy without relying on fungicides requires integrated approaches involving selection and breeding for resistance, and planting clonal mixtures. By designing mixtures to encourage the spread of the mycoparasite, the suppressive effects of mixtures on rust disease can be enhanced.

In all, there is a strong potential for a significant degree of biological control of rust by deploying the mycoparasite *S. filum* in SRC willow plantations. Compared with annual crops, perennial crops, such as SRC willows, which are harvested at 2–5-year intervals, are better suited for deploying mycoparasites for biological control, because they allow inoculum to be carried over from 1 year to the next.

Acknowledgements

This study was funded by the Department for Environment, Food and Rural Affairs (DEFRA), UK, and the European Union. Rothamsted Research receives grant-aided support from the Biotechnology and Biological Sciences Research Council of the UK.

References

Adams, J.F. (1920) *Darluca* on *Peridermium peckii*. *Mycologiga* 12, 309–315.

Bean, G.A. (1968) Growth of the hyperparasite *Darluca filum* on a chemically defined medium. *Phytopathology* 58, 252–253.

Bivona-Bernardii, A. (1813) *Stirpium rariorum* in *Sicilia provenienthium* descriptions. *Manipulus* 3, 12.

Calpouzos, L., Theis, T. and Buller, C.M.H. (1957) Culture of the rust parasite, *Darluca filum*. *Phytopathology* 47, 108–109.

Carling, D.E., Brown, M.F. and Millikan, D.F. (1976) Ultrastructual examination of the *Puccinia graminis–Darluca filum* host parasite relationship. *Phytopathology* 66, 419– 422.

Castagne, L. (1851) *Catalogue des Plantes qui Croissent Naturellement aux Environs de Marseille*. Supplément. Aix, pp. 52–56.

Eriksson, O. (1966) On *Eudarluca caricis* (Fr.) O. Eriks., Comb. nov., a cosmopolitan uredinicolous pyrenomycete. *Botaniska Notiser* 119, 33–69.

Eriksson, O. (1967) On graminicolous pyrenomycetes from Fennoscandia, 2. Phragmosporous and scolecosporous species. *Arkiv For Botanik* 6, 390–392.

Fries, F. (1823) *Systema Mycologicum* 2, 547.

Grey, S.M., Moyer, J.W. and Bloomfield, P. (1986) Two-dimensional distance class model for quantitative description of virus-infected plants distribution lattices. *Phytopathology* 76, 243–248.

Hau, B. and Kranz, J. (1978) Modellrechnungen zur Wirkung des Hyperparasiten *Eudarluca caricis* of Rostepidemien. *Zeitschrift für Pflanzenkrankheiten und Pflanzenschutz* 85, 131–141.

Hulea, E.E. (1939) Contributions à la connaissance des champignons commensaux des uredinea. *Bulletin de la Section Scientifique de l'Academie Roumaine* 22, 1–19.

Keener, P.D. (1934) Biological specialization in *Darluca*. *Bulletin of Torrey Botanical Club* 61, 475–490.

Keener, P.D. (1951) An ascigerous stage of *Darluca filum* (Biv.) Castagne. *Plant Disease Reporter* 35, 86–87.

Kranz, J. (1969a) Zur natülichen Verbreitung des Rostparasiten *Eudarluca caricis* (Fr.) O. Erikss. *Phytopathologische Zeitschrift* 65, 43–53.

Kranz, J. (1969b) Das Verhalten von *Darluca filum* (Biv.-Bern.) Cast. *in vitro* unter verschiedenen Versuchsbedingungen. *Phytopathologische Zeitschrift* 65, 325–331.

Kranz, J. (1972) Zur Infection mith *Eudarluca caricis* (Fr.) O. Erikss. *Phytopathologische Zeitschrift* 74, 13–20.

Kranz, J. (1981) Hyperparasitism of biotrophic fungi. In: Blakeman, J.P. (ed.) *Microbial Ecology of the Phylloplane*. Academic Press, London, pp. 327–352.

Kranz, J. and Brandenburger, W. (1981) An amended host list of the rust parasite *Eudarluca caricis*. *Journal of Plant Diseases and Protection* 88, 682–702.

Kuhlman, E.G., Mathews, F.R. and Tillerson, H.P. (1978) Efficacy of *Darluca* for biological control of *Cronartium fusiforme* and *C. strobelinum*. *Phytopathology* 68, 507–511.

Morelet, M. and Pinon, J. (1973) *Darluca filum* hyper-parasite du genre *Melampsora* sur peuplier et saule. *Revue Forestière Française* 25, 378–379.

Morris, R.A.C., Royle, D.J., Whelan, M.J. and Arnold, G.M. (1994) Biocontrol of willow rusts with the hyperparasite *Sphaerellopsis filum*. *Proceedings of the Brighton Crop Protection Conference – Pests and Diseases*, vol. 3. Brighton, UK, pp. 1121–1126.

Morris, R.A.C, Royle, D.J. and Whelan, M.J. (1995) Potential for biological control of willow rust with the hyperparasite *Sphaerellopsis filum*. Final Report of Project ETSU B/W6/00230/REP. Energy Technology Support Group, Department of Trade and Industry, UK.

Parker, S.R., Royle, D.J. and Hunter, T. (1993) Impact of *Melampsora* rust on yield of biomass willows. In: *Abstracts of the 6th International Congress of Plant Pathology, Montreal, Canada, 28 July–6 August 1993*. National Research Council Canada, Ottawa, p. 117.

Pei, M.H., Royle, D.J, and Hunter, T. (1995) A comparative study of stem-infecting and leaf-infecting forms of *Melampsora* rust on *Salix viminalis* in the UK. *Mycological Research* 99, 357–363.

Pei, M.H., Royle, D.J. and Hunter, T. (1996) Pathogenic specialisation in *Melampsora epitea* var. *epitea* on *Salix*. *Plant Pathology* 45, 679–690.

Pei, M.H., Ruiz, C., Hunter, T., Arnold, G.M. and Bayon, C. (2002) Quantitative relationships between inoculum of *Melampsora larici-epitea* and corresponding disease on *Salix*. *Plant Pathology* 51, 443–453.

Pei, M.H., Hunter, T., Ruiz, C., Bayon, C. and Harris, J. (2003a) Quantitative inoculation of willow rust *Melampsora larici-epitea* with the mycoparasite *Sphaerellopsis filum*. *Mycological Research* 107, 57–63.

Pei, M.H., Ruiz, C., Hunter, T. and Bayon, C. (2003b) Rust resistance in *Salix* induced by inoculations with avirulent and virulent isolates of *Melampsora larici-epitea*. *Forest Pathology* 33, 383–394.

Ramakrishinan, T.S. and Narasimhalu, I.L. (1941) The occurrence of *Darluca filum* (Bev.) Cast. on cereal rusts in South India. *Current Science* 17, 215–216.

Spegazzini, C. (1908) Fungi Aliguot Paulistani. *Revista del Museo de La Plata* 15, 22.

Sundaram, N.V. (1962) Studies on the parasites of the rusts. *Indian Journal of Agricultural Sciences* 32, 266–276.

Sutton, B.C. (1977) Coelomycetes. VI. Nomenclature of generic names proposed for Coelomycetes. *Mycological Papers* 141, p. 196.

Swendsrud, D.P. and Calpouzos, L. (1972) Effect of inoculation sequence and humidity on infection of *P. recondita* by the mycoparasite *Darluca filum*. *Phytopathology* 62, 931–932.

von Shroeder, H. and Hassebrauk, K. (1957) Beiträge zur Biologie von *Darluca filum* (Biv.) Cast. und einigen anderen an Uredi beobachteten Pilzen. *Zentralblatt für Bakteriologie Parasitenkunde Infektionskrankheiten und Hygiene II Abteilung-Naturwissenschaftliche-Mikrobiologie Der Landwirtschaft Der Technologie* 110, 676–696.

Whelan, M.J., Hunter, T., Parker, S.R. and Royle, D.J. (1997) How effective is *Sphaerellopsis filum* as a biocontrol agent of *Melampsora* willow rust? *Aspects of Applied Biology* 49, 143–148.

Yuan, J.Y., Yuan, Z.W. and Li, L.Z. (1988) Studies on the biological control of larch brown rust with a rust parasite – 1. Morphological and cultural characteristics of the rust parasite [in Chinese]. *Journal of Shenyang Agricultural University* 19, 17–22.

Yuan, X.Y. and Han, Y.J. (2000) Hyperparastism and biocontrol of *Hedysarum mongolicum* rust by *Sphaerellopsis filum* [in Chinese]. *Forest Research* 13, 103–106.

Yuan, Z.W. (1991) Observations on a pathogen parasitizing larch brown rust, *Triphragmiopsis laricinum* [in Chinese]. *Chinese Journal of Biological Control* 7, 61–63.

Yuan, Z.W., Pei, M.H., Hunter, T. and Royle, D. J. (1998) *Darluca caricis*, the teleomorph of the mycoparasite *Sphaerellopsis filum*, on blackberry rust *Phragmidium violaceum*. *Mycological Research* 102, 866–868.

Yuan, Z.W., Pei, M.H., Hunter, T., Ruiz, C. and Royle, D.J. (1999) Pathogenicity to willow *Melampsora epitea*, of the mycoparasite *Sphaerellopsis filum* from different sources. *Mycological Research* 103, 509–512.

Zadoks, J.C. and Schein, R.D. (1979) *Epidemiology and Plant Disease Management*. Oxford University Press, Oxford, UK.

Index

www.cabi-publishing.org/bookshop
Browse Read and Buy
ANIMAL & VETERINARY SCIENCES
BIODIVERSITY CROP PROTECTION
HUMAN HEALTH NATURAL RESOURCES
ENVIRONMENT PLANT SCIENCES
SOCIAL SCIENCES
CABI Publishing
A division of CAB International
Online BOOK SHOP
Subjects
Search
Reading Room
Bargains
New Titles
Forthcoming
Crop Pollination by Bees
Keith S. Delaplane and Daniel F. Mayer
A DICTIONARY OF Entomology
G Gordh and D H Headrick
Order & Pay Online!
MasterCard
VISA
AMERICAN EXPRESS
Principles of CATTLE PRODUCTION
C.J.C. Phillips
Seeds
THE ECOLOGY OF REGENERATION IN PLANT COMMUNITIES
2nd EDITION
Edited by Michael Fenner
FULL DESCRIPTION
BUY THIS BOOK
BOOK OF THE MONTH
Tel: +44 (0)1491 832111 Fax: +44 (0)1491 829292